K. Hirota (Ed.)

Industrial Applications of Fuzzy Technology

Translated by H. Solomon

With 217 Figures

Springer-Verlag
Tokyo Berlin Heidelberg New York London
Paris Hong Kong Barcelona Budapest

Kaoru Hirota, Ph.D.
Professor, Department of Systems Control Engineering, College of Engineering,
Hosei University, Koganei, Tokyo, 184 Japan

Harold Solomon
Keiyō Ltd., Tama, Tokyo, 206 Japan

ISBN 978-4-431-65879-5   ISBN 978-4-431-65877-1 (eBook)
DOI 10.1007/978-4-431-65877-1
Softcover reprint of the hardcover 1st edition 1993
Printed on acid-free paper

PREFACE

Fuzzy theory originated with a paper titled "Fuzzy Sets" by Prof. Lotfi Zadeh of the University of California, published in 1965. Practical applications, led by Japan, started in the 1980s. In 1990 there was a rush of applications to household electrical appliances, and the term "fuzzy" came to be heard even in tea parlors. As a result, people overseas who had been cool toward fuzzy technology began to change their attitude.

Thus, the application of fuzzy technology is a technological revolution which Japan can boast to the world of having led. It is highly significant and important for this work to be introduced not only in Japan but also to foreign countries. I was very pleased, therefore, when in the early summer of 1990 I was approached by the Tokyo office of Springer-Verlag, a publisher with an established international reputation, with a proposal to do this book.

As the value of fuzzy technology has come to be appreciated, a number of books on fuzzy theory have been published in Japanese, starting about 1988. Many introductory books and textbooks have been published, but this book, in contrast, has the purpose of providing easy - to - understand explanations centered mainly on concrete applications. This English edition was planned to appear soon after the Japanese edition to lay a foundation for the introduction of this technology of which Japan is proud to the rest of the world.

For this reason we selected authors who are among the leaders in this field and imposed on them to write the chapters in spite of their busy schedules. There are a total of 12 chapters. Consideration was given to making each chapter independently readable, but there is nevertheless a reason for the way the chapters are arranged.

A minimum of fuzzy theory that is needed to understand its practical applications is given in Chapter 1. Chapters 2 to 5 discuss hardware, including chips, and software tools used in constructing systems. Chapters 6 to 12 cover a series of practical applications. These include applications to industrial processes and plants, and to transportation systems, which were among the first applications; and applications to consumer products such as household electrical appliances, which produced the "fuzzy boom".

This book has been written to be understandable to anyone with a good undergraduate background in physical sciences or engineering. It is hoped that it will be read by a wide variety of people, from engineers and project managers working in plants to undergraduate and graduate students in universities. It is the common hope of the authors that this book will become a stepping stone for many readers to enter this field.

Finally, I wish to thank the authors who took time to write the chapters of this book in spite of their busy schedules, and the editorial staff members who have brought it to publication.

July, 1992
Kaoru Hirota, Editor

Editor's Note

Fuzzy engineering is still a young field. Not only has the terminology not yet been standardized, but there is still active controversy as to which terms should be used. The authors who contributed to the present volume have insisted that their own preferences be followed. For example, some chapters use "condition part" and "operation part" (or "conclusion") for what other chapters call "antecedent" and "consequent"; some use "degree of applicability" for what others call "grade". It is necessary to ask for the reader's indulgence in this matter.

Table of Contents

List of Contributors

Chapter 1
Kaoru Hirota

Department of Systems Control Engineering, College of Engineering, Hosei University, Tokyo, 184 Japan

Chapter 2
Takuya Ishioka

Industrial Electronics & System Laboratory, Mitsubishi Electric Corporation, Kyoto, 661 Japan

Chapter 3
Yohichi Ageishi

System Development Department, AdIn Research, Inc., Tokyo, 102 Japan

Chapter 4
Masayuki Otani

Research & Development Division, Mycom, Inc., Kyoto, 616 Japan

Chapter 5
Takao Ohta

IC Promotion Center, Kyoto Laboratory, OMRON Corporation, Kyoto, 617 Japan

Chapter 6
Osamu Itoh

Software and System Laboratory, Fuji Electric Corporate Research and Development, Ltd., Tokyo, 191 Japan

Kazuhiro Ozawa

Faculty of Economic, Hosei University, Tokyo, 194-02 Japan

Chapter 7
Takashi Ohnishi

Yokohama Research & Development Center, Mitsubishi Heavy Industries, Ltd., Yokohama, 231 Japan

Chapter 8
Jun-ichi Koizumi — Institute of Molecular and Cellular Biosciences, The University of Tokyo, Tokyo, 113 Japan

Yoshinobu Tsuchiya
Kazuo Suenari — Hiroshima Prefectural Food Technology Research, Hiroshima, 730 Japan

Shiro Nagai — Department of Fermentation Technology, Faculty of Engineering, Hiroshima University, Hiroshima, 724 Japan

Chapter 9
Shiro Hikita — Power and Transportation System Department, Industrial Electronics & System Laboratory, Mitsubishi Electric Corporation, Hyogo, 661 Japan

Chapter 10
Toshihiro Koyama — Control Systems R & D Section, Heavy apparatus Engineering Laboratory, Fuchu Works, Toshiba Corporation, Tokyo, 183 Japan

Chapter 11
Kuniaki Matsumoto — 1st System Department, Defense Division, Hitachi, Ltd., Tokyo, 101-10 Japan

Chapter 12
Noboru Wakami — Central Research Laboratories, Matsushita Electric Industrial Co., Ltd., Osaka, 570 Japan

Haruo Terai — Home Appliance Research Laboratory, Matsushita Electric Industrial Co., Ltd., Osaka, 661 Japan

Chapter. 1

THE BASIS OF FUZZY THEORY

1.1 Introduction

The importance of information processing, which is said to come naturally to human beings but to be difficult for computers, is at last becoming recognized by society in general. Perhaps it would also appropriate to say that its position is now well enough established that we can afford to raise questions about it.

It goes without saying that computer logic is binary Boolean logic, with 2 values, 0 and 1. The binary pair {0,1} can be equally well expressed in such terms as {no, yes}, {off, on}, {low, high} or {0V, 5V}. We describe such a universe of quantities which can be expressed as either of 2 values as a "crisp" universe. In general, it is sufficient for operations within a computer and communication between computers to be conducted in a crisp universe. However, in a man – machine system, there arises the problem of processing information with the "vagueness" that is characteristic of man. For example, a human will give as answers not only "yes" or "no", but also "almost yes" or "don't know". Such answers arise not only in the various forms of man — machine communication such as human interfaces, character recognition, and voice recognition/synthesis, but also in information exchange between human beings (man-woman communication?).

Prof. L. A. Zadeh of the University of California at Berkeley was one of the first to realize the importance of this concept, and in 1965 published his "fuzzy sets theory"[1]. This was the start of fuzzy theory. At first this theory was not well received, but in 1974 Dr. E. H. Mamdani of London University demonstrated the applicability of fuzzy inference to automatic steam engine operation[2]. Then, in 1980, fuzzy theory found its first practical application on the working level when the F.L. Smidth Company of Denmark used it in automatic cement kiln operation[3]. At present there is a quiet boom in fuzzy application with fuzzy theory being applied in close to 600 cases, centered in Japan, some examples being control of the Sendai municipal subway (developed by Hitachi Ltd.) and chemical injection in the Sagamihara water treatment plant.

In this chapter, a simple outline of fuzzy theory will be given, with emphasis on fuzzy set theory and fuzzy inference.

1.2 Fuzzy set theory

1.2.1 Review of crisp sets

The concept of a fuzzy set is an extension of the concept of an ordinary set (to be referred to here as a crisp set). Therefore, we will first review crisp sets. Crisp sets can themselves become quite difficult, as in axiomatic set theory, but here it is sufficient to take the point of view of naive set theory, which is an extension of the set theory learned in grade school arithmetic. However, the concept of characteristic functions, which is not learned in grade school, is necessary.

Let us discuss this concept in terms of a collection of 10 people, $x_1, x_2, \ldots\ldots,$ x_{10}, who form a project team. The entire object of discussion is

$$X = \{x_1, x_2, \ldots x_{10}\}. \tag{1}$$

In general, the entire object of discussion X is called a "universe of discourse", and each constituent member x_i (the fact that x_i is an element of X is written $x_i \in X$) is called an "element". Here, if team members x_1, x_2, x_3 and x_4 are female, the set of females in X (denoted by A) is

$$A = \{x_1, x_2, x_3, x_4\} (\subset X). \tag{2}$$

In grade school, such a condition is expressed by a Venn diagram such as the one in Fig. 1. 1.

This condition can also be expressed by the mapping χ_A from X into the binary space $\{0,1\}$ (this is called the "characteristic function" of A).

$$\chi_A : X \quad \rightarrow \quad \{0,1\}$$

$$\underset{x}{\cup} \quad \mapsto \quad \underset{\chi_A(x)}{\cup} = \begin{cases} 0 & x = x_5, x_6, x_7, x_8, x_9, x_{10} \\ 1 & x = x_1, x_2, x_3, x_4 \end{cases} \tag{3}$$

That is to say, the value $\chi_A(x)$ is 1 when the element x satisfies the attributes of set A, 0 when it does not. The characteristics of A are expressed by χ_A.

Next, supposing that within X only x_1 and x_2 are below age 20, we can define a set $B = \{x_1, x_2\}$ consisting of team members below age 20. In this case, $\chi_B(x)$ is 1 when x is x_1 or x_2, 0 otherwise. Thus, one can conceive of a variety of sets within X. Their totality is expressed as 2^X (called a "power set" of X). 2^X includes, besides A and B, the "null set" ϕ (the characteristic

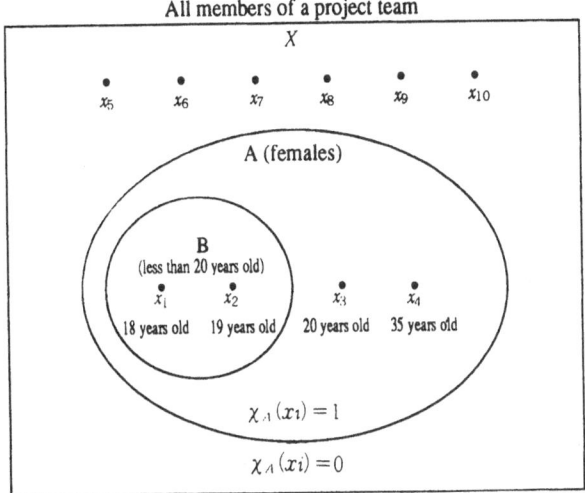

Fig. 1.1   The (crisp) set A of females among all members X of a project team.

function $\chi_\phi(x)$ is always 0) and the total set X (of which the characteristic function is always 1). (It is a question of how many different characteristic functions can be created; for each x_i it can have the value 0 or 1, so when there are 10 x_i, a total of 2^{10} can be created.)

Here let us review the operation 2^X. First there is the concept of "subsets". In the preceding example B is a subset of A; we write $B \subset A$. In general, if all elements of B are also elements of A, then B is a subset of A. Expressing $B \subset A$ in terms of characteristic functions, we have the inequality

$$\chi_B(x) \leqq \chi_A(x) \tag{4}$$

which holds for all elements x in X. The concepts of "complement", "intersection" and "union" are also important. These relationships can be expressed in the Venn diagram shown in Fig. 1. 2; the corresponding characteristic function relationships are given by equations (5) to (7).

Complement of A A^C

$$\chi_{A^C}(x) = 1 - \chi_A(x) \tag{5}$$

Intersection of A and B $A \cap B$

$$\chi_{A \cap B}(x) = \chi_A(x) \wedge \chi_B(x) \tag{6}$$

Union of A and B $A \cup B$

$$\chi_{A \cap B}(x) = \chi_A(x) \vee \chi_B(x) \tag{7}$$

Here, the symbols $\wedge$ and $\vee$ mean that the smaller value (min) and the larger value (max), respectively, is to be taken.

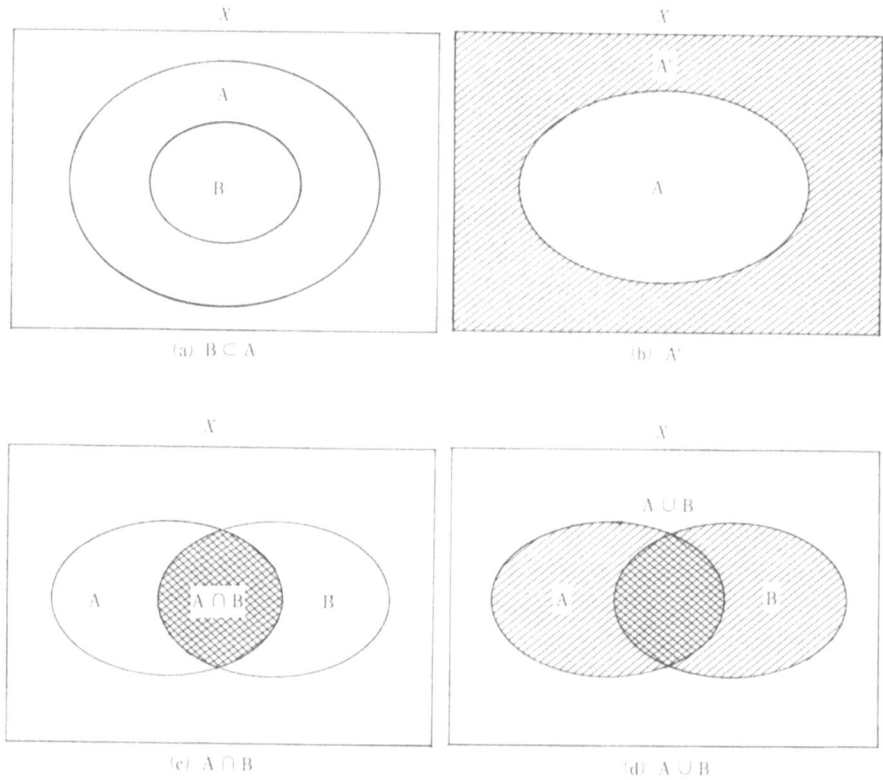

Fig. 1.2 Set inclusion relationships: a complement, an intersection, a union.

1.2.2 Fuzzy sets

Let us again consider the example of the project team X. We have considered the "set of females A" and the "set of minors B" in X; is it also possible to consider a "set of young females C"? If for convenience we consider the attribute "young" to be the same as "minor", then $C = B$; but in this case we have created a sharp boundary, under which x_2 who is 19 is still young ($\chi_C(x_2) = 1$), but x_3, who just turned 20 today, is no longer young ($\chi_C(x_3) = 0$). In just one day the value changed from yes (1) to no (0), and x_3 is now an old maid.

However, is it not possible that a young woman becomes an old maid only over a period of 10 to 15 years, so that we ought to be patient with her? Prof. Zaeh admitted values such as 0.8 and 0.9 that are intermediate between 0 and 1, thus creating the concept of a "fuzzy set". Whereas a crisp set is defined by the characteristic function (3) that can assume only the two values $\{0,1\}$, a fuzzy set is defined by a "membership function" that can assume an infinite number of values, any real number in the closed interval $[0,1]$.

With this definition, the concept of "young women" in X can be expressed flexibly in terms of the membership function m_C:

$$m_C : X \to [0, 1] \qquad (8)$$

The value of $m_c(x)$ is at least 0 and not more than 1. For example, it can be expressed as:

$$m_C(x) = 1/x_1 + 1/x_2 + 0.9/x_3 + 0.2/x_4 \qquad (9)$$

Here, $0.9/x_3$ means that $m_c(x_3) = 0.9$; / is called a "separator". + means "or". Normally, elements for which the membership function is 0 are omitted from the description.

Since $[0,1]$ incorporates $\{0,1\}$, the concept of a fuzzy set can be considered as an extended concept, which incorporates the concept of a crisp set. For example, the crisp set B of people under 20 can be regarded as a fuzzy set with the membership function:

$$m_B(x) = 1/x_1 + 1/x_2 \qquad (10)$$

(see Fig. 1. 3).

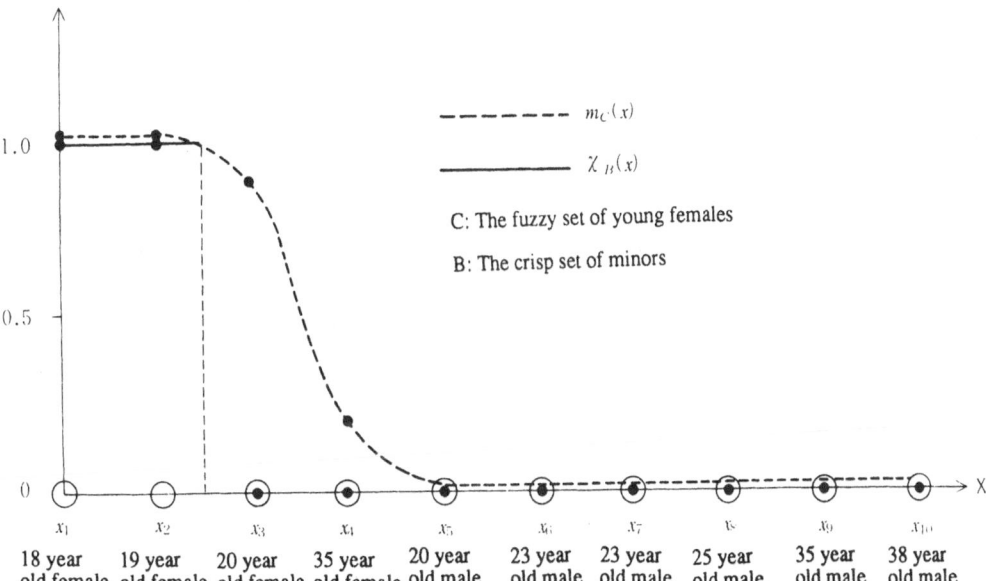

Fig. 1.3 Examples of a crisp set and a fuzzy set.

Next, let us explain the significance of underlining such terms as patient and flexibly in the above description. For example, in formula (9), $m_C(x) = 0.9$, but suppose that x_3 objects that "you are being unfair; I really ought to be a 1, but if you insist we can compromise at 0.95". A great deal of research has been done on this question, with a distinction now being made among type I (ordinary) fuzzy sets; type II fuzzy sets, in which the values are reevaluated to [0,1]; and more sophisticated fuzzy sets using vague "linguistic variables" such as "true", "almost true", "more or less true". However, in virtually all practical applications of fuzzy theory to date type I fuzzy sets are being used; even with this restriction, it has become possible to deal with many problems that could not be handled with only crisp sets. In the remainder of this chapter, only type I fuzzy sets will be considered.

As in the crisp case, the concepts of complements, intersections and unions are defined for fuzzy sets, with their membership functions satisfying relations (5) to (7) (Fig. 1. 4):

$$m_{A^C}(x) = 1 - m_A(x) \tag{11}$$

$$m_{A \cap B}(x) = m_A(x) \wedge m_B(x) \tag{12}$$

$$m_{A \cap B}(x) = m_A(x) \vee m_B(x) \tag{13}$$

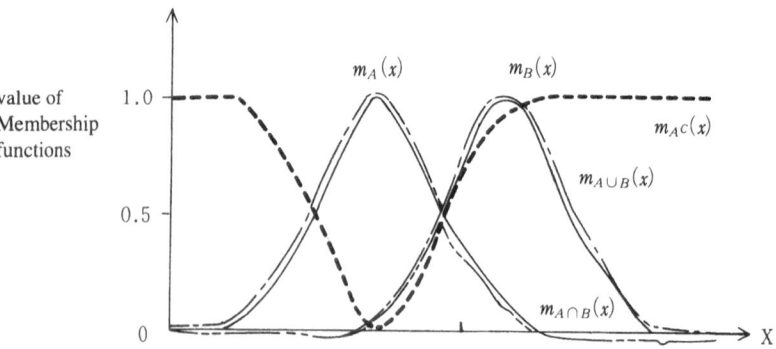

Fig. 1.4 Complement, intersection and union operations on fuzzy sets.

1.3 Fuzzy inference

The concept of fuzzy sets makes it possible to describe vague information, but description alone will not lead to the development of any useful products. Indeed, a good deal of time passed after fuzzy sets were first proposed until they were applied on the industrial level. However, eventually it became possible to apply them in the form of "fuzzy inference", and, as mentioned in 1.1, fuzzy theory has now become legitimized as one component of applied high technology. Many forms of fuzzy inference have been researched; but here we will restrict consideration to "approximate reasoning" (formerly called simply "fuzzy inference") by the Max — Min composition fuzzy production rule, which is the most important form of fuzzy inference in practical applications.

In this method of fuzzy inference, the knowledge of an expert in a field of application is expressed as a set of "fuzzy production rules" of the form IF-THEN, leading to algorithms describing what action should be taken based on currently observed information. (In fuzzy theory, nothing is done at random or haphazardly. Information containing a certain amount of vagueness is expressed as faithfully as possible, without the distortion produced by forcing it into a "crisp" mold, and it is then processed in a proper manner.)

Let us again explain a familiar example. Consider the fuzzy production rule

$$\text{"IF my lover is beautiful, THEN I will propose".} \tag{14}$$

("Beautiful" as used here is a typical fuzzy description; it might for example include not only physical beauty but also kindness of heart.)

The result of observation is usually that the lover is neither perfectly beautiful nor the complete opposite. If she leans somewhat toward being beautiful, what should one do? In a crisp world, this description could not be fitted to relation (14), so one would not know what to do. However, in fuzzy inference, it is possible to feed approximate information into this relation and come up with such an answer as "go out with her a few times and then decide".

Of course, even using crisp inference it is possible to add the rule "IF she is beautiful, THEN I will go out with her for a while and see how things go", but to respond flexibly in such a manner would require a great many rules. It is far more efficient to keep the number of substantive rules to a minimum and make it possible to respond flexibly using approximate values. This is one of the reasons that fuzzy inference has come to be widely used.

Now let us go into a bit more detail. First, we describe expert knowledge as a set of fuzzy production rules.

$$\{\text{IF (antecedent } i) \text{ THEN (conclusion } i)\}_{i=1}^{N} \tag{15}$$

Here N is the total number of rules. In general there are two or more antecedents and conclusions. For example, when a robot arm grasps an object to be moved, one could have a rule such as "IF the object to be grasped is moving fast AND it is a long distance away, THEN move the hand a long way in the direction of the object", which has 2 antecedents for 1 conclusion. In real industrial application systems, there may be for example 2 antecedents to 1 conclusion or 5 antecedents to 2 conclusions, and N varies from only a few rules to as many as 20 or 30.

To keep the present discussion simple, we restrict ourselves to the case of 1 antecedent and 1 conclusion, and use diagrams in the explanation. In this case relation (15) can be written

$$\{\text{IF}\, A \text{ is } A_i, \text{ THEN } B \text{ is } B_i\}_{i=1}^{N} \tag{16}$$

If we let X be the antecedent universe of discourse and Y the conclusion universe of discourse (for example X is the speed of an object and Y the estimated distance that it moves), the relation between the antecedent A_i and conclusion B_i can be described using fuzzy sets in terms of their membership functions, as in the upper part of Fig. 1.5. The values for different elements of X and Y having different subscripts can then be formed into a matrix. This matrix is prepared in advance as a knowledge base using knowledge obtained from an expert. In actual use, observed data A' are input as they are obtained. If sensor calibration is also considered, A' is input in the form of a thin bell as shown in Fig. 1.5. Next, each A_i is evaluated for A'; in general, the evaluations are approximate, as shown in Fig. 1.5. The matching degree is interpreted to be the maximum value α_i of overlap of A' and A_i; if we base the conclusion B_i on this value, then the i th rule gives the response B' with respect to A', as shown in Fig. 1.5. If the individual B_i' are then unified in parallel through maximum operation, the final conclusion B' in the lower right of Fig. 1.5 is obtained. That is to say, the conclusion B' from input information A' is output based on the N rules. This series of procedures by which B' is obtained from A' is called 1 fuzzy inference. In actual application it is necessary to limit the operation value from B' to 1 point; as shown at the lower right in Fig. 1.5, the "CG method" using the center of gravity of the membership function of B' is employed.

The series of procedures in Fig. 1.5 can be expressed by the following equations:

$$\alpha_i = \bigvee_{x \in X} \{m_{A'}(x) \wedge m_{A_i'}(x)\} \tag{17}$$

$$m_{B_{i'}}(y) = \alpha_i \wedge m_{B_{i'}}(y) \tag{18}$$

$$m_{B'}(y) = \bigvee_{i=1}^{N} m_{B_{i'}}(y) \tag{19}$$

$$CG = \int_Y y \cdot m_{B'}(y) dy / \int_Y m_{B'}(y) dy \tag{20}$$

and then programmed.

In actual application, CG is found from A' and the procedure executed, then CG is found from the next A' and so on. In the Sendai municipal subway, 10 fuzzy inferences per second (10 flips) are carried out for 24 rules using 1 8-bit microprocessor, and automatic operation is then performed based on

the anticipated condition of the train 3 seconds later. For applications in
which even faster operation is required (for example in control of airplanes
and rockets), VLSI chips called "fuzzy inference AI chips" which are capable
of several hundred thousand to 10 million flips have been developed, and are
starting to appear on the market.

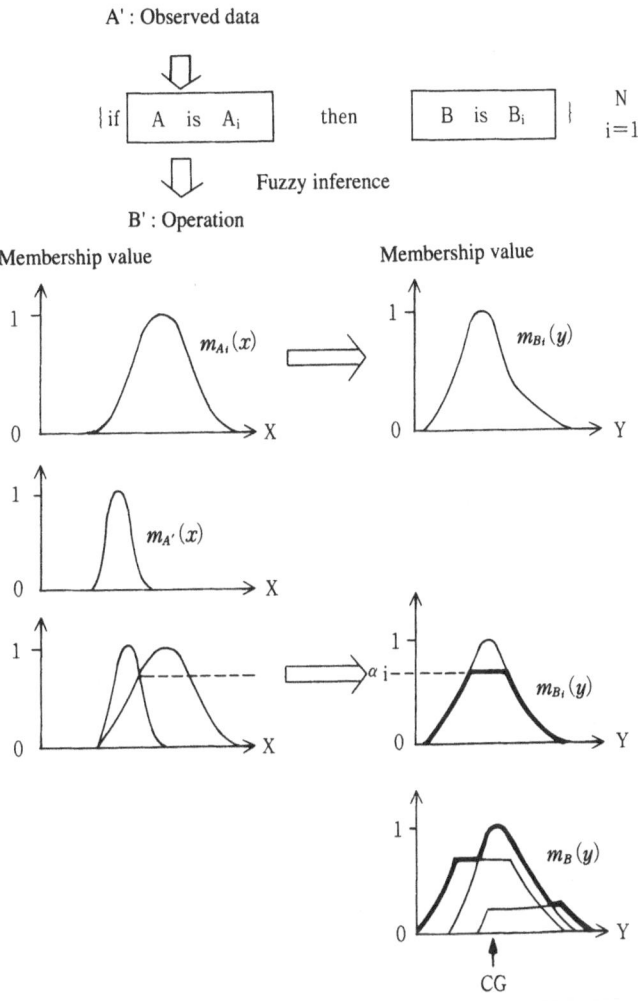

Fig. 1.5 Fuzzy inference using a fuzzy production rule.

What is shown in Fig. 1.5 is called max-min composition (refer to equations (17) to (19)) fuzzy inference, which is the standard method in use today,
but of course there are many variations (according to one estimate over 100).
In any case, fuzzy inference is now widely used as a means of saving labor in
high technology.

1.4 Conclusion

The basic concepts of fuzzy theory have been introduced. The explanation has centered on fuzzy sets and fuzzy inference using rules, hopefully providing the reader with an adequate minimum of knowledge needed to understand the applications to be discussed in succeeding chapters. The reader who wishes to pursue the study of theory further is referred to references 4 and 5.

References

1. Zadeh LA (1965) Fuzzy sets. Int J Information Control 8:338–353

2. Mamdani EH, Assilion S (1974) An experiment in linguistic synthesis with a fuzzy logic controller. Int J Man-Machine Stud 7:1–13

3. Holmblad LP, Ostergaard JJ (1982) Control of cement kiln by fuzzy logic. In: Gupta MM, Sarchez E (eds) Fuzzy information and decision processes NorthHolland pp389–399

4. Terano T, Asai K, Sugeno M (eds) (1989) Introduction to applied fuzzy systems chapter 2 (by Hirota K) (in Japanese). Ohm Sha, pp9–48

5. Hirota K (ed) (1989) Fuzzy CAI (Fuzzy Expert System) (CAI learned on a PC98 series or FMR series personal computer) (in Japanese). Corona Sha

Chapter. 2

ERIC — A Shell for Real-time Process Control

2.1 Background

When performing operation in an actual process, it becomes necessary to deal with complicated characteristics and environmental changes that cannot be described by a numerical model. For this reason, in addition to conventional control methods, it becomes indispensable to perform operations in a way that makes use of an operator's experience[1,2]. Expert systems and fuzzy control systems are now attracting attention as ways to computerize such control operations.

In order to create such an intelligent control system, a system that performs data processing that is typical of all of the control operations is developed; then, by giving instructions that are characteristic of each process with respect to this system, it becomes possible to reduce the software development cost and improve the ease of system maintenance.

We have developed ERIC (Extended Rule-based system for Intelligent Control) as a tool for the purpose of investigating knowledgeable control methods and incorporating them into real-time computerized control systems[9,10]. ERIC is an expert shell for use in process control, based on a working memory with data in frame format and an if...then...-format rule base.

ERIC's rule base has a framework that combines a boolean logic format with a fuzzy logic format. It does not merely consist of descriptions of an operator's linguistic operations as fuzzy control rules, but fuzzy control rules can be structured using boolean logic rules. This makes it possible to combine fuzzy control rules to permit flexible processing, with for example switching among control rules according to the control operation being performed and the operating conditions.

ERIC has been designed with consideration to its suitability for installation in a real-time computer system and for use in real-time processing. ERIC can be installed in a variety of computers and used for real-time control.

In this chapter, ERIC will be explained with emphasis on fuzzy processing.

2.2 The design of ERIC

In developing ERIC, design emphasized 2 points: highly sophisticated functionality for flexible processing in control operations, and suitability for use in process control computers. The principal items dealt with in this development were the following.

1. Information processing necessary in control operations

   a) Logical processing, for example evaluation of situations
   b) Quantitative evaluation for operations
   c) Real-time processing

2. Use in control computers

   a) Real-time computer processing
   b) Portability in a process computer
   c) Memory capacity limits

2.2.1   Information processing for control operations

In order to describe control operations, such as those performed by an operator, as a "rulebase", it is important for design to consider the type and structure of information processing[1].

When thinking of process operations, evaluation of whether the present process status is satisfactory or not and decision making to select an operation method based on the present status can be thought of as a type of logical processing. In contrast, processing to evaluate the process status and to determine operation quantities is quantitative evaluation related to process status quantities.

In order to perform these 2 types of information processing, in ERIC the rule base uses both boolean logic rules (logic judgments) and fuzzy rules (quantitative judgments) together. Numerical operation descriptions are also included in both rule descriptions, so that both linguistic quantitative evaluation and rigorous numerical evaluation can be used as appropriate.

When considering a total operation, the number of states that can occur in a process is extremely large; in order for computer processing to be done efficiently, the rule base must be structured. In ERIC, the descriptions of judgments in various situations are modularized; this structuring makes it possible for various process states to be recognized. This modularization is in the form of a tree structure that makes it possible to deal with anything from gross judgments on the overall status of the process to specific control operations. These modules are described as sets of if...then...rules, called "rulesets"; the rule base is structured through ruleset-calling relationships.

In addition, in order to permit complicated status judgments to be made, it is important to also use a mental model of the process to perform inference processing. Therefore, in the frame data used to express process devices in module form, slots that describe the connections between frames are added to the general slot variables, making it easy to create a model based on the device topology of the devices being controlled. This permits conditions to be judged and control operations to be performed based on the process component structure.

2.2.2 Installation in a real-time computer system

When considering installation in a computer control system using a real-time operating system, the constraints on computer resources must be considered[6,7].

1. Constraint on amount of memory that can be used
2. Real-time computer processing
3. Interface with sensor information

Languages such as lisp and prolog are well suited for describing logical processing, but they have the defects that they require a great deal of memory and that computing must be stopped for "garbage collection". To increase the speed of inference and reduce the amount of memory required, ERIC was developed in the C language. This makes it easy to install ERIC in a computer that is supported by the C compiler. In order to handle large quantities of data such as process information, ERIC is provided with input/output tables that use shared memory, to facilitate exchange of information between the working memory and the outside world.

When considering the operation format of ERIC as a control shell, the working environment in the knowledge base development stage, including verification and maintenance, is important. Considering such ease of operation, inference processing is performed in an interpreter format in which knowledge base processing is performed in a readable form. (Here the term "interpreter format" is used in contrast to an approach in which execution modules are created by compiling rule descriptions.) For this reason, ERIC provides the following advantages in operation.

1. The entire knowledge base is described in character text; corrections and maintenance are easy.

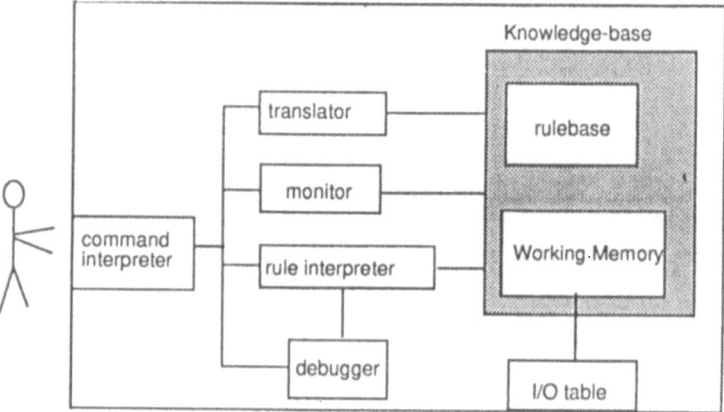

Fig. 2.1   Internal composition of ERIC.

2. Only those parts which are necessary are re-input when correcting the knowledge base.
3. Additional tools such as a compiler are not necessary when correcting the knowledge base.

In addition, ERIC is structured so that all of its operations are performed through an internal command interpreter; the command interpreter operates with input and output based on a character stream, so that the control and diagnosis system can be structured to match a wide variety of operating environments, from character terminals to engineering work stations.

2.3   Internal composition of the shell

ERIC internally consists of the following modules, which correspond to the various types of processing which it performs.

1. Command interpreter
2. Translator
3. Monitor
4. Rule interpreter
5. Rule debugger

These modules are constituted so as to be able to cope with the full range of work stages, from knowledge base development through execution. They have been designed to be able to operate independently of one another. It is also possible for the different parts of ERIC to be divided up and installed in a way that matches the computer environment.

The command interpreter is the interface between ERIC and the user; it initiates action of the other modules in accordance with commands from a keyboard or other input device. The input operation consists merely of reading in a character string; input can be read in from a command description file, by FIFO, by message communication, etc. For this reason, it is easy to link a user interface to ERIC by communication between processes, making it possible to construct a system that matches the computer environment.

A knowledge base description in the form of character text is read into ERIC and converted to internal data structure by the translator. Working memory variables in the rule base description are converted to pointers, flags, etc. in the conversion process, which speeds up the execution during inference. These conversions can be performed at arbitrary times for each working memory type and rule set, making it easy to correct the data base.

The monitor is used for display of the data base presently being read into ERIC and for performing settings. It is also possible to create a data file by redirecting this output to a file.

The rule interpreter consists of 2 inference engines, one for production rules and one for fuzzy use. The system automatically switches between these engines, depending on the rule set, so they can be viewed as a single inference engine that has 2 different functions.

When rule base action is verified, step execution and break point operation can be performed by operating the debugger at the same time. This is specified by means of an environment variable or a command operation parameter. During step execution, every time the rule condition section makes a judgment, when break point is specified, after the corresponding rule condition judgment is made the rule interpreter processing is interrupted, and the command interpreter is called. This permits working memory values to be checked and adjusted via monitor commands during rule execution. In addition, it is also possible for the rule interpretation process to be output onto the screen so that the rule logic can be checked.

2.4 ERIC's knowledge expressions

2.4.1 Working memory

The working memory describes the various devices being controlled in frame format, to make it easy to describe the total structure of the plant being controlled or diagnosed. This frame data structure is described as a working memory type for each device; then working memory elements to describe the actual data are created based on that format. For example, in describing a piping system such as the one shown in Fig. 2. 2, the basic structure of pipes, valves, pumps, etc. is defined as a working memory type.

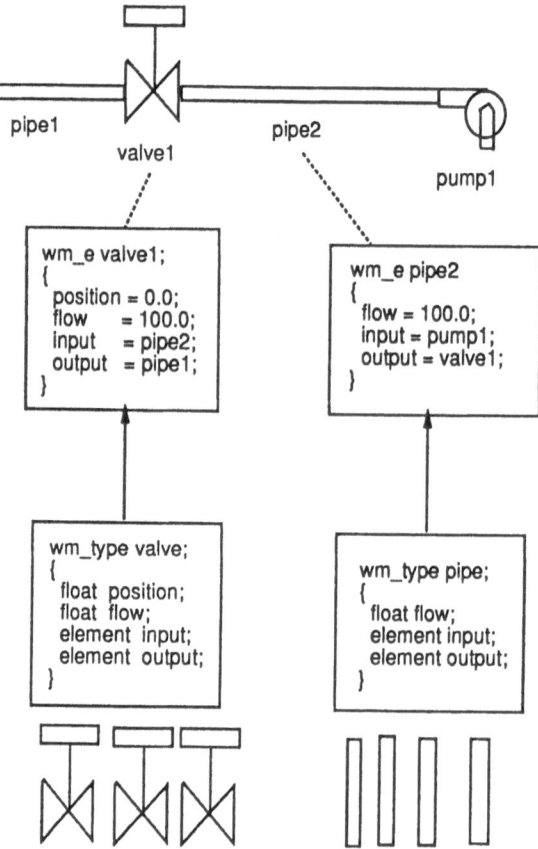

Fig. 2.2 Example of descriptions in working memory.

Based on this type definition, the various devices which comprise the process are described as working memory elements; then the system can be described by setting the connecting structure among devices as data.

This frame structure is described in the following format:

wm_type [*type name*] ;

{

      [*variable type specifier*] [*attribute variable name*] ;

      [*variable type specifier*] [*number of attributes name*] ;

      . . .

}

The 5 types of [variable type specifier] which can be set are given in table 2.1; they can also be set in a matrix.

Table 2.1 Types of variables

int	Integer
float	Floating point
f_string	Fixed length character string
strings	Character string
element	Pointer to element in working memory

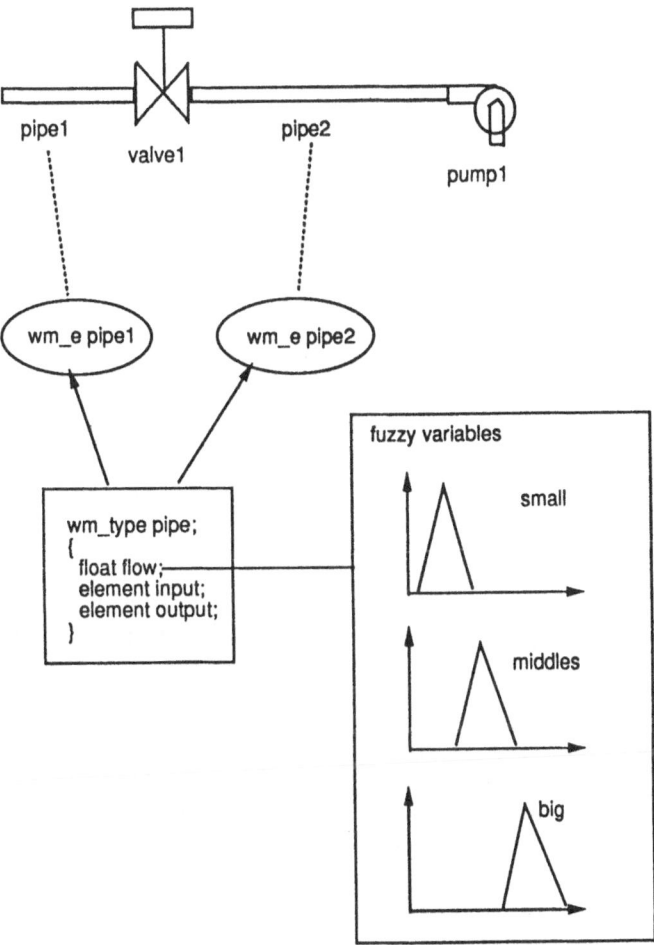

Fig. 2.3 Descriptions of fuzzy variables.

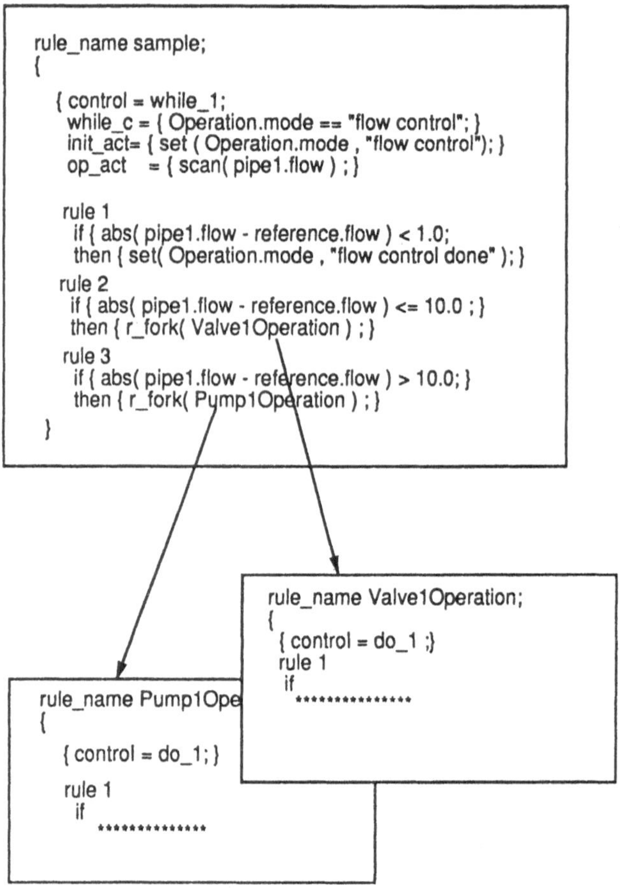

```
rule_name sample;
{

   { control = while_1;
     while_c = { Operation.mode == "flow control"; }
     init_act= { set ( Operation.mode , "flow control"); }
     op_act  = { scan( pipe1.flow ) ; }

   rule 1
     if { abs( pipe1.flow - reference.flow ) < 1.0;
        then { set( Operation.mode , "flow control done" ); }
   rule 2
     if { abs( pipe1.flow - reference.flow ) <= 10.0 ; }
     then { r_fork( Valve1Operation ) ; }
   rule 3
     if { abs( pipe1.flow - reference.flow ) > 10.0; }
     then { r_fork( Pump1Operation ) ; }

}
```

```
rule_name Valve1Operation;
{
   { control = do_1 ;}
   rule 1
     if ••••••••••••••
```

```
rule_name Pump1Ope
{
   { control = do_1; }
   rule 1
     if ••••••••••••••
```

Fig. 2.4   Description of a rule set.

The last "element" (used as key word in ERIC) is set for the purpose of expressing the device topology of the device composition, etc. of the process being controlled; the process can be modeled with connecting structure data in working memory.

Fuzzy membership functions used in fuzzy inference are defined by fuzzy variables with respect to the working memory type attribute variables (Fig. 2.3).

By using membership functions in common for one type of attribute, the treatment of fuzzy variables can be modularized. Membership functions are described by line segments, for ease of description and tunability.

2.4.2 Rule sets

Using the above process model description using working memory, the logic judgments and variable manipulations for operations are described by a rule base in "if⋯ then⋯ " format. When handling a large-scale process and when describing a wide range of operations, it is necessary to consider a very large number of conditions, and a large-scale rule base must be developed. Considering the efficiency of developing such a rule base and its ease of maintenance, in ERIC not necessary rules that deal with one situation are described in modules; the rule base is structured in units of rule sets.

For example, a set of device operations in which the operation to be performed differs depending on conditions can be described as in Fig. 2. 4.

The rule set "Sample" switches from one operation to another depending on error magnitude; specific device operations are described by "Pump1Operation" and "Valve1Operation" which are called depending on the value of r_fork( ). The respective rule sets can be described independently; rule sets can be freely added and corrected via the translator.

One rule set is described in a format such as the following:

rule_name [*rule set name*] ;

{

    {

        control = [*rule execution method*]

        while_c = [*repeat condition*] ;

        init_act = [*initialization procedure*] ;

        op_act = [*updating procedure*]

    }

    if

        [*condition that initiates use of rule*]

    then

        [*executed procedure*]

    . . .

}

A selection is made from among mutually independent inference procedures.

"Rule execution selection" selects the rule initiation procedure; the 5 descriptions in Table 2. 2 are available.

Table 2.2 Firing rule execution options

do_1	Execute 1 succeeded/matched rule.
do_all	Execute all succeeded/matched rules.
while_1	Execute 1 succeeded/matched rule repeatedly.
while_all	Execute all succeeded/matched rules repeatedly.
order	Execute succeeded/matched rules in the order of rule description.
fuzzy	Execute fuzzy inference.

These can be specified for each rule set; a selection can be made corresponding to the rule. The 5 procedures from do 1 to order specify processing with respect to the production rule; "fuzzy" specifies fuzzy inference execution. In ERIC, switching between production rules and fuzzy rules is done only with this specification; the inference processing is automatically switched according to this description at the time when rule set processing starts (for example on "r_fork()").

In the example in Fig. 2. 4, "while_1" is specified; rule set execution is repeated until the value of "Operation.mode" specified by "while_c" is other than "flow control". The value of "Operation.mode" is set to "flow control" in the initialization procedure "init_act"; it is rewritten by rule 1 when the error becomes sufficiently small. Also, the value of "pipe1.flow" is read in as process data in the initialization procedure "op_act" using "scan()"; it is updated every time the rule processing is repeated.

The "rule initiation condition" and "repeat condition" are described as the logical sum or logical product of logical expressions. These logical expressions compare variables in working memory and are described in the form [variable] [comparison operator] [variable]. Each variable can be operated on by a working memory variable or constant on either side; using these the results of arithmetic operations and arithmetic functions are described (Table 2. 3).

Table 2.3 Comparison operators

$==, !=, >, >=, <, <=$	Comparison of variable values
is, is_not	Fuzzy logical sum
$->$	Substitution of intermediate variable

This makes it possible to describe standardization, data processing, etc. in the rule base. In the example in Fig. 2. 4, abs() is used to compare absolute values.

"Execution procedure" and "updating procedure" describe the operations to be performed when a certain condition is fulfilled, or on each repetition. The principal descriptions used are given in Table 2. 4.

Table 2.4   Execution procedures

set()	Substitution of working memory variable
scan()	Input from input/output table
out()	Output to input/output table
r_fork()	Call of rule set
fset()	Fuzzy inference operation part

Process input and output are performed by scan() and out() functions through the input/output table. This prevents values in working memory from being accidentally changed during rule set processing, and makes it possible for input/output management to be explicitly described by rules.

The "variables" that are available are given in Table 2. 5.

By combining these, it is possible to describe variable specifications corresponding to complicated pattern matching and data structures with respect to variables.

Table 2.5   Variable descriptions in rules

Direct reference description	Working memory variable value
Constant	Value such as number or character string
Indirect reference description	Variable used for pattern matching
Temporary variable description	Intermediate variable
Fuzzy variable description	Fuzzy variable reference

ERIC is constructed so as to be able to deal explicitly with mixtures of production rule sets and fuzzy rule sets. For example, a fuzzy operation during production rule processing is generally a contradiction and causes an error, but logical conditions which come up during fuzzy inference are processed in boolean form. In addition, normally calling a production rule set from a fuzzy rule set is an error, but a fuzzy rule set can be called from a fuzzy rule set. (Boolean can be called fuzzy but fuzzy can not be always called boolean.) In this case, the conclusion of the fuzzy rule set that is called is processed as a fuzzy proposition of the rule that called it.

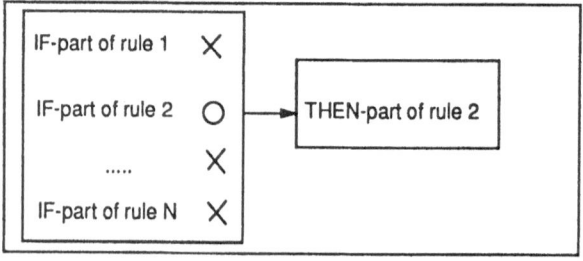

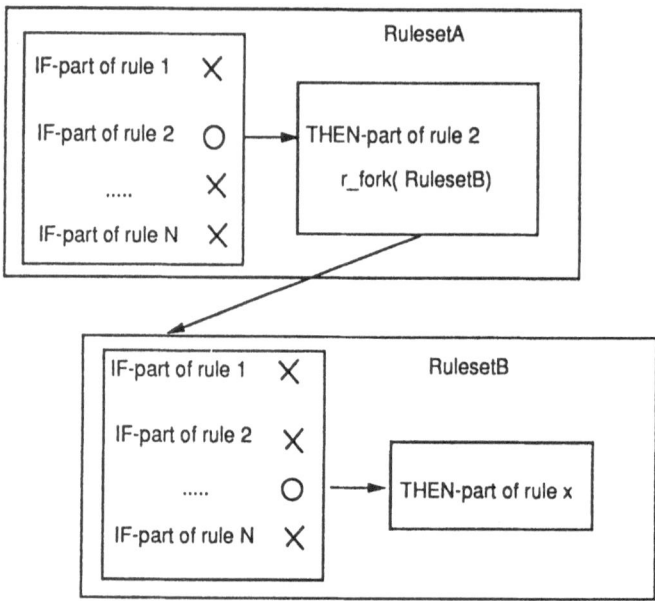

Fig. 2.5 Boolean logic rule processing.

2.5 An overview of ERIC inference processing

2.5.1 Rule setting processing

Rule interpreter inference action is based on repeated processing of inference processing units of the form [condition section judgment] implies [conclusion section judgment] for each rule set, as shown in Fig. 2. 5. For this reason, when hierarchical processing of rule sets is performed, there arise situations in which a condition judgment for a higher ranking rule set is incompatible with execution processing of a lower ranking rule set. However, in order to maintain the compatibility of condition judgments in the rule set hierarchical structure during execution, it is necessary to, for example, constantly be checking the condition judgments up to the present; and in some

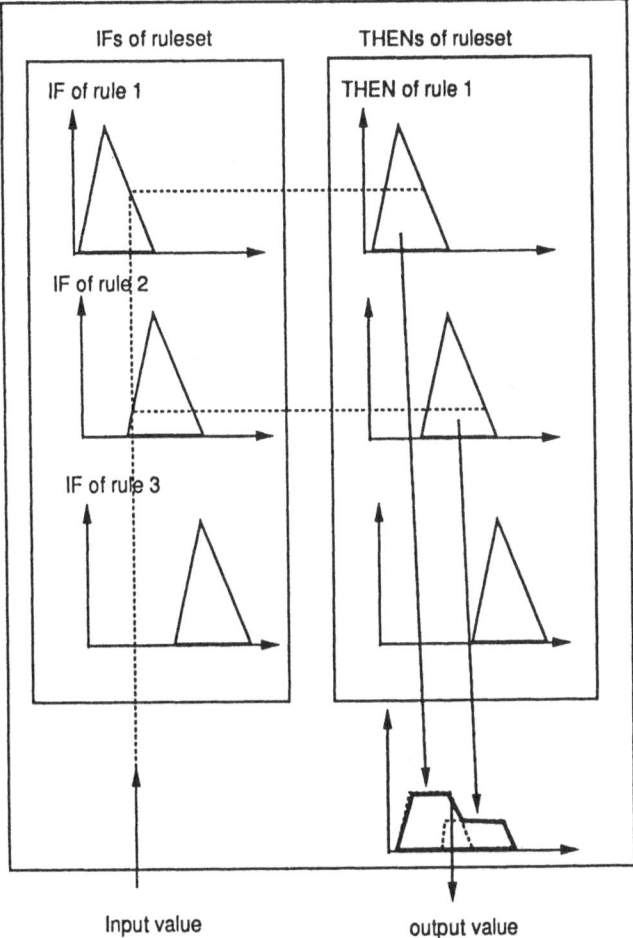

Fig. 2.6 Fuzzy rule processing.

cases it might be necessary to restore a variable to its former value. In ERIC, processing speed is emphasized, so it was decided to perform rule condition judgment processing based on the status of working memory at the time the rule set is started.

In addition, in fuzzy inference, as well, in place of performing condition section judgments, production rules are processed in basically the same way except when a grade value is evaluated (Fig. 2. 6). This inference processing is for the purpose of deriving control operation quantities, so fuzzy inference is performed using the process status quantities set in the working memory as input; then the membership function resulting from the inference is converted to numerical values and the processing set in the WM variable is consistently executed.

In this processing, fuzzy inference (including defuzzying operations) input and output are limited to crisp values. Therefore, fuzzy rule sets can be processed as special functions called from boolean logic rules.

2.6 Fuzzy processing in ERIC

2.6.1 Fuzzy logic, which is suited for process control

Membership function expressions

A fuzzy set X which expresses vague quantities such as "large" and "small" is expressed by the membership functions $\mu_X(x)$:

$$\mu_X \rightarrow [0, 1] \tag{1}$$

$\mu_X(x)$ is a function which expresses the degree to which the elements x belong to the set. A fuzzy set, like an ordinary set, can assume a variety of forms such as a continuous set or a discrete set. When dealing with process control, it is necessary for fuzzy inference to process continuous fuzzy sets in order to express continuous quantities, such as process quantities and operation quantities. At this time, the following items relating to the method of expression of $\mu_X(x)$ must be considered from the point if view of suitability for installation in a control device.

1. Support of the fuzzy set
2. The shape of the membership function
3. The membership function description precision
4. The ease of fuzzy logic operations

For example, when expressing water temperature in daily life, the range from 0 to $100°C$:centigrade $\leftrightarrow °F$ is sufficient. However, depending on the purpose of water use, there are a variety of linguistic expressions for water temperature. For example, the term "hot" can mean different things depending on whether the water is to be used for bathing or cooking; a variety of applicable temperature ranges can be conceived of. In addition, it is necessary to express membership functions in a variety of forms, and it is necessary to describe the precision required for the control objective.

To directly describe linguistic expressions relating to process operations, and to make it easy to correct the membership functions, we have decided to use the following line segment expressions.

$$Array\text{-}\mu : [(x_1, y_1), (x_2, y_2), \cdots, (x_n, y_n)] \tag{2}$$

x_i : Value of an element of a set
y_i : Grade value corresponding to x_i
(x_i, y_i) : i'th break point
n : Total number of break points

$$\mu(x) = \begin{cases} y_1 & , x < x_1 \\ y_i + \dfrac{(x - x_i)(y_{i+1} - y_i)}{x_{i+1} - x_i} & , x_i \leq x \leq x_{i+1} \\ y_n & , x \geq x_n \end{cases} \tag{3}$$

When this method of expression is used, sets in arbitrary ranges can be described, and operations can be freely performed on membership functions by adding, moving and deleting break points.

Fuzzy relationships

Fuzzy inferences are calculated using fuzzy relationships R which express fuzzy propositions:

$$Y = X \bigcirc R_{xy} \tag{4}$$

X : A fuzzy set relating to x
Y : A fuzzy set relating to y
R_{xy} : A fuzzy relation from x to y
$\bigcirc$: Fuzzy composition

In the case of fuzzy inference using IF-THEN rules, which are often used as fuzzy control rules, a rule such as:

$$\text{rule: IF } x \text{ is } A_1, \text{ THEN } y \text{ is } B_1 \tag{5}$$

corresponds to the fuzzy relation R_1 from A_1 to B_1. The fuzzy relationship for all of the rules corresponds to the sum of the fuzzy relationships for the individual rules, expressing a single fuzzy relationship such as the following:

$$R = R_1 \cap R_2 \cdots R_m \tag{6}$$

Here R_{rulek} is the fuzzy relationship obtained from the fuzzy proposition expressed by the individual IF-THEN rules:

$$R_k = A_k \times B_k \tag{7}$$

where:

$$\mu_{R_k}(x, y) = \mu_{A_k} \wedge \mu_{B_K} \tag{8}$$

μ_{A_k} : Membership function for A_k
μ_{B_k} : Membership function for B_k
μ_{R_k} : Membership function for R_k

Fuzzy inference with respect to non-fuzzy quantities

In process control, signals, for example from sensors, are non-fuzzy quantities such as integers and real numbers. When fuzzy inference is performed based on a numerical value x_0, μ_{x_0} is treated as a membership function such as the following.

$$\mu_{x_0} = \begin{cases} 1 & , x = x_0 \\ 0 & , x \neq x_0 \end{cases} \tag{9}$$

At this time, if the product between a fuzzy set and a non-fuzzy quantity is to be found using operations such as min and algebraic products, we have:

$$\begin{aligned} \mu(x) &= \vee \mu_{x_0}(x) \wedge \mu(x) \tag{10} \\ &= \mu(x_0) \\ &= g(= constant) \end{aligned}$$

Then the fuzzy inference using R_{rulek} can be transformed into a form such as:

$$\begin{aligned} \mu_k(y) &= \mu_{x_0}(x) \bigcirc R_k(x, y) \\ &= \mu_{x_0}(x) \bigcirc (\vee \mu_{A_k}(x) \wedge \mu_{B_k}(y)) \\ &= \vee(\mu_{x_0}(x) \wedge (\vee \mu_{A_k}(x) \wedge \mu_{B_k}(y))) \\ &= \vee(\mu_{A_k}(x_0) \wedge \mu_{B_k}(y)) \\ &= g_k \cdot \mu_{B_k}(y) \tag{11} \end{aligned}$$

so that the processing can be performed simply as scalar products of fuzzy sets. Similarly, a fuzzy inference using the fuzzy relation R expressed IF-THEN rules can be calculated in the form:

$$\begin{aligned} \mu_k(y) &= \mu_{x_0}(x) \bigcirc R(x, y) \\ &= (\mu_{x_0}(x) \bigcirc R_1(x, y) \\ &\quad \vee(\mu_{x_0}(x) \bigcirc R_2(x, y) \\ &\quad \cdots \\ &\quad \vee(\mu_{x_0}(x) \bigcirc R_n(x, y) \\ &= g_1 \cdot \mu_{B_1}(y) \\ &\quad \vee g_2 \cdot \mu_{B_2}(y) \\ &\quad \cdots \\ &\quad \vee g_n \cdot \mu_{B_n}(y) \tag{12} \end{aligned}$$

Fuzzy inference processing with multiple input/output

We can conceive of a general form for a fuzzy rule such as: Rule K

$$\text{IF } X_1 \text{ is } A_{1k} \text{ and } x_2 \text{ is } A_{2k} \cdots X_n \text{ is } A_{nk}$$

$$\text{THEN } y_1 \text{ is } B_{1k} \text{ and } y_2 \text{ is } B_{2k} \cdots y_m \text{ is } B_{mk}$$

describing a logical relationship between n preconditions and m postconditions. At this time, if $x_1, x_2, \cdots, x_n$ and $y_1, y_2, \cdots, y_m$ are independent variables, respectively, this rule can be thought of as the following $m \times n$ fuzzy relationship.

$$R = (A_{1k} \times A_{2k} \cdots A_{nk})$$
$$\times (B_{1k} \times B_{2k} \cdots B_{mk}) \tag{13}$$

$$\mu_R(x_1, x_2, \cdots, x_n, y_1, y_2, \cdots, y_m)$$
$$= \mu_{A_{1k}}(x_1) \wedge \mu_{A_{2k}}(x_2) \cdots \wedge \mu_{A_{nk}}(x_n)$$
$$\wedge \mu_{B_{1k}}(y_1) \wedge \mu_{B_{2k}}(y_2) \cdots \wedge \mu_{B_{mk}}(y_m) \tag{14}$$

If, in addition, $x_i(i = 1, 2, \cdots, n)$ and $y_j(j = 1, 2, \cdots, n)$ are continuous fuzzy sets, the fuzzy inference becomes very complicated, but if the x_i are non-fuzzy quantities, then simple processing becomes possible. The fuzzy inference results for the non-fuzzy input quantities $(x'_1, x'_2, \cdots, x'_n)$ are, as in the case of 1 input and 1 output, given by:

$$\mu_k(y_1, y_2, \cdots, y_m)$$
$$= \mu_{A_{1k}}(x'_1) \wedge \mu_{A_{2k}}(x'_2) \cdots \wedge \mu_{A_{nk}}(x'_n)$$
$$\wedge \mu_{B_{1k}}(y_1) \wedge \mu_{B_{2k}}(y_2) \cdots \wedge \mu_{B_{mk}}(y_m) \tag{15}$$

This equation can be transformed into:

$$\mu_k(y_1, y_2, \cdots, y_m) = (gk \cdot \mu_{B_{1k}}(y_1)) \wedge (gk \cdot \mu_{B_{2k}}(y_2))$$
$$\cdots \wedge (g_k \cdot \mu_{B_{mk}}(y_m)) \tag{16}$$

and:

$$g_k = \mu_{A_{1k}}(x'_1) \wedge \mu_{A_{2k}}(x'_2) \cdots \wedge \mu_{A_{nk}}(x'_n) \tag{17}$$

Since the y_i are mutually independent, the membership function for y_i can be given as $g_k \cdot \mu_{B_{i'k}}(y_i)$. This permits processing of the above equation in inference based on 2 or more fuzzy rules to be processed as one logical sum for each y_i, as follows:

$$\mu(y_i) = g_1 \mu_{Bi1}(y_i) + g_2 \mu_{Bi2}(y_i) \cdots + g_L \mu_{BiL}(y_i) \tag{18}$$

Defuzzification

The result of fuzzy inference is given as a fuzzy variable or variables; defuzzification is necessary before it (they) can be used in control operation. In ordinary fuzzy control, the areal bisecting point, peak point and center of gravity point in relation to the membership function are used.

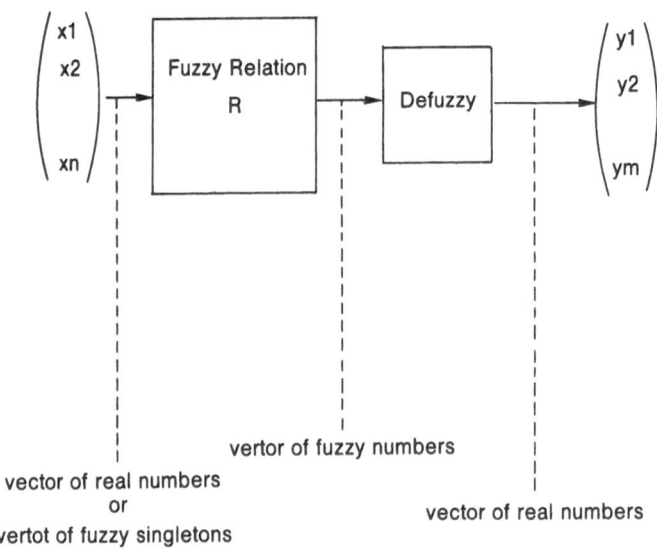

$$\begin{pmatrix} x1 \\ x2 \\ \\ xn \end{pmatrix}$$ → Fuzzy Relation R → Defuzzy → $$\begin{pmatrix} y1 \\ y2 \\ \\ ym \end{pmatrix}$$

vertor of fuzzy numbers

vector of real numbers
or
vertot of fuzzy singletons

vector of real numbers

Fig. 2.7 Flow of fuzzy inference.

When thinking about fuzzy control, fuzzy inference based on non-fuzzy quantities and defuzzification of an inference result can be thought of as a single processing operation. The fuzzy rule can be thought of as an $n \times m$-dimensional function from the vector $(x_1, x_2, \cdots, x_n)$ to $(y_1, y_2, \cdots, y_m)$. The defuzzification operation can be thought of as (Fig. 2. 7):

$$
\begin{aligned}
y_i &= f_d(\mu_{A1}(x_1') \cdot \mu_{A2}(x_2') \cdots \mu_{An}(x_n') \cdot \mu_{Bi}(y_i) \\
i &= 1, 2, \cdots, m
\end{aligned}
\tag{19}
$$

2.6.2 Fuzzy inference processing in ERIC

The fuzzy inference processing system in ERIC has been constituted based on the items explained above. The variables and fuzzy sets used in inference are all organized as the working memory, and fuzzy rule sets are described based on it. These fuzzy rule sets describe the procedures which perform fuzzy inference and defuzzification; the fuzzy calculation methods and the defuzzification method can be specified as fuzzy processing conditions. Fuzzy inference is processed according to the following procedure.

1. The grade values in the precondition section of each rule are evaluated.
2. The logical products of the grade values of each rule and the postconditions are taken, then the logical sums of the results are calculated for each y_i.
3. Defuzzification processing is performed for each y_i.

In the processing of preconditions, it is sufficient to calculate only the grade values of the preconditions. For this reason, the number of precondition section propositions can be arbitrarily chosen. For a rule which does not have any propositions relating to x_i that are used in other rules, by taking $\mu_{A_{ik}}$ to be 1.0 it is possible to use a different number of x_i with each rule. As in the case of the preconditions, by taking $\mu_{B_{jk}}$ to be 0.9 similar processing becomes possible.

In addition, fuzzy inference with respect to the result if unit conversion, etc. is also possible; the fuzzy rule precondition section can then be described in a format such as:

$$x_i * 100.0 \text{ is } A_i \tag{20}$$

$$log(x_i) \text{ is } A_i \tag{21}$$

Coupling and adjustment of fuzzy inferences

ERIC processes inferences in units of 1 or more rules (rule sets). The input variables to the inference, and the output variables which are the results of the inference, are all variables in working memory, and can be used in common with arbitrary rule sets. Two or more fuzzy inference rules can be linked using variables in working memory, and a variety of fuzzy inference structures can be formed corresponding to the fuzzy inference execution order. For example, by processing 2 fuzzy relationships $R_1(x \rightarrow y)$ and $R_2(y \rightarrow z)$ described by fuzzy rule sets in that order, multi-stage fuzzy inference processing via the non-fuzzy operator $f_d()$ can be performed:

$$Z = f_d(f_d(X \bigcirc R_1) \bigcirc R_2) \tag{22}$$

Also, by storing the membership functions of 2 fuzzy inference results as variables and taking their logical sum, parallel fuzzy inference such as:

$$Y = X \bigcirc (R_1 \cap R_2) \tag{23}$$

can be performed. Such combination of fuzzy inferences is accomplished by starting a fuzzy rule set from a boolean logic rule. It is also easy to change the fuzzy inference which is started by a boolean logic rule according to conditions, making it possible to construct a variety of fuzzy inferences.

Fuzzy variables are also processed as working memory variables; substitution and operations can be performed just as with ordinary variables. The principal operations that are performed on fuzzy variables are operations on membership function break points and calculation of membership function characteristic values, as shown in Fig. 2. 8.

By combining these capabilities, it is possible to evaluate fuzzy inference results in detail. Also, by adjusting the membership functions used in inference using binary inference rules, flexible fuzzy inference adjustments are possible.

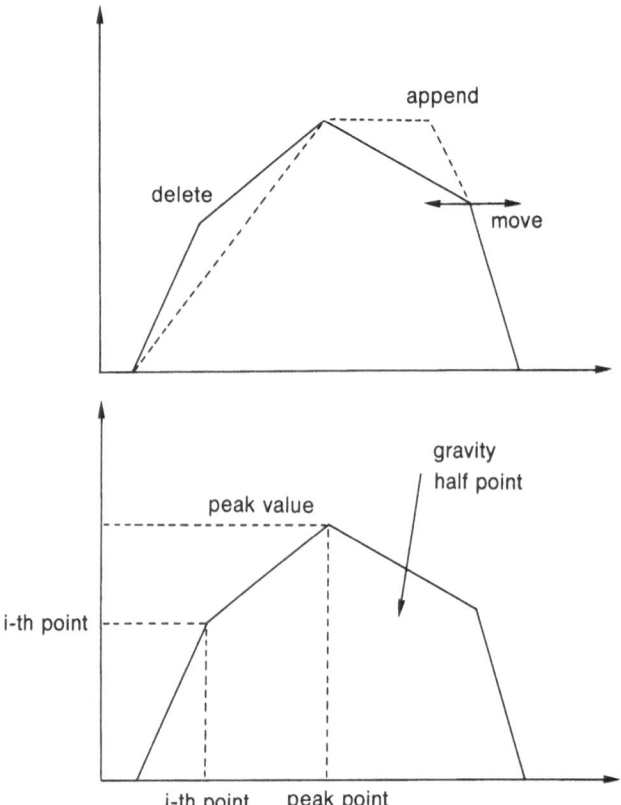

Fig. 2.8 Membership function handling.

2.7 Functions for real-time control

The status of a process changes with time in accordance with control operations, changes in environmental factors, etc. In order to process control operations on a computer, the quantities that indicate the process status must be constantly monitored and operation quantities determined. Such processing is a kind of sample value control algorithm; in order for stable control to be achieved, process input and output must be performed at fixed time intervals. In addition, to obtain smooth response those intervals must be sufficiently shorter than the time scale of the process dynamics. Thus, computational processing for control operations is required to be done in real time.

When an operator's knowledge-based operations are to be performed by a computer, many rule base controls and fuzzy controls using rules of the form "if...then " are required. Since the conditions that are dealt with by computerized processing are represented by continuous quantities, the calculations which are repeated at fixed intervals take the form "operation quantity

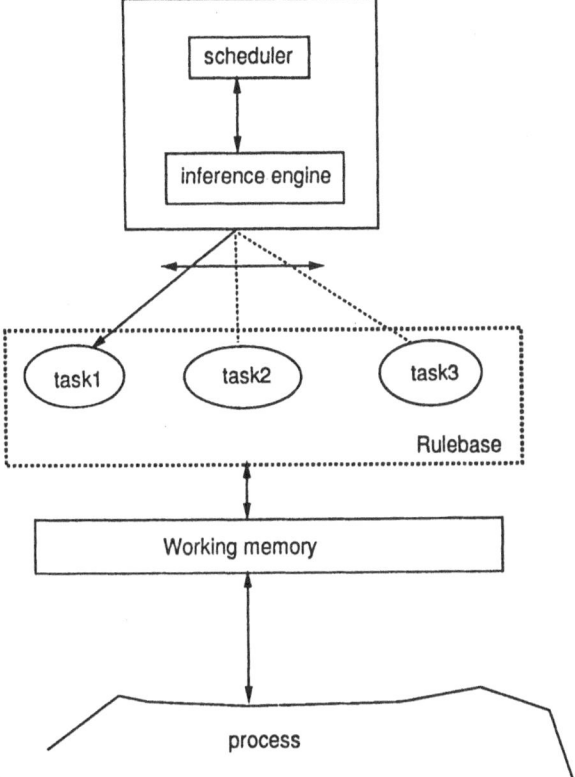

Fig. 2.9   Shell used for real time control.

calculated from status judgment". As a means of performing such real-time control processing, we have incorporated a scheduler into the inference engine (Fig. 2. 9). In this method, the calculation for each cycle is performed by the inference engine, and the scheduler keeps track of the cycles to make it possible to add the real-time capability to the processing of knowledge-intensive information. In addition, by processing rule sets in a form that can be done on a time-sharing basis, rule sets can be processed in parallel.

This method replaces a calculation algorithm such as sample value control by the entire inference processing based on process status quantities at certain times. This approach generalizes ordinary fuzzy control system processing to a rule base system.

Schedule information is set in the WM variables, so by manipulating the variable values in the WM elements which give schedule information, from any arbitrary rule an arbitrary rule set can be stopped, or the starting period or time limit can be changed. By using this capability, it is possible to describe a rule set that will do high priority processing in a concentrated manner in an emergency.

2.8   In conclusion

Since ERIC was designed with emphasis on expandability and transferability, it can be applied to a variety of control systems in combination with many different applications.

Applications of fuzzy control systems include applications to product manufacture such as the Fuzzy Control System [12] and fuzzy control of electrical discharge machining [13], and use as a tool in research [11, 14].

In addition, ERIC can also be used in fields other than fuzzy control by using its transferability and rule base modular structure. Applications which have been developed include a transformer facility maintenance support system[15], transformer station trouble diagnosis system[16] and signal interpretation system[17].

References

1. Rasmussen J (1983) Skill, rules and knowledge, signals, signs, and symbols and other distinctions in human performance models. IEEE Trans Syst Man Cybern SMC-13 : 257–266

2. Åström KJ, Anton JJ, Årzén K-E (1986) Expert control. Automatica 22 (3) : 277–285

3. Zadeh LA (1969) Fuzzy sets. Information and Control 8 : 338–353

4. Mamdani EH (1974) Application of fuzzy algorithms for control of simple dynamic plant. Proc IEEE 121 (12) : 1585–1588

5. Tong RM (1977) A Control engineering review of fuzzy systems. Automatica 13 : 559–569

6. Mitsumaki T, Kuwahara H (1986) Real-time processing technologies in process computers (in Japanese). Corona

7. Stankovic JA, Ramamritham K (1988) Hard real-time systems. IEEE Computer Society Press, pp1–37

8. Stefik MJ, Bell AG, Bobrow DG (1983) Rule-oriented programming in LOOPS. Knowledge-based VLSI D.G. Technical Memo KB-VLSI-82-22, Jan, 1983

9. Ishioka T, Takegaki M, Ohi T (1988) ERIC—the shell for industrial control system (in Japanese). Automation 33 (6) : 17–21

10. Takegaki M, Ishioka T (1990) Intelligent control system (in Japanese). Kaibundou

11. Ishioka T, Takegaki M (1989) Design of a knowledge-based model of process dynamics (in Japanese). Trans ISCIE 2 (1) : 1–9

12. Kobayashi K, Takagi M, Takeuchi Y, Takegaki M, Ishioka T (1989) Development tools for real-time fuzzy control system (in Japanese). Mitsubishi Denki Gihou 63 (3) : 248–251

13. Morita A, Imai S, Noda T, Maruyama T, Kobayashi K (1989) Fuzzy controller for EDM (in Japanese). Technical Report of EDM 13 (41) : 25–31

14. Maeda Y, Takegaki M (1988) Collision avoidance control among moving obstacles for a mobile robot on the fuzzy reasoning (in Japanese). J Robotics Soc Jpn. 6 (6) : 50–54

15. Hamanaka S, Fujimura H, Tada S, Sugio T, Shiota K (1989) Maintainance support system for power system (in Japanese). Mitsubishi Denki Gihou 63 (6) : 491–495

16. Goda T, Kyomoto S, Takegaki M (1990) A method for locating section by using sequential information at substraction (in Japanese). Trans IEE Jpn 110-B (6) : 475–484

17. Inushima H (1990) Signal interpretation system (in Japanese). Mitsubishi Denki Gihou 64 (5) : 432–435

Chapter. 3

MODEL BASE FUZZY INFERENCE

3.1 General concepts

3.1.1 Introduction

Intelligent Systems (referred to below as ISs) have created a need for new ways to solve the problems that arise in constructing them.

Not all systems that use artificial intelligence can be considered intelligent systems. Here we consider autonomous dispersive systems, and define an intelligent system as one that "achieves a human-like goal in a human-like way". To put it simply, it is a system that understands and can explain, that can "acquire knowledge" and "see through things", and has a high level of "ability".

At present there are many artificial intelligence paradigms including parallel processing; fuzzy inference is a basic technique to unify them.

Major problems in IS are how to respond to changes in the environment surrounding the system and how to balance conflicting requirements. More flexibility is demanded of systems today than ever before.

In FMS and CIM systems, it is necessary to constantly make changes in material inputs and load distributions in response to predictions of environmental conditions in order to achieve an overall objective. Similar systems exist outside of industrial fields. For example, in institutional investment, efforts are being made to develop systems that will go beyond qualitative rules to support decision-making based on market dynamics.

Today there is great demand for multi-purpose intelligent systems[1]. Intelligent systems blend into society and support human beings, smoothing the man-machine interface. Here we will discuss the means of constructing such systems through model-based fuzzy inference, and explain the new trends in system engineering (SE).

3.1.2 Multi-purpose systems and intelligence

An IS is required to take human desires (safety, economy, time to completion) into account and support human beings taking many things into account. In addition, the system is expected to cope with many kinds of changes in the situation that it faces, make accurate predictions and decide on appropriate action.

A multi-purpose system requires a separate evaluation function (such as an effectiveness function) for each condition that it has to deal with. These can be in subjective, non-linear or compound form. In order to "obtain" such knowledge with "insight", it is necessary to approximate concepts using a specialist's knowledge framework.

In modeling human know-how, starting with formulation of an evaluation function, it is necessary to have a grasp of the processing structure. A skilled worker's knowledge - based information processing can be analyzed to obtain a hierarchical structure. The knowledge and inference processes used in forming a hierarchy and in modeling are different for different intellectual levels, and a different approach is needed in each case.

A subsystem of a dispersed system has to be able to respond to situations that it encounters by changing its own schedule, solving the problem and reporting to a higher rank system. The higher rank system, in turn, must predict the load and make dynamic schedule changes, and apportion tasks among lower rank systems in pursuit of an overall multi-dimensional objective.

Thus, roles in the hierarchy are clearly defined. Let us use a self - propelled robot as an example to explain hierarchical processing. A self - propelled robot executes orders from a higher rank system while modifying its own motion, for example to avoid obstacles. Degrees of intelligence are determined for example by the way in which knowledge is described, inference methods, ways of acquiring and updating knowledge, types of missions performed and conversational means of explaining the inference process. Here we will judge based mainly on the instructions which a robot can carry out and the range within those instructions can be altered.

[Intellectual level]   [Instructions and changes]

Level 1 (reaction) :Travels accurately to the indicated position. Speed can
                    be varied.

Level 2 (Rule)   :Detects obstacles and takes whatever action is necessary
                    to avoid them.

Level 3 (Target)   :Judges the cause of an obstacle and decides on an alter-
                    nate course to reach the goal.

Level 4 (Objective):Anticipates obstacles and alters instructions accordingly.

Level 5 (Morality) :Decides on order of priority of objectives.

Looking at these, types of instructions and changes, it is clear that there are great differences among the levels of intelligence. If there is an overall system of which the above robot is a subsystem, the combined intelligence level of the two is also important. An IS can be a system which includes a human being. In such a case, it is essential that an adequate explanation be given to the person in charge who has overall responsibility for making decisions. In order that conversation will proceed smoothly, it is desirable to have a screen and/or voice that will adequately explain the situation, so that high level instructions can be given with symbolic manipulations. The requirements for such an intelligent man-machine system are as follows.

1. The system is a man - machine cooperative system.

   a. Understanding of instructions: Symbolic conversation
   b. Quality of explanation: Explanation of inference close to human thought process
   c. Judgment of situation: Grasp of overall situation and explanation of problems.

2. The system has a means of acquiring knowledge (easy input, learning).

   a. Knowledge processing: Using experience rules
   b. Adaptation capability: Subjective acquisition of on-the-job knowhow

3. Human objectives are achieved approximately the way that humans would achieve them.

   a. Overall plan: Balance among several objectives is determined.
   b. Decision making: Strategy and action are decided based on one's own thought.
   c. Coordinated work: Coordination among subsystems.

3.1.3 Human problem-solving processing

No matter how automated a system is, inevitably a human being becomes involved; the higher the intellectual level, the more numerous the occasions on which coordination with the human becomes necessary. In a man-machine cooperative system in which the machine processes information and prompts the human in high-level problem solving, a human interface through which the machine explains the anticipated problem and method of solution to the human in easily understood terms is required. The manner in which the human grasps problems is hierarchical; achieving interaction through which the overall situation can be grasped improves the overall performance.

In the field of cognitive engineering, J. Rasmussen has analyzed the plant

operator's information processing process, divided problem solving into 3
levels, and conceived models for the information processing flow at each level:
identification, decision and planning. His work can be briefly outlined as
follows:

[Intellectual level]   [Type of action]

1. Skill-based   : Reactive, correction of deviations

2. Rule-based   : Judge trend and plan optimum action

3. Knowledge-based : Analytically diagnose the basic problem.

The 3 levels of this human problem solving model correspond to the above
- mentioned intellectural levels 1, 2 and 3 of a robot. Taking the example
of an electrical generating plant, intellectual level 1 information processing is
to measure values with sensors and indicate whether or not there is danger
on control panel meters and lamps. In this case the human assumes the
intellectual level 2 and higher roles, grasps the situation, makes predictions
and gives instructions to the system. In an intellectual level 2 system, changes
of conditions are shown, for example by graphs, so the human can concentrate
on intellectual level 3, determining the cause. In a large plant such as a
nuclear power plant it is difficult to determine the cause of trouble. Systems
of intellectual level 3 are gradually being introduced so that human judgment
and operation errors will not cause major accidents;such systems support
human thought and operation.

3.1.4 IS intellectual levels

J. Rasmussen investigated the extent to which human objectives are focused; in a problem solving model, knowledge-based inference is used for the
highest level of processing. However, not all system operations can be determined from logical knowledge. The present author has been expanding a
problem solving processing model suitable for higher level IS processing, up
to the level on which human desires are taken into account. In order that
knowledge at the level of a specialist can be acquired, problem solving processing level 3 has been subdivided into 3 sublevels. The dotted line in Fig.
1.3 is the part that corresponds to J. Rasmussen's problem solving models.

The sublevel nearest to intellectual level 1 is a region that is closely
bound to social and natural phenomena. The closer to intellectual level 5,
the deeper the relation to human objectives and expectations; the capability
to understand one's own role becomes necessary. The intellectual levels are
subdivided into the following processing categories.

I (Identify)   : Observation, identification, grasp of conditions and prob-
                     lems

D (Decision)   : Thought; determination of strategy

S (Scheduling)   : Action pattern planning; action

To make the above intellectual levels easier to understand, we consider the example of control of an air conditioning facility. Intellectual level 1 is the ability to maintain sequential temperature levels from temperature sensor observations (I1). Intellectual level 2 involves using level 1 sensor information together with fuzzy state information (I2) in effective combination (D2) to plan a temperature gradient (S2). In intellectual level 3, the relation between distribution and capacity affecting air conditioning over a wide area is described in a frame as an object model; then inference (D3) according to a rule is performed so that the model status will be correct (I3) and control target procedures for each floor are set up (S3). In intellectual level 4, additional knowledge (effect function) for evaluating control objectives is introduced, so that control targets can be set and changed freely (D4). The optimum control for air conditioning over a wide area varies depending on changes in climate, season, time zone and heat sources. If a model that can emulate human sensation (what is desirable) is described, control that many people will feel is optimum in any conditions can be achieved.

3.2 Model - based fuzzy inference

3.2.1 Outline of model-based fuzzy inference

Suppose we have a model that structurally describes the objectives (of a problem to be solved). We input observed information into the model, and obtain corrections and instructions for actions as output. This is model - based inference.

The use of hierarchical models in fuzzy is still rare. It is believed that this is because fuzzy has had so many successes that there has been little incentive to expend the necessary effort.

The advantages of using fuzzy theory include the ability to run a system with few empirical rules and the ability to express feelings in rules. It is true that people have been expectantly waiting for such convenience, but just because such convenience has been achieved does not mean that "models are not needed in fuzzy theory". The first reason for this is that it is clear that performance is increased by predicting behavior. In addition, in the area of diagnosis and control systems other than numerical models, in objects such as expert systems (ESs) and inference rules on meaning nets are used, but in a numerical model it is difficult to make a model nonlinear without considerable analysis. AI paradigms can be described nonlinearly, but (the same is true for fuzzy rules) suffer from the defects given below. This is the second reason. First let us give the advantages of AI:

1. Empirical knowledge of experts can be handled: Qualitative judgment with regard to unique aspects of the problem and changes in conditions is effective.

2. Knowledge can be expressed naturally: Empirical knowledge can be described clearly in natural language, and easily updated.

In contrast, there are the following disadvantages:

1. With rigorous symbolic logic, there is weakness among crisp logics.
2. Production rules are qualitative; quantitative inference is difficult.
3. Description of multiple-input, multiple-output relationships for the rules is difficult.
4. Complicated description is possible using frame-type knowledge base ; but dynamics cannot be described.

Inference by rules belongs to intellectual levels 1 and 2, but suffers from defects (1) and (2). Rules obtained from specialists are highly effective, but reaction to phenomena that have not been experienced before is totally unpredictable. The reasons that fuzzy is so effective are:

1. The problem of weakness in disadvantage (1) can be covered.
2. With regard to disadvantages (2) and (3), if fuzzy rule mapping is performed considerable improvement is obtained. Formation of a state-space makes it possible for this theory to deal with even more complicated problems. In addition, when the problem is well-structured, improvement can be achieved with respect to problems (3) and (4) by using a hierarchical model. It is believed that fuzzy theory has room to incorporate other AI paradigms on the model concept.

Of course, there are also many difficult problems with having a model in intellectual level 1. Even when the characteristics of the object are unclear, in order to make system construction by inductive learning possible it is necessary to use a different method for each intellectual level, for example by using a neural network.

3.2.2 Instructions and reporting among intellectual levels

Strategic information systems (SISs), CIMs and intellectual robot systems have the evaluation knowledge of intellectual level 3 or higher shown in Fig. 3.1, and can be considered systems which execute information processing on intellectual level 2 and higher. In order to handle instructions of intellectual level 3, there can be a division of labor between the autonomous subsystem and higher level system while using an expert system (abbreviated as ES below) as a support system. The total system including the support system can be said to be an IS. Let us now list the instructions and reports

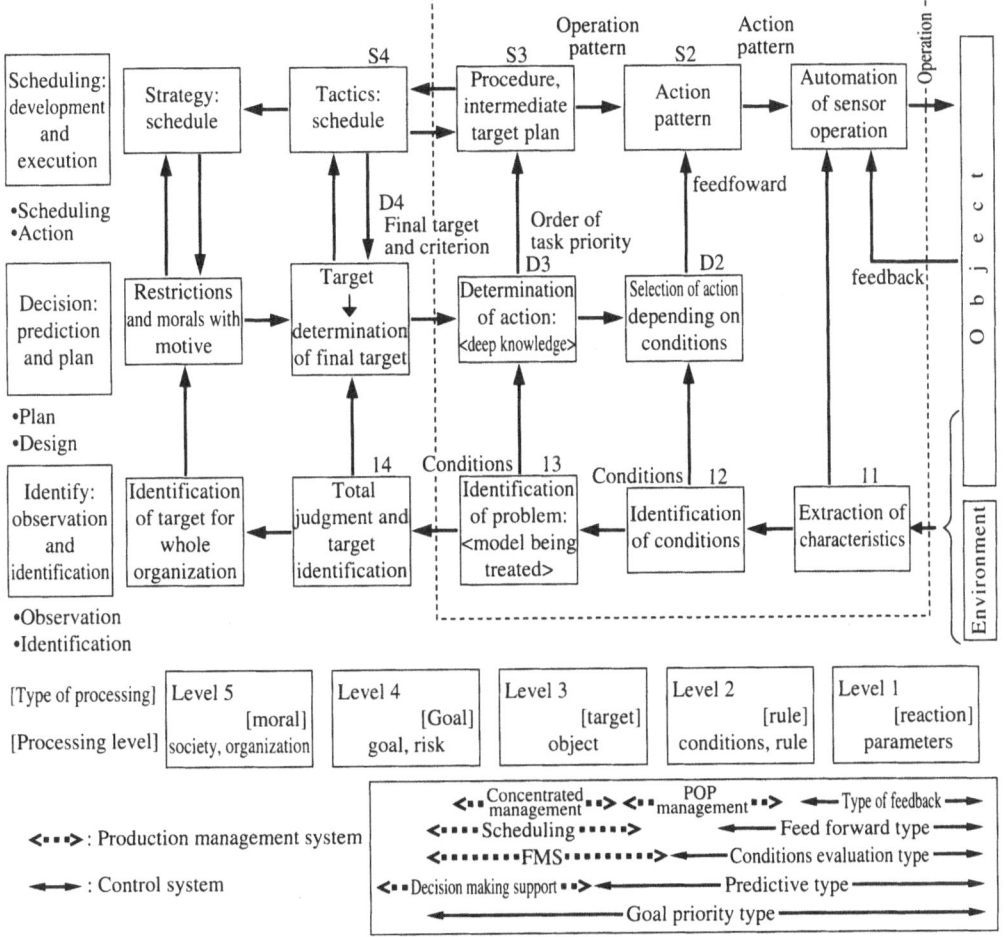

Fig. 3.1 Classification of problem solving models.

among intellectual levels that are needed for the division of labor.

<Reports and recognition>

[→ 1 → 2] Index data for the object behavior are observed, and error and deviation signals are recognized and reported.

[1 → 2 → 3] Erroneous and abnormal data strings are received; trends, characteristics and status signs (index values) are discerned and reported.

[2 → 3 → 4] Indices and characteristics of phenomena are received; causes are inferred from information about the object (such as cause - and - effect relationships) and reported in symbolic form.

[3 → 4 →] Recognition of the cause of the problem is received; possible objectives are determined, the mode is analyzed, and the effective function is updated.

<Transmission of instructions>

[→ 4 → 3] A search for a highly evaluated objective is conducted with the effective function, a new objective is determined, and instructions are given.

[4 → 3 → 2] The optimum action pattern (tasks, intermediate objectives) is determined from the functional characteristics of the object and the new objective, and instructions are given.

[3 → 2 → 1] Action patterns are received and expanded in partially corrected action patterns according to conditions; and instructions are given.

[2 → 1 →] Action patterns are transformed into action sequentially, and errors are minimized.

3.2.3   Differences between IS intellectual levels

Intellectual level 1

Information processing of intellectual level 1 is the automatic control component based mainly on feedback. An action pattern given from the next higher level (intellectual level 2) is converted to control quantities and implemented so as to have an external effect. The deviation from the pattern is determined from observational information, and the operation is corrected to minimize the error. The operation can be controlled with the unique quantity called error, but this uniqueness is lost if there is interference with another system.

Examples of interfering systems are heat conduction in a plant, and the mutual interaction between currency and stocks. Even when there is a relatively large effect from the natural world (external disturbance), in many cases fuzzy control in accordance with the sense of a skilled worker is effective. In such cases calibration by statistical hypothesis is necessary; recognition and learning from patterns which model the process are effective. The same is true for systems in which information is insufficient because sensors have been omitted to cut the cost.

Many mechanical systems can maintain linearity, and/or control on the level of frequency correction. Fuzzy control is effective in improving AI defect (2) in systems having a certain amount of interference or wasted time. For full - fledged removal of interference, a neuro system to improve AI defect (3) and level 2 fuzzy state-space are necessary.

The difference between intellectual levels 2 and 3 (tasks and objectives)

At present many production systems (example: concentrated management devices) and control systems (example: numerical control devices) are separated horizontally at intellectual level 2. If intellectual level 2 information collection is strengthened for total information management, greater system detail is obtained but the quality of problem solving is not improved. In level 2, adaptation improves. Some intellectual level 3 roles, such as optimum path creation and optimum processing condition settings from the processing objective, can now be expected.

Intellectual level 3 is a special level, created with the logic of conventions, law and design in human society. It is possible to say that this is why ES frames exist. An intellectual level 3 IS divides up loads and does scheduling based on a grasp of how the production system is organized. If a plan is changed and task instructions are given to a lower level system, then coordination with other robots (intellectual level 2) becomes possible, making the system more flexible.

The difference between intellectual levels 3 and 4 (objectives and goals)

Intellectual levels 3 and 4 make it possible to construct multi-purpose decision - making mechanisms. Intellectual level 3 performs object management and decides on tasks in pursuit of specific objectives. Intellectual level 4 achieves multiple purposes. It is easy to confuse the terms "objective" and "purpose". When speaking of intellectual processing, the two should be clearly distinguished. An objective is a point to be reached when hypothesizing a certain action. It is the "what" in "what to do to reach multiple purposes"; what can be changed is the "how" (method) in "how to proceed". An objective is expressed by a control function or symbol that the system can understand; the object is diagnosed, an action schedule organized, and progress monitored so that the final objective will be reached. All of these processes can be handled with the knowledge and inferences up to intellectual level 3.

A purpose is a socially recognized benefit such as safety or comfort. It corresponds to the "motive" in "what to do from what kind of motive"; the objects are separate pieces of knowledge that come under the concept of intellectual level 4. It is impossible for all purpose functions to be optimized. In a multi-purpose system, two or more purposes compete with each other; depending on conditions, tradeoffs might become necessary. Human desire becomes a standard for evaluating what is optimum; objectives can be changed depending on the motive. A multi-purpose system reconstructs groups of multiple purposes (effectiveness functions such as riding comfort and safety) that are achievable in certain conditions; evaluates objectives, decides which objectives are optimum and generates an action pattern by giving instructions to intellectual level 3.

3.3   Model-based fuzzy inference in intellectual level 2

3.3.1   Roles and observations (I2)

Roles

With regard to roles, it is normally thought that most problems can be solved by correcting errors through feedback (intellectual level 1). However, if the system is systematically generating errors, local description by the control rules of intellectual level 2 becomes necessary. Even in combinations of linear systems, singular points exist, so non-linear rules become necessary. A characteristic of intellectual level 2 information processing is that it is possible to construct a feed - forward system in which changes of state can be identified (locally recognized)and an action pattern which based on experience is expected to approach the goal can be generated.

Observations

Also regarding observations, if a reliable prediction model is available, a better control result is obtained from feed-forward control than from mere feedback control. Counting and integration make it possible to obtain overall observations of a system; these are effective in maintaining stability. Regardless of which AI paradigm is used, delay of observation and/or response makes prediction of the effect of the effector unreliable; there is also a limit as to the suppression of external disturbances, so that the system becomes unstable.

It is often said that "in fuzzy control, it is not necessary to construct a model; a skilled worker's experience is expressed as qualitative rules, so that good results are obtained". It is true that in an uncomplicated case in which cause - and - effect relationships of objects (such as of a water purification plant) are not complex, status monitoring and operation planning capabilities are sufficient.

Construction of fuzzy application systems can be started quickly, but when the characteristics are complicated, empirical rules alone are insufficient and other rules, for example from analysis, must be added. In this case, if rules are added carelessly, the rule-base will lose consistency and the system structure will gradually become ill-structured. If constructing a state-space model is necessary, make judgment of conditions which need to be satisfied. The use of indicators using differential and integral functions in intellectual level 1 is effective in judging conditions.

3.3.2 Intellectual level 2 dynamics

Description of 1-to-1 relationships by numerical formulas (deep knowledge) does not provide sufficient detail to solve dynamic problems. A skilled worker operates a system skillfully; the basis of AI is to incorporate his knowledge into the inference process. However, are humans always that skillful? Can that knowledge be incorporated into a system without due caution? There is room for doubt.

Putting a human's condition judgments and understanding into rules is effective for initial rules, but is often far from sufficient to form the state-space of an applied system. In particular, in multiple input systems, and in systems involving differential equations of 2nd or higher order, humans cannot make dynamic forecasts. A fuzzy membership function is visually easy to understand, but when an attempt is made to apply it to a high order nonlinear situation, it becomes hard to understand. Even intuition is necessary for empirical interpretation. It is dangerous to emphasize intuition in control rules; is there are many situations in which the dynamics exceed human intuition.

It is common sense that control is handled quantitatively, but in the case of qualitative rules with 1 - to - 1 correspondence, this is impossible. It is difficult to prove whether or not the system is effective with regard to all of the changes taking place around it, but in general, an applied system requires 2 or 3 more decimal places than a sample problem.

A frame-type knowledge base in ES must go beyond the initial meaning of the term, which is simply a semantic network, and move to an object oriented knowledge base which describes a relation in demon and method of frame base system. This makes it possible to describe complicated relationships involving the object. If the sequence is simple, it becomes possible to describe it, but the description does not extend to dynamic relationships of the object.

The various processes which are going on around us form a multi-input multi-output system. In systematizing them, it is necessary to describe their mutual dynamic relationships in parallel, and to process observations and predictions "simultaneously". To do this it is necessary to perform operations on either rows or columns in parallel.

The C language edition high - speed AI tools which are presently available give performance of several hundred times per second in an inference test on EWS; a high - speed edition gives on the order of 1,000 times per second.

In addition, the following are necessary in constructing an applied system:

1. Grouping of rules into clusters, and use of multi- stage inference
2. Multivariable input-output rule descriptions
3. Parallel operations in inference processing

Among these, (2) and (3) will be realized in the state-space to be discussed in the next section.

3.3.3 State-space (I2, D2)

Just as an IS must acquire knowledge, effort is required to improve its performance. State-space corresponds to the observer in control theory. This space has the role of sensing a dynamic status, following an action pattern according to instructions from a higher level, and creating an action sequence; and, before that, deciding on changes and insertions in local action patterns.A matrix - type format is necessary to make the multi-input multi-output relations of the status space model easy to describe. Status is divided into a number of modes; in addition, there are many cases in which it is desirable to distinguish between judgments and decisions, so the matrix comes to have multiple linkages. If the rule preconditions are applied to the matrix operations and fuzzy functions are taken as components, the matrix becomes a fuzzy relation matrix. In addition, the matrix components can be thought of as fuzzy membership functions in constructing a multi-input multi- output state-space model (Fig. 3.2)[3].

One effective method of using a state-space model is to have a status observation sensor that works by sensor fusion. Sensor fusion performs information processing simultaneously with a combination of input information obtained from sensors. In pre-analysis of system behavior, the correlation between this information from a combination of sensors and the status is extracted; the fuzzy membership function is given. Fuzzy has the advantage of reducing the number of rules and making nonlinear descriptions possible, but more important, describing rules for multidimensional combinations of state-space models makes it possible to construct dynamic models that could not be described in a conventional ES.

In image understanding of a dynamic, irregularly - shaped object (an animal), there are several tens of characteristic quantities (image processing outputs) at level 1. The number of combinations for two or more is nearly infinite. A human gives rough rules, the calculation load is lightened by some means such as linear discrimination, and characteristic axes are extracted[4]. Several tens of thousands of characteristic axes are produced; among them, it is believed that there are several thousand cognitive rules that are effective. Thus, an experiment with rough rules which one thinks up is valueless. If the goal is limited and learning is done by experimentation, there is also a

Occurrence/cause	Temperature 1	Temperature 2	Amount of dust and smoke
Medium - scale disaster over whole region			
(Median value)	60	60	60
(Dispersion value)	30	30	40
(Degree of vagueness)	10	10	10
Small - scale disaster in region A			
(Median value)	50	20	30
(Dispersion value)	30	30	40
(Degree of vagueness)	10	10	10

(a) Matrix expression

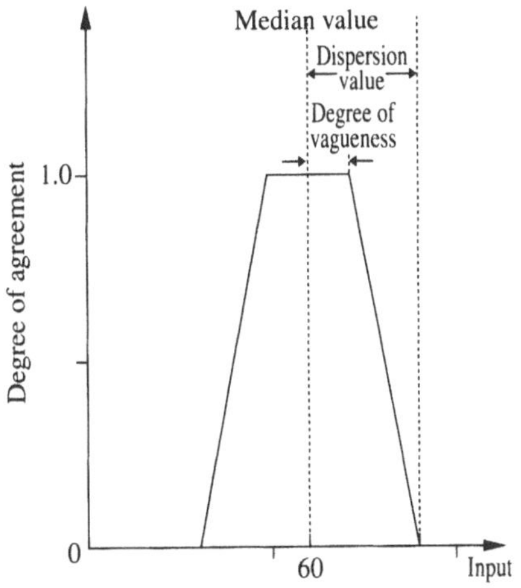

(b) Membership functions of components

(Example: Occurrence: medium - scale
disaster over whole region; Cause: temperature 1)

Fig. 3.2 Description of fuzzy condition space.

possibility of neuro application, but in constructing concurrent systems with multiple goals, effective rules are narrowed down by testing several tens of thousands of rules with the fuzzy state-space.

The state-space inference processing speed is important. At present, with a CPU capable of several MIPS, with non-object type membership function rules an inference speed of 14,000 times/second has been reached [3]. The speed of this status space inference tool can be increased dramatically by using CPUs in parallel or fuzzy chips.

Intellectual level 2 extracts natural and social mechanisms by scientific means, clarifies the means by which humans deal with them and constructs a system explained in terms of rules and numerical formulas. If a state-space is constructed from the clarified mechanisms, compliance control of robots with multiple degrees of freedom and construction of arm avoidance action capabilities can be learned on-line.

3.3.4 Action plans (D2, S2)

In some cases it becomes necessary to perform fuzzy modeling of vibration elements, etc.; but often there are many advantages to coordinating fuzzy with a linear control system such as conventional PID control. Such a hybrid method does not raise the intellectual level of the control, but more skillful control can be expected. This method is also effective as an easy method of raising the intellectual level of level 1 control.

Rules based on experience make it easy to respond to phenomena which are typical of intellectual level 2. State-space models can basically be described with multiple inputs and outputs. However, in cases requiring high - level control, for example strategies for solving problems and their search procedures, and in systems which include discrete and/or nonlinear elements, it is not easy for either a rule or status space to evaluate an action pattern for the purpose of forecasting conditions or pursuing an objective; rather, an intellectual level 3 knowledge system directed toward an object is more likely to be effective.

Whereas observations using a state-space proceed in the direction [signal to state], an action plan performs inferences in the direction [state to change of action pattern]. It is sufficient to describe the action pattern change rule in a matrix table and construct a state-space.

If there are external disturbances that are difficult for the system to predict, if status judgment and control prediction can be performed then it is not difficult to determine the optimum control (task) from 2 or more control methods. Several prediction control plans can be created from status space model (I2, D2) inference. Control plan prediction values are compared by an evaluation rule, and the action pattern (task) that gives the optimum evaluation value is decided upon. It can be said that this is an objective evaluation type system in which the objective values are evaluated. Since this objective evaluation type system has evaluation rules that correspond to intellectual level 3, they are positioned between intellectual level 2 status control and intellectual level 3 objective control.

3.4 Adaptive systems

3.4.1 Learning

It can be conjectured that shortening of the lifetime of industrial products and consumable materials has spurred the application of AI. The work of starting production as soon as possible is called concurrent engineering (CE). It involves a parallel approach for related processes including products and production technology. Also, with regard to a product, if knowledge of the conditions under which it will be used can be increased, its robustness with regard to particular changes, for example in the environment, can be increased.

Using empirical knowledge (know-how) in a natural form (natural language) for inference is an ideal method of CE for products of short life cycle. In particular, learning - type adaptation technology appears likely to become an essential technology since it permits tuning to be done while the system is operated.

Human feelings and areas related to nature are strongly fuzzy and chaotic. Problems in intellectual level 1, 5, or a world of external levels cannot be judged, or cannot even be grasped from the distribution, in such an area inductive learning by itself is effective; and learning by neural networks and fuzzy multi-dimensional weighted matrices is applied.

Neural networks

Neural networks are used in sensory and technological level processing such as audio and video, and time sequence information. Evaluating output and assigning potentials to network linkages (corresponding to a state-space) is effective as an adaptation capability to intellectual level 1 and 2 status judgments (I1, I2). Incorporating a neural network into fuzzy inference observations is helpful in initial system construction.

However, the term "neuro" does not apply well to an image. In understanding an irregularly shaped moving image, there are cases in which the different objectives do not have a same feature potential and the evaluation cannot be performed. The objects of action naturally assume a variety of forms. If an objective is fixed for each such form, there will be no limit to the number of objectives. Also, if other objects resembling the various objects are caused to act, the potentials are buried in the variety of forms. Basically, separation functions are determined for each of the similar forms, but it is preferable to begin operation first, then add objects of recognition. If an attempt is also made to change the recognition goal, neuro learning will not be able to cope with the change. In addition, it is difficult for a neural network to explain the obtained potential patterns and the processes of and reasons for inferences. However, there are many objects which it is difficult to model, and their importance is steadily increasing. We hope for future progress in this area.

3.4.2 Fuzzy adaptation

Adaptation using a script

Machines such as arc welding robots, wire cut electrical discharge machine tools, and plastic extrusion and molding machines face different demands in each job regarding materials, work methods and finishes. For each combination different high level control is applied. Since there are many parameters, selection of machining conditions is left to the operator's experienced operation. Since there are an infinite number of possibilities (Paleto's optimal solution), it is necessary to design the optimum control process from document knowledge. The method of approaching this problem is to prepare a script that appears likely to reach the goal, then select the optimum solution. There are 2 selection methods.

The first, which involves adaptation in a broad sense, is called an open - loop adaptation system. Multiple goals are inferred from conversation, then the optimum machining conditions are selected (S2). In the future this will be developed into a man - machine conversational optimization method, in which the decision maker's effect function is extracted, and the optimum solution is obtained from a variety of objectives (S3). In addition, it is also possible to change the control parameters (D2) from environmental measurements in order to stabilize the machining.

The second method is closed loop adaptation. As the first stage of closed loop adaptation control, the machining status is observed, and machining conditions are selected (S2) so that the status will become optimum (D2). Optimization of the observer himself (I2) is also conceivable.

Memory learning type adaptation systems

In full-scale closed loop adaptation, the model is updated by information that can be newly used at each sampling time (continuous identification); this new model is then used to update the controller parameters[5]. The problem is that if the system is complicated (conditions change, control structure characteristics change), then it is necessary to perform probabilistic adaptation control to reduce uncertainty while proceeding toward the goal. If changes in the control object and environment are complicated (parameter fluctuations), then the effectiveness of the fuzzy adaptation learning control capability becomes pronounced.

Let us explain a learning type adaptation system, using a large diameter muddy water pressurized type shield digging machine as an example. In shield construction, every time a segment is inserted the plant returns to its initial state and the control direction changes. Linear control has already been applied, but a large diameter shield digging machine produces only one item; the action characteristics are only design conditions, and it has been difficult to find the correlation between the instructions and the control results.

Control ability is determined by the ability to cancel external disturbances. Unknown external disturbances include changes in the item being machined (earth layer, soil properties) and wear or breakage of the bit. The learning procedure is given below.

1. Default data are obtained: Design prediction characteristics are set in the multi-dimensional weighted matrix as default values.
2. The index of performance is measured: The IP is measured by laser. When averaging processing such as parameter tuning is performed in a discrete system, true dynamical characteristics are lost, so memory type probabilistic processing using known external disturbance timing is used.
3. Comparison: What are obtained last are the total pattern of control char-

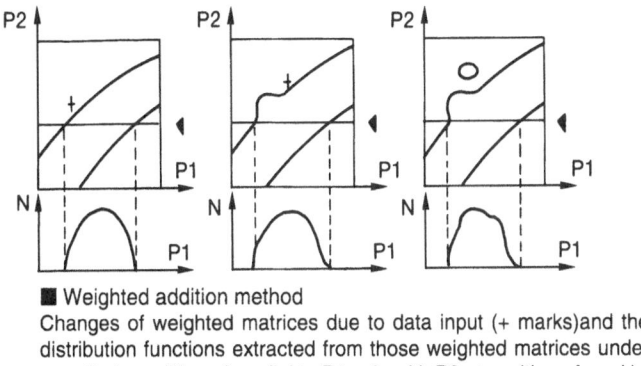

■ Weighted addition method
Changes of weighted matrices due to data input (+ marks)and the
distribution functions extracted from those weighted matrices under
specified conditions (parallel to P1 axis with P2 at position of mark).

Fig. 3.3 Examples of operations on multi - dimensional weighted matrices

acteristics and its subpatterns. In the comparison section, it is stored as a
probability distribution in several - dimensional characteristics space; then,
after time constant processing, it is formed into a pattern and finally cut
up into subpatterns.

The characteristic axes are mutually different collections of indices; in this
example the statuses shown by the IPs can all be considered errors with
respect to the given action pattern. If the indices have the same meaning,
then the subtraction capability and declining function (a kind of averaging
processing of the time constant) have significance. Averaging is performed
using a weighted addition method as shown in Fig. 3.3. The distribution
effect function (when the number of samplings is small) has an effect on
the surroundings, and there is a population management capability so that
the patterns are not concentrated in only the new observation section.

4. Output: A rule which performs processing on normalized functions and
   multi-peak functions. The distribution is managed as a macro pattern;
   from it, membership functions (subpatterns) are cut out and output.
5. Intellectual level 2 inference section state-space change: When a new fuzzy
   membership function is transmitted, state-space components are changed
   in real time to match the control cycle, and provided to inference.

In an intellectual level 2 memory learning type adaptation mechanism
such as described above, a system can change its own state-space. On the
work site, the actual machine in operation is the object of control, so a
test signal cannot be generated. Considering safety, it is recommended that
adaptation be carried out through the following steps.

1. Initial value (assumed distribution) input: Design
2. Pre-measurement data off-line tuning: Design
3. Open loop on-line tuning: On site

4. Complete closed loop measurement, control: On site
5. Counting data analysis and new IP extraction: Design

In this example, one month after an initial explanation of shield construction is received system design is started; half a year later the system is installed on site.

If the system is not a multi-purpose system, it is not necessary to raise the intellectual level, and effort can be concentrated on increasing the IP and improving control characteristics. If an IP cannot be determined at the start, knowledge is obtained as follows:

1. An index having high correlation is selected by principal component analysis.
2. From the nonlinear distribution of characteristics space, a pattern that has significance (nonlinear correlation, separation rate) is automatically cut out.
3. The multiple dimensions are visualized and a fuzzy function is created and visually confirmed.

The AI-DNA model X removes the characteristics space dimension limitation, so the processing load for these analyses can be set to match the facilities, such as CPU capacity or memory space [4, 6].

For example, it is possible to first analyze the correlation of characteristics from past accumulated data, then narrow down to the characteristic axes (indices), then (on the experimental site) perform real time learning for only a certain specified index. Alternatively, it is also effective to create a state-space, change the weights of the indices in accordance with the success contribution weight, and perform parameter tuning.

3.5   Intellectual level 3, 4 model - based fuzzy inference

3.5.1   Goal - oriented inference systems [7]

Introduction

First, the roles of level 3 are to clarify whether the functions/performance/ conditions are normal or not and to logically clarify problems, to plan the division of labor at lower levels and to give instructions in the form of action patterns. ES systems share the common problem that in logical inference objectives cannot be changed in response to changed conditions. It is necessary to have partial changes of action plans, changes of intermediate objectives, and interruption (including abandonment) of missions. In order to dynamically change objectives and schedules, it is necessary to accurately judge changes of conditions and their causes, but it is difficult to do this with rules. It is necessary to clearly indicate the structural relations of the world that are the objects of control and to have a dynamic prediction capability. If the focus of the multiple goals is narrowed down from judgment of conditions, then objectives can be changed. If a self - propelled robot has a prediction capability, a new action control command string can be created from predictions of residual energy and of breakdowns, so that stoppage of action can be prevented. These are the roles of intellectual level 3.

As with intellectual levels 1 and 2, effectiveness and knowledge of goals in intellectual level 4 have a strongly fuzzy character. Meta knowledge which expresses multiple goals is difficult to express logically. We will explain "goal - oriented inference" including intellectual level 4 goal functions. In the following explanation the parts that duplicate ordinary ES will be briefly summarized, while emphasis will be placed on the parts that are unique.

Multi-stage inference procedure

In goal - oriented inference, objectives are changed in response to conditions by a goal function group. With the object's structural relations clearly indicated, conditions are appropriately judged and the focus of multiple goals is narrowed down so that objectives can be changed and multi-stage inference combining intellectual levels 3 and 4 performed. The inference procedure is as follows (Fig. 3.4).

Inference step 1   : Judgment of conditions [I2, I3]

Inference step 2   : Evaluation weights [I4]

Inference step 3   : Constraints on goals [D4]

Inference step 4   : Selection of goals [S4]

Inference step 5   : Final objective [D3]

Inference step 1 is to judge conditions. Inference step 2 is to determine the constraint limits of the effect function (for the purpose of performing multi-purpose evaluation). Inference steps 3 and 4 are intellectual level 4: tradeoffs and balance among multiple goals. Inference step 5 is an intellectual level 3 capability, to select objectives; for processing after that, it is necessary to have detailed scheduling capabilities S3, S2, etc. to create intermediate goals. The above procedure is capable of dealing with complicated problems, but in cases such as small - scale plants, where one does not have to deal with multiple goals, it is sufficient to proceed from inference step 1 to inference step 3 to inference step 5.

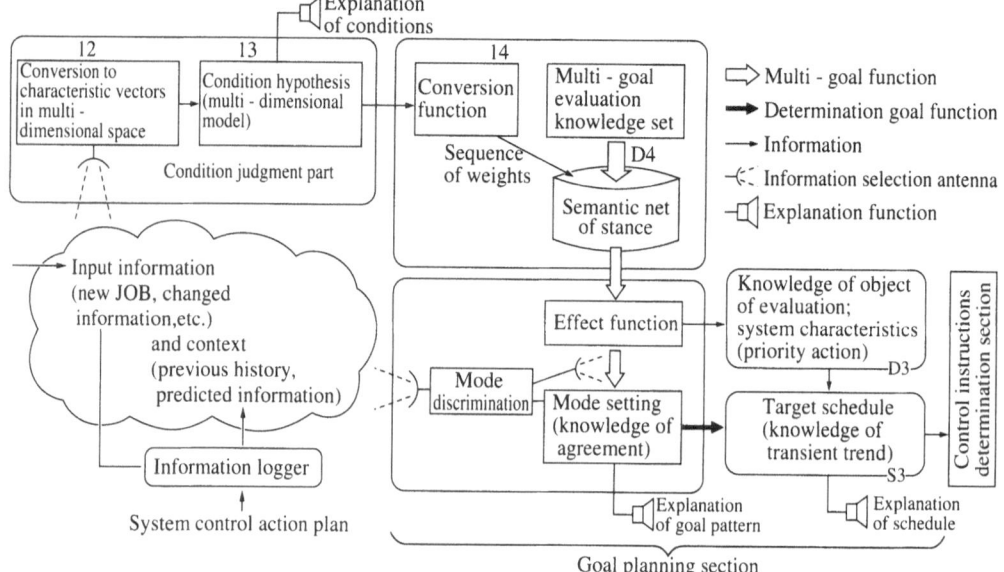

Fig. 3.4 Inference primarily directed at a goal.

3.5.2 Intellectual level 3 condition judgment

Inference step 1: Judgment of conditions [I2, I3]

[State input, state hypothesis] to Judgment of [conditions]: Reverse inference has promoted diagnostic type ES, but if there are many combinations of goals and input information, the number of search branches increases out of control, destroying the practical utility of this method. In order to dynamically change the schedule, it is necessary to perform appropriate judgments on changes in conditions and their causes in a short time. As for the speed with which a status is observed and identified, intellectual level 2 state-space models are much faster than other methods.

If a difficult problem is anticipated, it is desirable to switch to a new objective that minimizes the risk. In order to select or change an objective, it is necessary to evaluate proposed objectives, having first applied a constraint to eliminate goals that are incompatible with conditions. I2 rapidly judges the trend of status; I3 judges fluctuations in conditions from a status and conditions model, and reports the judgment. I3 is an ordinary diagnostic type ES. The cause is searched for by tracing the compound relationships of objects (is-a, has-a) described in the frame.

When intellectual level 4 knowledge, such as logical diagnostic knowledge about the object, goals and risks, effectiveness, etc. is mixed in with other information, condition judgment knowledge becomes unclear. If the intellectual level of the model is made clear, it will not invite a poor structure. Indeed, if it has been updated sufficiently, it is sufficient to treat the goal as fixed D3 knowledge.

[Inference step 1 Description of frame: I3]

When an AI paradigm other than fuzzy theory is used in coordination with other methods, the possibilities of IS are expanded. In intellectual level 3, the problem is treated as an object, and grasped on a frame type knowledge base [8]. For example, an intellectual CAD system uses the transmission of constraints to maintain specifications, rules and safety standards; when a change is made, it is possible to return to the higher level concept of the design model and redesign. In addition, if a complicated map is described in a self - propelled robot frame, another route can be searched. If there is no highly reliable model for deciding strategy, an application system with fuzzy analogies reasoning and search of action patterns from a script by case based reasoning, etc. are effective.

An intellectual level 2 system does not change the action plan itself; all it can do is detour around obstacles. In order to change an action plan, it is necessary to judge whether or not the new plan is appropriate on the plan object model. If the object is static, the frame alone is sufficient.

In addition, in a dynamic problem, if a state-space model is constructed in intellectual level 2; a complicated, static object structure and relationship are (by demon and method) described in an object - oriented type frame; and state-space output is extracted; the dynamic behavior can be grasped with level 2 and 3 coordinated inference. Dynamic prediction capabilities that are available include differential/ integral type condition models coupled to the status model [3] and mathematical models.

3.5.3 Intellectual level 4 decision making

[Inference step 2 Determination of evaluation weights: I4]

(possibility of omissions according to the problem):

From conditions that have been judged, the system investigates whether or not a certain specified pattern exists, and determines weights (selection) to constrain goals that are incompatible with conditions (evaluation). The result makes it possible to reconstruct the group of multiple goals (effect function: riding comfort and safety) in inference steps 3 (D4) and later. In a problem that is not complicated, it is sufficient to directly constrain the goal function from the index that expresses the condition.

[Inference step 3: Goal constraint D4]

[Weight string, meaning net] to weighting of [effect functions]:

In the next inference step, the objective that is most suitable for present and near future conditions is selected from candidates (scripts) for objectives (hypotheses made up of scenes and intermediate objectives). For this purpose, it is necessary for the relation between conditions and goals and the order of priority to be made clear on the meaning net.

Intellectual level 4 knowledge has a strongly fuzzy character; meaning net weights, not crisp logic constraints, are processed.

Fuzzy theory is superior for subjective descriptions; in addition, constraints are transmitted in both directions as evaluation functions. Other descriptions are also possible, for example the distribution pattern of center of gravity traces for the purpose of non-linear description. When a system uses this kind of evaluation function and models highly specific knowledge and know- how, it becomes possible to determine action in a way that approximates human decision making.

Examples of schedule effect functions are shown in Fig. 3.5. Fig. 3.5(a)

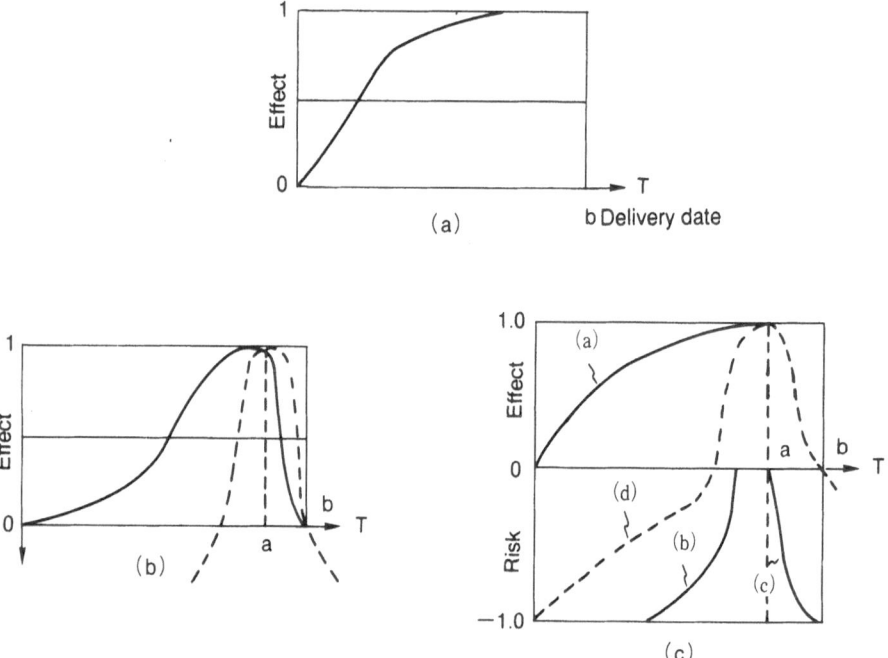

Fig. 3.5 Time - varying effect functions.

shows an ordinary effect function. Fig. 3.5(b) and subsequent figures show effect functions based on the concept of fuzzy center of gravity traces. A center of gravity trace is a set of centers of gravity when an infinite number of fuzzy membership functions are considered. Conversely, an index value can be input and infinitely many membership functions cut out from the center of gravity trace [7].

The solid line in Fig. 5(b) is the part for which the attributes are near delivery date (b). However, the increase is not monotonic; there is a decrease immediately before the delivery date (a). This is not due to external condition fluctuation; even as time passes, the risk is changing. Fig. 5(b) can be thought of as a case involving on-the-job know-how, in which employees consider the possibility of delayed delivery as a risk.

If the function in Fig. 5(c) is split into an effect function (a) and risk functions (b and c), the nonlinear function (d) can be regarded as the sum of them. One of the functions from the above multi-purpose group can be understood as a set of effect functions. Although the concept is level 4, it is possible to choose whether the accumulation of knowledge is placed in a frame or indexed. The evaluation function may be a fixed value, but if possible it is desirable to describe the quantified effect function in a meaning net (hierarchical frame) so that the balance of goals can be changed.

[Inference step 4: goal selection S4](possible to omit depending on the problem)

[Goal function, pre-goal, knowledge of agreement] to (mode analysis) to selection of [string of goals]:

A mode is a characteristic of a group of goals. Mode analysis is performed with weights assigned to the effect function that makes the goal concrete; the central goal is enhanced, the goals that can be traded off mutually are lowered, and the focus of goals is made clear.

The simultaneous pursuit of multiple goals is equivalent to simultaneously maximizing a number of goal functions. If an attempt is made to increase the overall sum of amounts of improvement of goal functions, it is necessary to improve certain specified goal functions; inevitably this results in improvement or worsening of other goal functions. It sometimes happens that goals contradict one another, so it is not possible to automatically decide to abandon another goal. In order to improve a certain specified goal function, knowledge corresponding to a so-called strategy is necessary. When comparing the pattern of weights assigned to a group of weighted effect functions with the previous time's weighting pattern, the goal functions are determined by rules such as [maintain the present state], [select what appears to be most rational], [select whichever is better since the effect on the whole is small], etc. with respect to the degree of change (a vector string).

3.5.4   Intellectual level 3 goal determination

[String of goals, candidate objectives] to selection of [objective]:

Inference step 5 (D3) is where the final objectives are selected; information processing after inference step 5 waits for scheduling such as S3 (intermediate objective planning), D2 (adding objectives so that a task can be passed through as favorably as possible), S2 (action pattern), S1 (action pattern), etc., decision information processing, and action.

IS preserves a mission in response to the appearance of an obstacle, and selects whether to continue action, change it, or switch to a different objective. If change is selected, depending on the extent of the obstacle a trade-off between main mission and sub-mission(s) is performed, and the policy (schedule, objective) is changed. Here the mission is described in intermediate language format, and the goal is handled in almost the same sense as that mission converted to an effect function.

An objective is a collection consisting of a principal scene and a group of intermediate objectives.

Depending on the objects that comprise the IS, it sometimes plays an extremely important role in intermediate transitional states, as in a plant or strategic information system. In many cases, as when it is sufficient for a self - propelled robot to reach its final objective, there is a great difference between the ways the objective and intermediate objectives are formulated. When there is a policy change, or it is judged that there has been a failure and the mission is to be aborted, the system switches to the new objective of retreat; but there are cases in which it is impossible to switch smoothly from the objective which is presently being pursued. For a smooth transition between objectives, it is necessary to first switch to a different intermediate objective. In addition, even in a case in which it is sufficient to only change the action pattern, it is sometimes even necessary to change the intermediate objective. In order to change the action plan flexibly, it is desirable that either the intermediate objective on which the action plan is based be described in fuzzy form, or a means of fuzzification be included in the method of formulating the action plan.

In order to compute evaluation of an objective, attribute values of the objective are applied to the above - described effect function. An evaluation value of one objective is computed as the (for example) weighted average of the degree of agreement that is obtained; then ordering objectives are selected. However, it often happens that the group of objectives lie on the same plane and the evaluation cannot be done. A method that is easier to understand is to first prepare a group of scripts based on conditions, then repeat the selection of the final objective from among them; but the number of cases in which the optimum selection is not obtained is increased by this method. Better results are obtained by describing conditions as attributes of the objective and various intermediate objectives, and the selection conditions; then executing a fuzzy search in attribute space (multi-dimensional characteristic space).

In the case of a self - propelled robot, the information that is received and passed on as the final objective of the action plan could be [object: name, characteristics, quantity, work], [location: position, path], [action: action pattern, intermediate objectives], [intermediate objective: intermediate role], [associated information: degree of importance, priority], etc.

The above output is provided in support of decision making, or used as input to an existing goal - oriented type control system.

3.6   In conclusion

It is said that artificial intelligence (AI) is not a special scientific technology; rather it is an attitude of clarifying problems. This chapter explains applied methods of fuzzy inference which are effective in multi-purpose system construction, centering on hierarchical models, from the point of view of SE (System Engineering).

Adding intelligence, as in AI, has the purpose of increasing the robustness of the system with regard to all of the changes that might occur in the situation that the system faces. A computer system, like a skilled worker, must acquire empirical knowledge and have an inductive learning mechanism[6] in order to make appropriate predictions and act accordingly.

A concept that will make multi-purpose systems and coordinated man - machine systems a reality is obtained from clarification of the problems of considering intellectual level. The inference mechanism obtained from a hierarchical model, which uses a human information processing model obtained from clarification of plant operation problems as an analogy[2], is useful for transferring a specialists's concepts without distortion.

The present author and colleagues have investigated an autonomous system using a hierarchical information processing model[1, 7], and used it in designing the autonomous processing structure of a self - propelled robot.

From the operational point of view, it is necessary to have a structure in which it is difficult for instruction mistakes and misunderstandings to occur at the man - machine interface. In a hierarchical model in which goals (intentions) are at the highest level, so as to make higher level conversation possible, explanations can be given based on effectiveness; such models can be used in many fields such as autonomous systems incorporating decision making capabilities[1, 10].

A specialist's know - how and knowledge cannot always be put into language; formats also differ from one level to another. In a processing system that cannot use simple AI tools, it becomes necessary to use fuzzy theory that can unify both symbol information and non- language sign information. In a knowledge processing structure such as the above (even if it is hierarchical), it becomes necessary to achieve coordinated processing such as parallel processing[1],[3] of different types of AI paradigms (frames, productions, blackboards, neural networks, etc.).

A hierarchical model can be built up from partial structures to make it easier to construct a multi-goal system. In addition, it can be easily combined with existing technology. I will be pleased if the methods in this chapter contribute to the clarification of problems in system construction.

References

1. Ageishi Y, Sasaki K (1990) A new approach to intelligent control (in Japanese). J Jpn System Control Information Soc 33 (6) : 324–331

2. Rasmussen J, Kaiho H, Kato T, Akai M, Tanabe F (1990) Interface recognition engineering (in Japanese). Keigaku Shuppan.

3. Nakamura K (1990) Explanation of the AI-RNA Model 01 (in Japanese). AdIn Research Institute.

4. Nakamura K (1990) Explanation of the AI-RNA Model X (in Japanese). AdIn Research Institute.

5. Landau ID, Tomizuka M (1981) Applied control system theory and practice (in Japanese). Ohm Sha.

6. Nakaya J, Nakamura K, Ageishi Y (1990) Knowledge acquisition in fuzzy inference systems (in Japanese). Collected Papers of the 1990 Nationwide Conference of the Artificial Intelligence Society, pp645–648

7. Ageishi Y (1988) Autonomous control systems using fuzzy inference (in Japanese). Collected Papers of the 2nd Nationwide Conference of the Artificial Intelligence Society, pp269–272

8. Ueno H (1986) Knowledge expression based on the concept of an object model, study of electronic communications (in Japanese). Papers of the Artificial Intelligence and Knowledge Processing Soc AI86–4

9. Nakamura K, Oyamada S, Horikawa R, Ageishi Y (1989) Searching for special vectors by fuzzy operations (in Japanese). Collected Papers from the 5th Fuzzy Systems Symposium, pp449–454

10. Ageishi Y, Nakamura K, Yaegashi K, Oyamada S, Murakami Y (1990) Achievement of artificial feeling using a human interface used in human problem solving models (in Japanese). Collected Papers of the 5th Fuzzy Systems Symposium, pp389–394

Chapter. 4

FUZZY DEVELOPMENT STATIONS AND FUZZY INFERENCE PROCESSORS

4.1 Background to development

With the incorporation of fuzzy control into household appliances, "fuzzy" has become a household word. The number of fuzzy applications is increasing every year, providing great incentive to researchers engaged in fuzzy control development.

However, the household appliances that have brought the term "fuzzy" into common use do not contain fuzzy chips that perform fuzzy inference; most perform inference using software that runs on general-purpose microprocessors. Naturally, the cost of fuzzy chips is a deterrent to their use in such equipment, but when a general-purpose microprocessor has to perform fuzzy inference while managing the whole system, the control is inevitably inadequate.

Fields which are considered suitable for application of fuzzy control are mainly those involving complicated control applied by a skilled worker. Inference by software operations is a burden to management of the whole system because of the time required for inference. Consequently, it is desirable to perform fuzzy inference with a fuzzy chip that will not interfere with management of the system as a whole.

Against this background, a Virtual Paging Fuzzy Inference Chip has been developed jointly by Prof. Hirota's laboratory at Hosei University and Mycom K.K. In addition, a number of companies and research institutions have announced and are marketing fuzzy chips that perform fuzzy inference[7, 11].

In addition, unless one has tools to support the construction of fuzzy chips and fuzzy chip application systems, effective and smooth application systems cannot be constructed no matter how high performance the chips that one has. Support tools are necessary to obtain the maximum performance from fuzzy chips. What is important in fuzzy control is how the knowledge base for controlling the controlled object is incorporated into the membership functions and fuzzy production rules. In other words, the software is more important than the fuzzy chip hardware. These tools are the hardware needed to obtain the maximum performance from the software.

Therefore, we decided to develop a fuzzy development station that will obtain the maximum performance from the Virtual Paging Fuzzy Inference Chips. In this development work we had the cooperation of Prof. Hirota of Hosei University and Prof. Sakawa of Hiroshima University (then at Iwate University).

This chapter emphasizes the specifications and method of operation of the Virtual Paging Fuzzy Inference Chip and the Mycom Fuzzy Work Station, which is a support tool for fuzzy inference development. Characteristics of the Virtual Paging Fuzzy Inference Chip are also explained.

4.2 Characteristics of the Mycom Fuzzy Work Station

The Fuzzy Work Station has the following characteristics.

4.2.1 Simulation

This function judges what kind of fuzzy inference operation result is to be derived by the fuzzy production rule and membership function that have been constructed. At present there is a relative lack of simulations and commands, but it is expected that in the future there will be a more complete range including an automatic learning function.

At present, there are 3 types of simulation available, making evaluations from a variety of points of view possible. The purpose is to evaluate the constructed rules, etc. with this function before they are connected to the target system.

Also, special specifications can be provided on the user's request, such as linking with the user's evaluation software and performing closed loop simulation.

4.2.2 Flexibility

Compared to crisp logic implication operations, a number of operations such as t-norm and s-norm have been introduced. For this reason there are a number of fuzzy operation methods. Even under the same conditions (rule etc.), when the operation method is changed, the inference result will be somewhat different. In other words, it is necessary to support as many operation methods as possible, even to discover which fuzzy operation method is best suited to the target system. Consequently, in this Work Station it has been made possible to choose from among 3 representative operation methods, and this number will be increased in the future.

Also, several different membership functions are supported, providing the flexibility to match the user's approach.

4.2.3 Easy to operate

As stated at the start of this chapter, fuzzy control makes it easy to introduce a skilled worker's thought into the control process. Consequently, for skilled on-site workers who are not used to using a personal computer, menus in a tree structure are used; only the menu needed for operation at a given time is displayed, increasing the ease of operation.

Since operation is so easy, anyone who has read the reference manual through once can operate the Work Station from only the displays that appear on the screen.

4.2.4 The fuzzy inference engine

When fuzzy control is being constructed, it is not known how many bits to assign for the support sets of the condition part and the operation part. In addition, although the Mycom fuzzy chip is of the digital type, in many cases the condition part or the operation part is of analogue type. If a problem such as that mentioned above occurs in such a case, it becomes difficult to select the resolution of the analogue-to-digital converter or the digital-to-analogue converter.

With the aim of solving this kind of problem, the capability for the Fuzzy Work Station itself to act as an emulator is provided as an option. It also incorporates analogue-to-digital and digital-to-analogue converters, and is also capable of processing analogue signals.

Since a target is connected and actual fuzzy control is performed, the trouble of first producing hardware for use with fuzzy chips is eliminated. Also, even if, for example, a rule is changed, operation including the target can be performed immediately, greatly reducing the time required for tuning.

At present, Work Stations are being supplied with specifications tailored to each user, but it is planned to release a general purpose direct inference version in the near future.

4.3 Configuration of the Mycom Fuzzy Station

A standard specification Mycom Fuzzy Work Station is shown in Fig. 4.1, and the fuzzy inference engine configuration in Fig. 4.2.

Fig. 4.1 System composition with standard specifications.

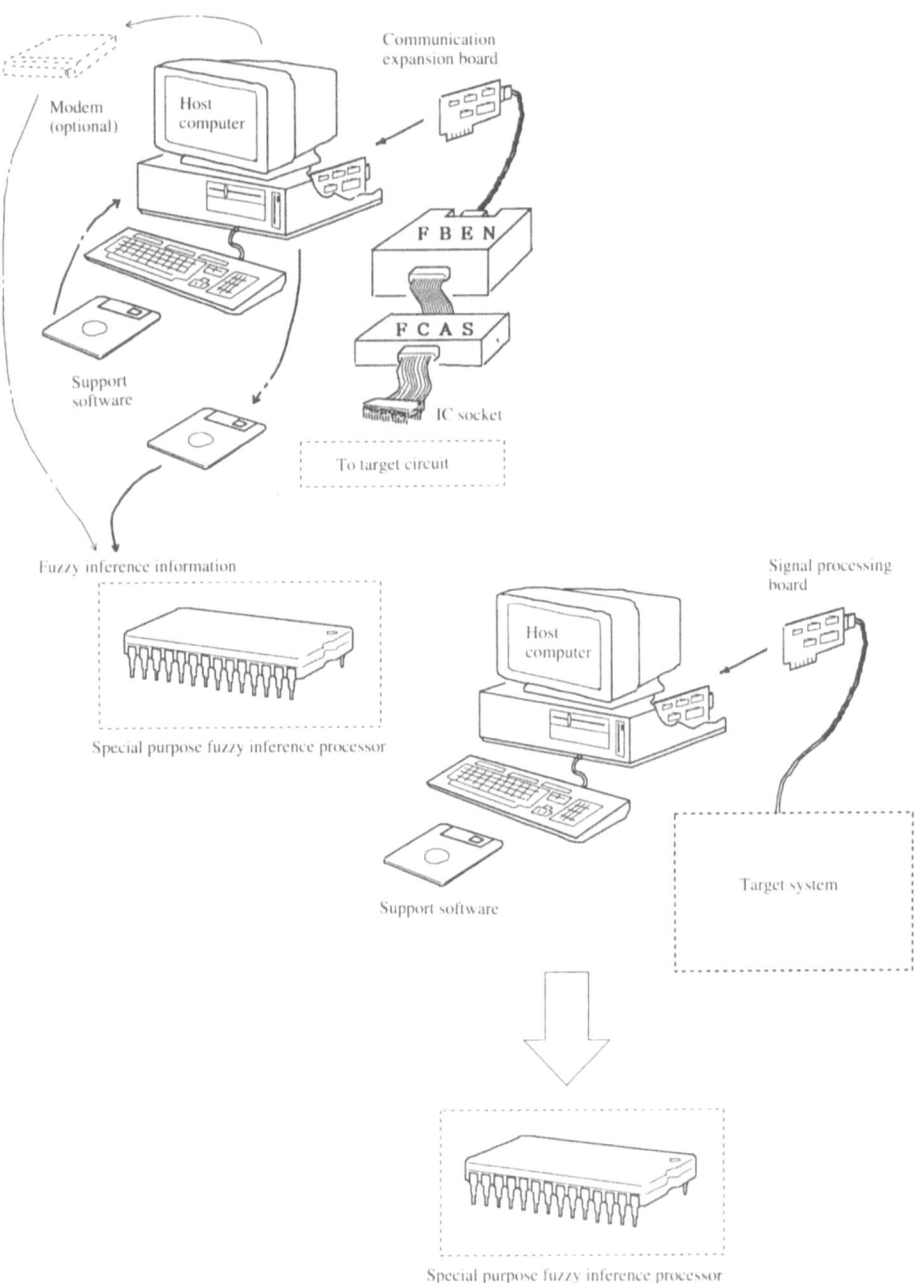

Fig. 4.2 System composition with inference engine

4.3.1   Host computer (Table 4.1)

Table 4.1   List of driver functions

Model	IBM PC/AT compatible
CPU	80386 (20MHz)
	with 80387 (20MHz)
Hard disk	40MB
Floppy disk	5.25″ 1.2MB
	3.5″ 1.44MB
RAM disk	1.04MB
Display	VGA mode

4.3.2   Development support software (Table 4.2)

Table 4.2   ICE specifications for use with FP-3000

OS	MS-DOS V5.00
Key input method	Conversational method
Display	Graphic display
Fuzzy variable definition	Free form definition
Rule format	Maximum 4 conditions 2 conclusions (standard specification)
	Maximum 8 conditions 2 conclusions (engine)
Fuzzy composition method	3 methods
Defuzzification method	Area method
Membership Function	5 shapes
Fuzzy support set	Dispersed function corresponding to number of set bits(256 in case of 8 bits)
Membership values	Real floating point operations
Language values	selected from 3/5/7/9/11

4.3.3   Communication expansion board

This is a RS422A serial interface conforming to the EIA standard, designed for use with the IBM PC/AT expansion throttle. Its function is to transfer transmitter files created with development support software (standard specifications) from the host computer to the FBEN (to be discussed below).

4.3.4   Fuzzy controller (FBEN, FCAS)

The FBEN (Fuzzy Best Emulation Navigator) and FCAS (Fuzzy Controller & Advanced Simulator) can be separated; they perform fuzzy inference with respect to the target in place of a fuzzy chip.

Although it is perhaps superfluous, we note that in Japanese the names FBEN and FCAS can be thought of as representing a BEN (lunch box) in the size of a CAS (cassette); the workers developing these devices referred to them so often as "f-ben" and "f-cas" that it was decided to make these their official names, and later with great effort we came up with English names to match the acronyms.

The FBEN receives transmitter file data from the host computer through a communication board, while the FCAS implements fuzzy inference algorithms.

Only the FCAS is connected to the target. The FBEN remains disconnected in use. The place where a fuzzy chip is to be installed is an IC socket; by plugging the cable from the FCAS into that socket the same operation as with a fuzzy chip can be obtained.

4.3.5 Fuzzy inference processors for special uses

These are interchangeable with the above-mentioned FCAS as far as inference speed is concerned, but rules cannot be changed as they can with the FCAS; these can only be used for specific purposes (change not possible). These inference processors have functions that are not supported by the development support software, and are expected to be useful with new operation methods that will be developed in the future.

These processors are made by our company, based on transmitter file data.

In the present version, these are only for special uses (in other words cannot be changed), but a general purpose fuzzy inference chip is under development, and it is possible that this chip under development will be supported by the Work Station.

4.3.6 Signal processing board

When the Work Station functions as a fuzzy inference engine, this board is used. It inputs and outputs analogue and digital signals.

Analogue signals are the principal targets; in the case of a digital signal, a board with many processing bits is added.

Condition part and operation part signals are input and output through this board; then fuzzy operations are performed according to the conditions set in the host computer.

Table 4.3 List of inference chip terminals

Analogue-to-digital conversion section	CH	14CH
	Resolution	12bit
Digital-to-analogue conversion section	CH	2CH
	Resolution	12bit
Digital input section	TTL level	8bit
Digital output section	TTL level	8bit

Also, in the standard specification unit this board can be installed, and target system analogue data sampled.

4.4 Fuzzy Work Station functions

The Mycom Fuzzy Work Station support software configuration is shown in Fig. 4.3.

The main menu has 6 items; each item selects a detailed menu, in a tree structure.

By starting the support software on the host computer, the main menu is displayed (Fig. 4.4, 5).

The contents which are presently set are displayed inside the frame at the bottom of the screen.

Next, the individual items will be explained.

4.4.1 Define Membership Function

Contents of input and output items, and membership functions, are defined.

When this menu appears, the choice is among 1. Edit 2. Load 3. Save. Load and Save are used to read and write files created with Edit, respectively.

When Edit is selected, the display shown in Fig. 4.6 appears.

Shape of Function

This defines the shape of the membership function. The choice is among 5 shapes: triangular, trapezoidal, sinusoidal, squared sinusoidal and normal distribution curve. This shape is not selected for each input and output item, but is defined for all items. The shape cannot be changed for each item.

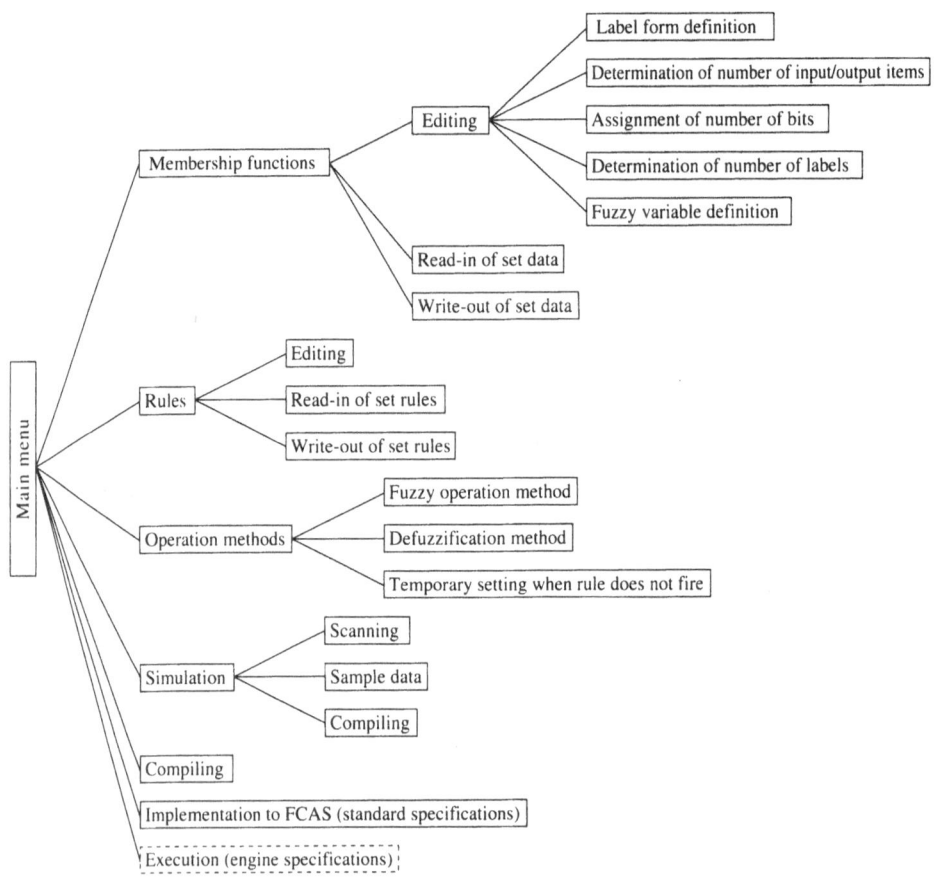

Fig. 4.3 Menu composition.

Number of Items

The number of input and output items is defined. In the case of standard specifications (a special purpose Mycom Fuzzy Inference Processor), the limits of normal input and output are 4 items and 2 items, respectively, so the upper limits of these settings are 4 inputs and 2 outputs. In the case of engine specifications, up to 8 inputs and 2 outputs can be specified.

Exchange Bit-map

The numbers of bits for each item are assigned. The Mycom Fuzzy Inference Processor uses digital input and output, so it is necessary to assign

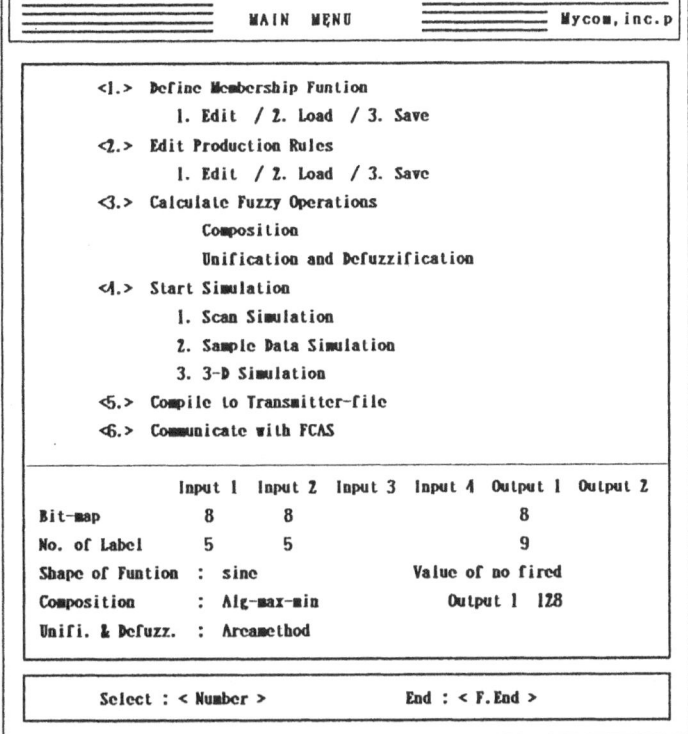

Fig. 4.4 Main menu (standard specifications).

the numbers of bits.

In the case of standard specifications, the total number of bits assigned to input items must be 16 or less, while the number assigned to output items must be 8 or less. In the case of engine specifications, the number of bits for each item must be set to 12 or less.

Exchange Number of Labels

The language value for each item is set. The choice is from among 5 numbers of labels: 3/5/7/9/11. In creating an ordinary rule, it is sufficient to have 11 labels; conversely, if there are 11 labels, rule creation becomes difficult.

Move Membership Function

The membership variables for each item are corrected. In the initial condition the default option (equal intervals) is set; the correction is performed while watching the graphic display.

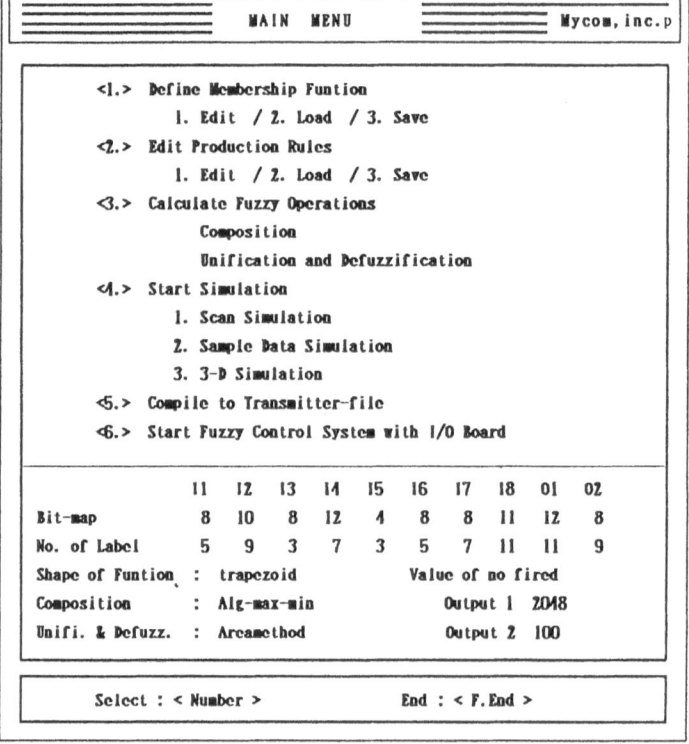

Fig. 4.5   Main menu (engine specifications).

When this menu appears, all of the membership functions (standard specifications) for each item are displayed as shown in Fig. 4.7. When the item which it is desired to correct is selected from among them, the display of the selected item is enlarged and the correction can be made (Fig. 4.8).

The display is graphic, so the membership function image is easily grasped.

4.4.2   Edit Production Rules

Fuzzy production rules are created and edited.

When this menu appears, the choice is among 1. Edit 2. Load 3. Save. Load and Save read and write files that have been created and edited with Edit, respectively.

When Edit is selected, the display shown in Fig. 4.9 appears, and input is performed in accordance with the label number for each item.

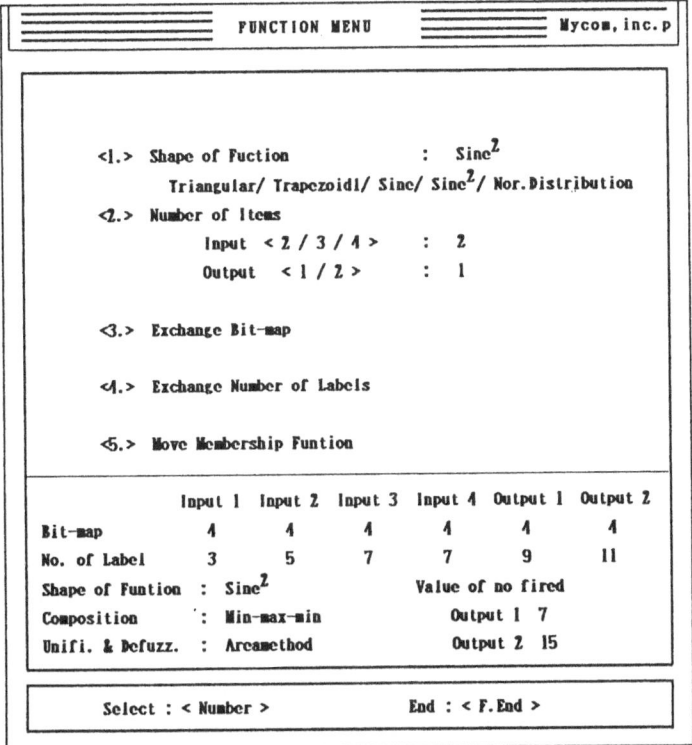

Fig. 4.6 Function menu.

When a rule is created and "∗" is input in place of the number corresponding to the label, this item is ignored. For example, the rule "IF A=Z(2) & B=∗ THEN E=Z(2) & F=∗" becomes "IF INPUT A IS Z, THEN SET OUTPUT E TO Z".

As the number of input items becomes large, the creation of rules for all input items becomes difficult. From a normal human perspective rules can be created for up to about 3 or 4 input items, so "∗" is a convenient option.

Also, the membership functions for the input items are displayed at the same time so the shapes of the membership functions that are created can be checked as the rules are created.

Normally up to a maximum of 2,000 rules are supported.

4.4.3 Calculate Fuzzy Operations

The fuzzy inference operation method is selected, and operational values at no-fire are set.

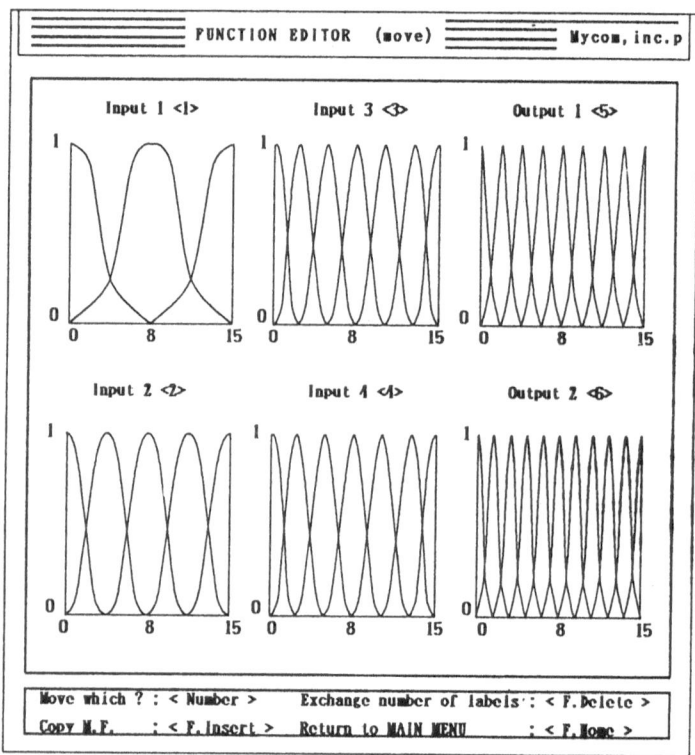

Fig. 4.7 Membership function display.

This menu is shown in Fig. 4.10.

The synthesis operation method is set. As was stated at the start of this chapter, many operation methods have been announced. Among them, 3 representative operation methods are supported.

Also, expressions which ,do not appear in other manuals are used. They are almost the same as existing operation methods, but are expressed in a bit more detail.

Three types of operations are explained below. The contents of each operation method are divided into 3 items; the order of operations is right to left. If the preconditions of the rule are A, the observed (fuzzy) data are A', the conclusion of the rule is B and the inference result is B', the operation can be expressed by the following formulas:

$$
\begin{array}{llllll}
* * * * * * * * & - & * * * & - & \min & : \quad \alpha_i = m_A(X) \cap m_{A'}(X) \\
* * * * * * * * & - & \max & - & * * * & : \quad \alpha = \cup \alpha_i \\
\text{Minimum} & - & * * * & - & * * * & : \quad m_{B'} = \alpha \cap m_B \\
\text{Algebraic} & - & * * * & - & * * * & : \quad m_{B'} = \alpha \cdot m_B \\
\text{Bounded} & - & * * * & - & * * * & : \quad m_{B'} = \alpha \odot m_B \qquad (1)
\end{array}
$$

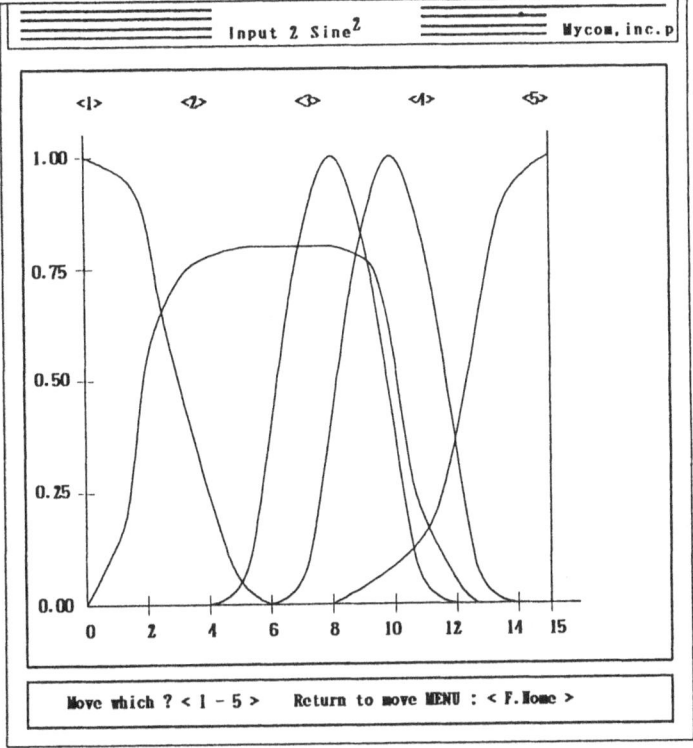

Fig. 4.8 Membership function correction.

where:

$$i = 1 \sim n$$

$\cap$: logical product
$\cup$: logical sum
$\cdot$: algebraic product
$\odot$: bounded product

"$* * * * * * * * * - \max - \min$" is the same for all 3 operation methods. As shown in Fig. 4.11, logical products are performed for all of the observational (fuzzy) data, and $(\alpha)1$ and $(\alpha)2$ are obtained. Next, the logical sum of α_1 and α_2 is taken, determining the degree of (α) of m_B.

The parts of the 3 operations that are different are at the left end. The membership m_B of the degree of (α) and the conclusion of the rule is derived by the specified operation (Fig. 4.12).

Unification & Defuzzification

The method of defuzzification is determined from this menu; at present only the area method is supported. The center of gravity (CG) method is

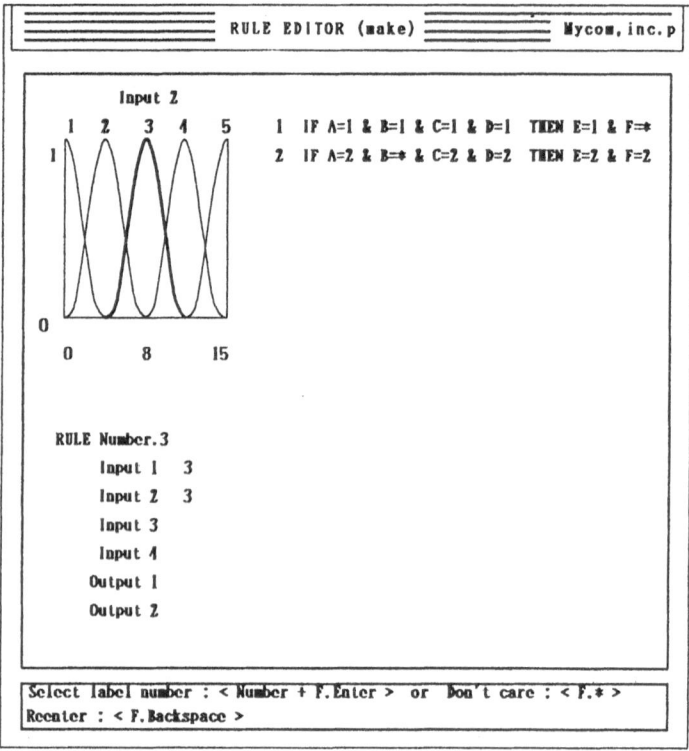

Fig. 4.9 Creation of a fuzzy production rule.

in general use, but we believed that the area method is more appropriate for control, and decided to use it. In future versions the center of gravity method will be included so that other defuzzification operations will also be supported.

Whereas, in the conventional center of gravity method, when conclusion labels are superimposed the center of gravity of the outer circumference shape is calculated by logical sums, in the area method the center of gravity of each label is calculated, then finally the center of gravity of all of the labels is found. This way the areas that are superimposed for 2 or more labels count 2 or more times toward the conclusion. Depending on the fuzzy variable settings and the synthesis operation settings.

The conclusion is completely contained in the label of another conclusion; in the center of gravity method even if there is a rule that has fired it sometimes happens that it is not reflected in any conclusion. The area method has the advantage that this kind of problem does not occur.

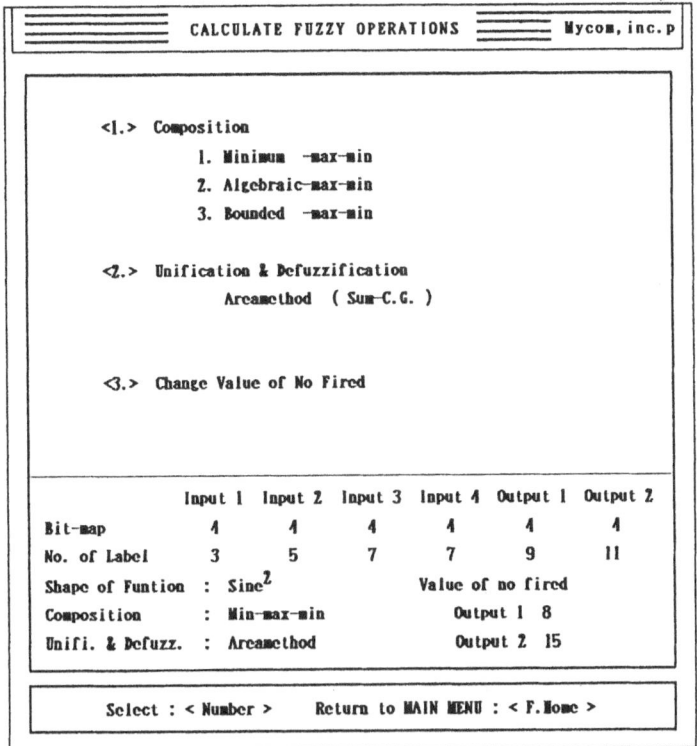

Fig. 4.10 Fuzzy operation methods.

Change Value of No Fired

If there is no set fuzzy production rule that has fired, normally it will be in the middle of the support set. However, the fact that there is no rule that has fired normally means that something is abnormal; depending on the system, the value that should be output in an abnormal condition is not necessarily in the middle of the support set.

Consequently, it has been made possible to freely set the conclusion to be output when no rule has fired.

4.4.4 Start Simulation

This menu performs fuzzy inference on the membership functions and fuzzy production rules that have been created, based on the set input conditions, and evaluates the inference results.

At present there are 3 simulations that can be performed on this menu.

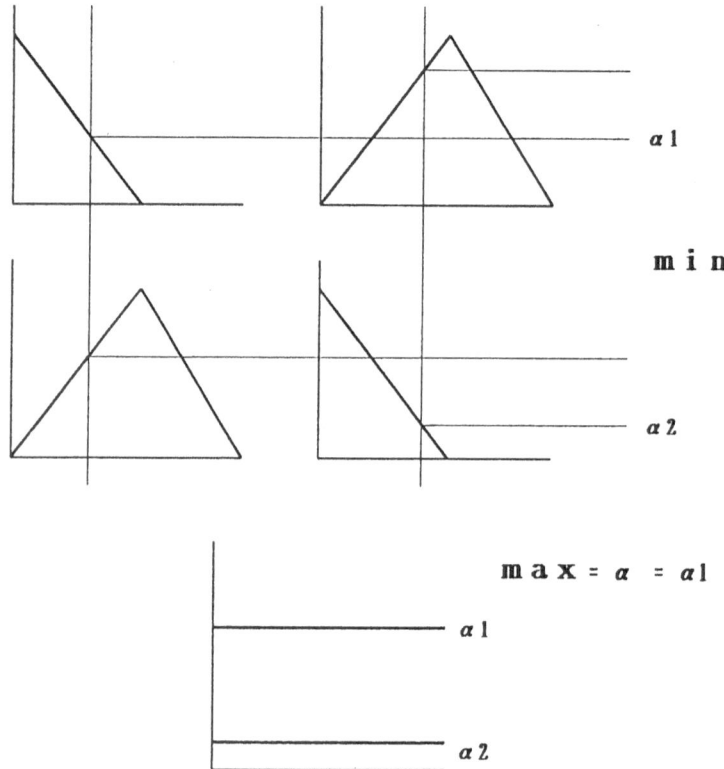

Fig. 4.11   Synthesis operations 1.

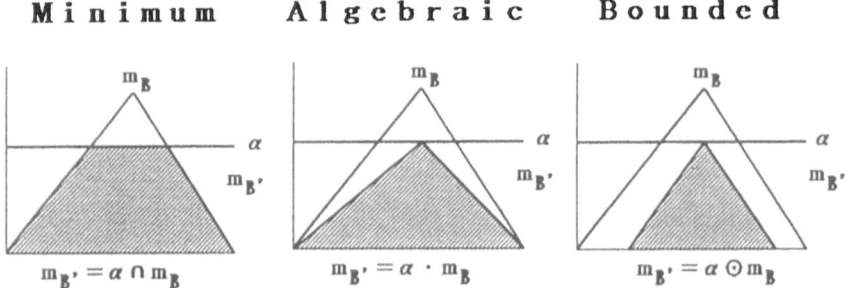

Fig. 4.12   Synthesis operations 2.

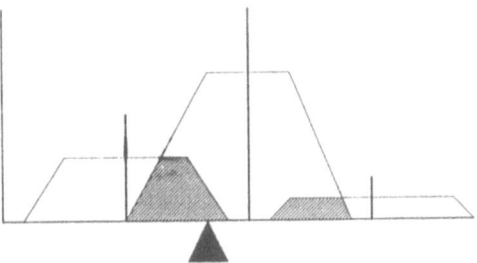

Fig. 4.13 Area method.

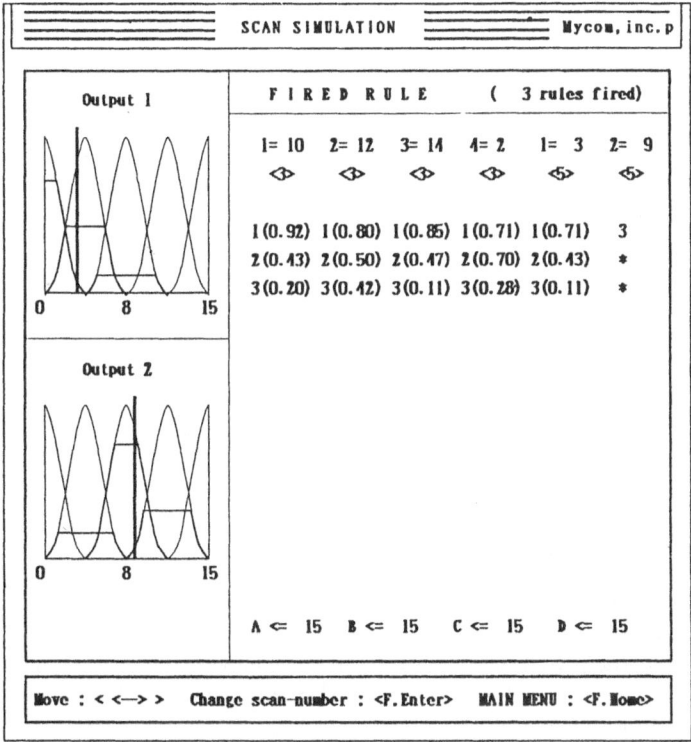

Fig. 4.14 Scan simulation.

Scan Simulation

The results of fuzzy inference performed on the set inference conditions are displayed (Fig. 4.14).

This scan simulation is for the purpose of verifying the details of fuzzy inference results with respect to 1 condition.

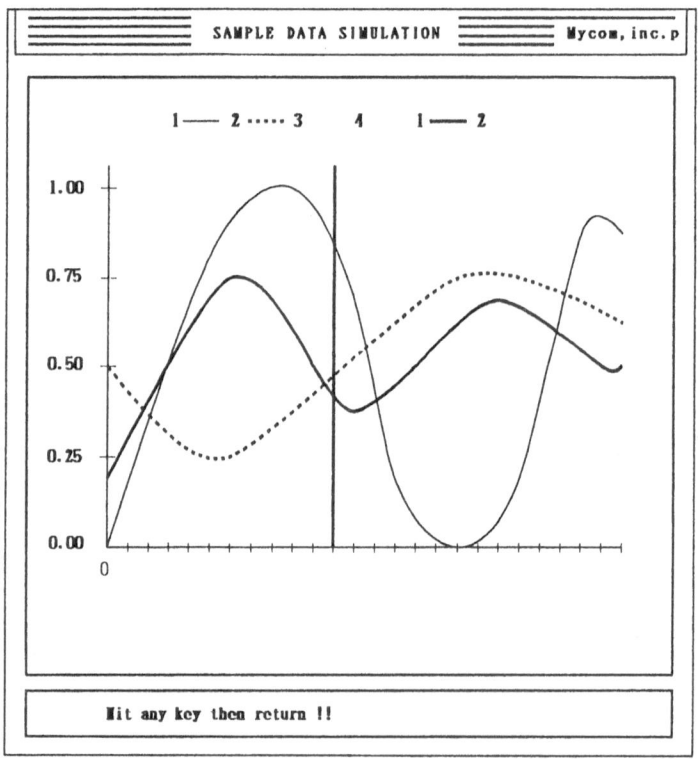

Fig. 4.15 Sample data simulation.

In Fig. 4.14, all of the rules that fired on OUTPUT 1 are shown in the frame at right. In this example there are 3 rules that fired. The input conditions (1=10 means that the INPUT 1 condition datum is 10) and the output inference results (1=3 means that the OUTPUT 1 result is 3) are shown. Below them the group of rules that fired are shown. The value $(\alpha)i$ that was explained in section 4.4.3 is shown in the parentheses following the number corresponding to the input label, and the value (α) in the output parentheses. The synthesized fuzzy inference and defuzzification result (vertical bars) are shown in the frame at left.

The considerable detail which can be obtained from this display helps greatly in membership function correction and rule correction when tuning.

Sample Data Simulation

Input conditions are modeled, data set and the transition of the fuzzy inference result is shown (Fig. 4.15).

Fig. 4.15 shows the case of 2 inputs and 1 output. Data are set in the order of input condition; the fuzzy operation result is shown in the same way. In this simulation, the fuzzy inference result with respect to input change can be seen, so the convergence status etc. can be judged at a glance.

In addition, the input conditions and output conditions of the system that has already been operating (including systems other than fuzzy control) are counted; then the output results can be compared with and evaluated against those of the existing system. Fuzzy production rules, membership functions, operation methods, etc. can be changed under the same input conditions, and the output results compared and evaluated. By using this function it becomes possible to do tuning up very efficiently.

In this figure vertical a bar is shown; by moving this bar right and left the inference under the conditions at the positions of the vertical bar can be transferred to the Scan Simulation mode display. This makes it possible to see the detailed contents of the fuzzy operation at the set location, in turn making it possible to judge more clearly at which locations corrections are needed when tuning up is done. When the check is completed it is possible to return to Sample Data Simulation mode.

3-D Simulation

3-D Simulation is a mode in which all fuzzy inference results for 1 output in 2 operation parts are shown (Fig. 4.16).

In the case of 2 inputs and 1 output, all fuzzy inference results can be seen, but in a case with 3 or more inputs the number of variable parts is limited to 2, so the remaining conditions are displayed as fixed data. Consequently, in such a case it becomes difficult to judge the overall image.

However, in the variable part the overall image can be grasped, so it can be judged of which part the operation quantity should be raised or lowered at the time of tuning up. In such a case, in what way the rule or membership function will be affected when the operation quantity is raised or lowered must be left to the operator's judgment. If it becomes possible to do tuning automatically, for example by a learning function, the capability of the Work Station will be increased. At present, even in Mycom research is underway on learning in neural networks.

4.4.5 Compile to Transmitter-file

A transmitter file is created for the purpose of implementation to the FCAS, based on the membership functions, fuzzy production rules and fuzzy operation methods that have been set (Fig. 4.17).

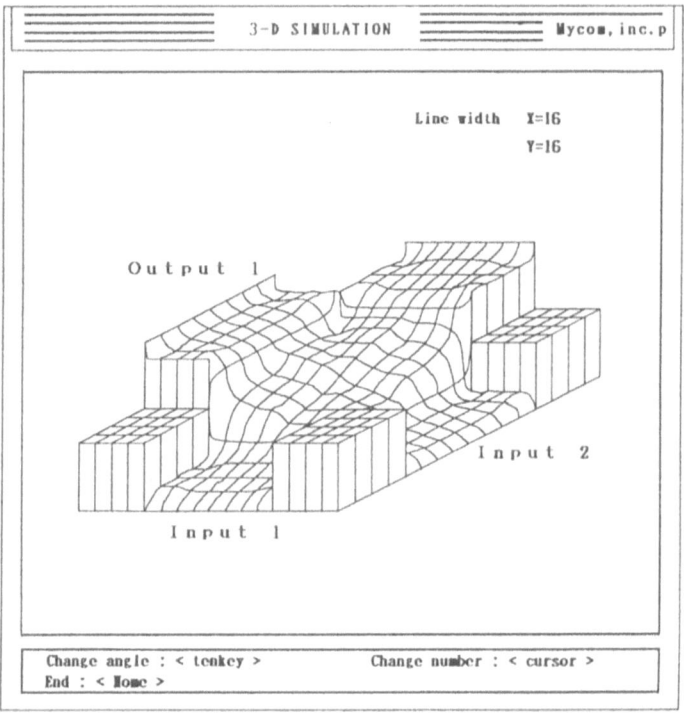

Fig. 4.16 3-dimensional simulation.

A special use fuzzy inference processor is created from these compiled data.

4.4.6 Implementation to Emulator (Communicate with FCAS) (standard specifications)

The transmitter file data created in section 4.4.5 is implemented in the FCAS through the FBEN (Fig. 4.18). After implementation, the FBEN and FCAS are disconnected, and only the FCAS is connected to the target system. After that fuzzy control is applied and the system is evaluated.

The compiled data stored in this FCAS are backed up by a battery, and normally will last for at least 2 months.

4.4.7 Execution (engine specifications) (Start Fuzzy Control System with Input/Output Board)

Fuzzy control is executed in the membership functions, fuzzy production rules and fuzzy operation methods that have been set (Figs. 4.19, 4.20).

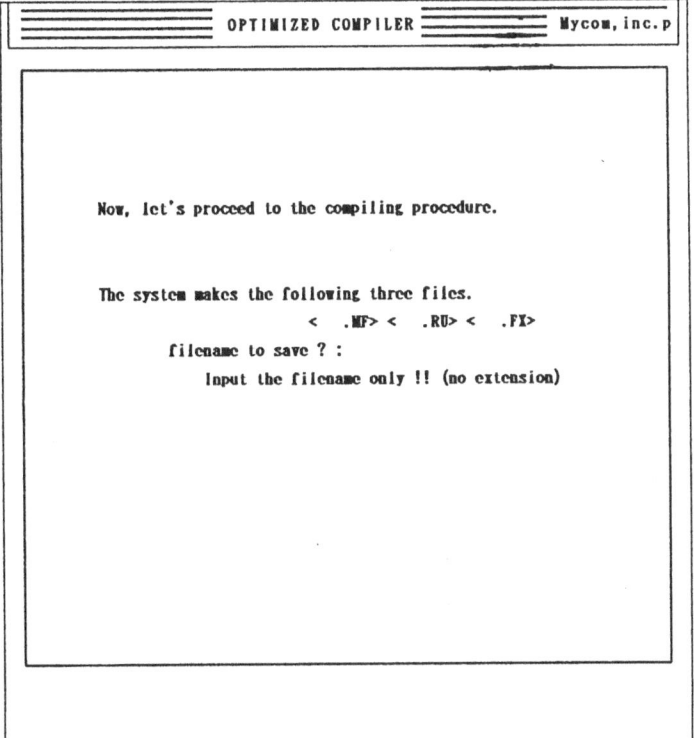

Fig. 4.17 Compiling.

In this case, whether or not the transmitter filter data in section 4.4.5 are used differs depending on the version. At present this is matched to the convenience of each user, so the standard specification has not yet been decided. It is planned to release the standard specification in the near future. Consequently, the figures merely show examples of the Execution mode.

In Fig. 4.19, the transmitter filter data created in section 4.4.5 are read in. In Fig. 4.20, the number of items and numbers of bits for each are shown in the middle of the screen; the fact that execution is in progress is indicated below these.

While execution is in progress the input data and inference results at that instant can be displayed (not shown in Fig. 4.20).

4.4.8 Verification of differences among fuzzy operation methods with the Fuzzy Work Station

As was stated at the beginning of this chapter, even if the same membership function and fuzzy production rule are used, if the operation method is changed there will be a subtle difference in the result.

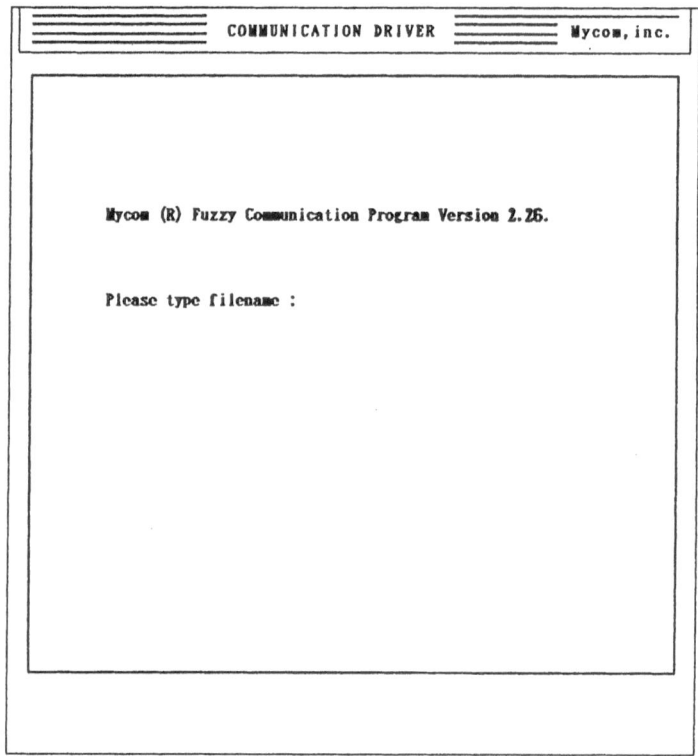

Fig. 4.18   Implementation to emulator.

Accordingly, let us use the same membership function and fuzzy production rule and compare 3-D simulations with 3 different fuzzy operation methods. We use 2 inputs and 1 output to make the 3-D simulations easy to compare.

As shown in Fig. 4.21, the membership function has a trapezoidal shape; each item is performed with a language value of 5 labels. As shown in Fig. 4.22, 9 fuzzy production rules are compared. Proportional rules are used.

The differences among the 3-D simulations with these settings are shown in Figs. 23(a) to 23(c).

It is seen that there are subtle differences among them. Since the rule itself is proportional, it is believed that the algebraic product operation shown in Fig. 4.23(b) is the most appropriate, but which operation method is optimum can differ depending on the target system.

Next, we verified the manner in which the conclusion changes when the shape of the membership function is varied. In the Fuzzy Work Station, 5 different shapes are supported, but if the fuzzy operation method is fixed at logical product (Minimum-max-min) and the membership function shape is

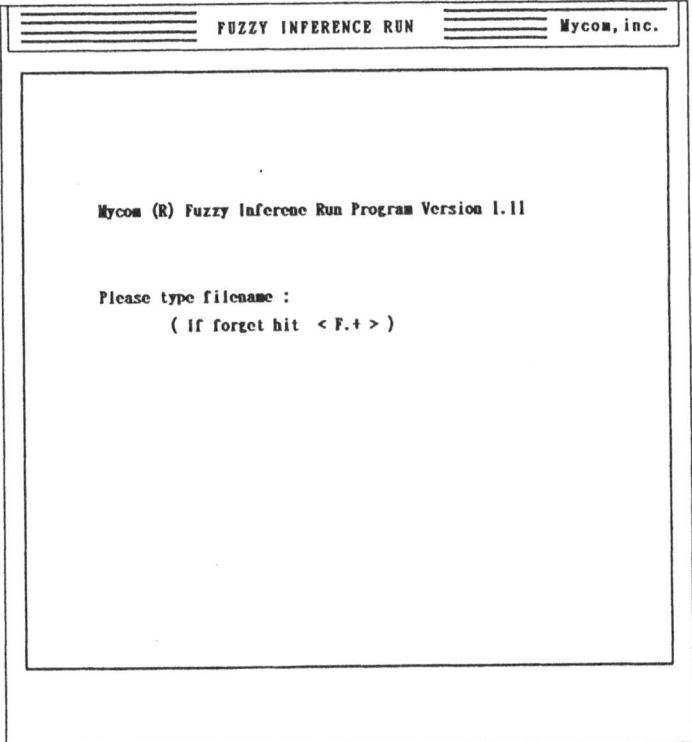

Fig. 4.19 Execution display 1.

varied among triangular, sinusoidal and squared sinusoidal, there is almost no change from Fig. 4.23(a); but if a normal distribution curve is used (Fig. 4.24), then, as shown in Fig. 4.23(d), the inference result becomes greatly different.

From this fact, it is seen that as long as the membership function shape is not set to normal distribution curve, the inference result hardly changes at all. If fuzzy inference is done with software operations rather than special purpose hardware, it is believed that the operation time factor will have a great effect on the system. When the operation is performed, if a curve-shaped membership function is used each operation takes time. In spite of this, if the operation result is not greatly different from that obtained with a linear membership function, there is no advantage to going to the trouble of using a curve-shaped membership function.

In software operations, it is recommended that a linear-type membership function be used.

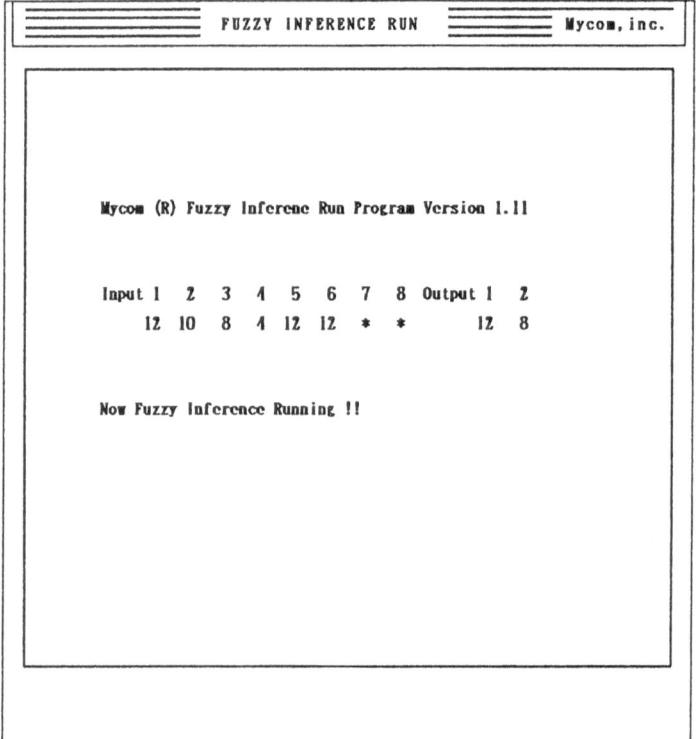

Fig. 4.20   Execution display 2.

4.4.9 Miscellaneous

When "Mycom, inc.p" appears in the upper right of the screen while operation is in progress, hard copy of the screen display can be obtained. This is convenient in obtaining records of graphic displays of membership function shapes, simulations, etc.

Also, co-process can be started by key operations, so MS-DOS command.com can be called at any time.

4.5 Characteristics of the Virtual Paging Fuzzy Inference Chip

This Fuzzy Inference Chip is a processor based on a memory element, and has flexibility with regard to, for example, operation method.

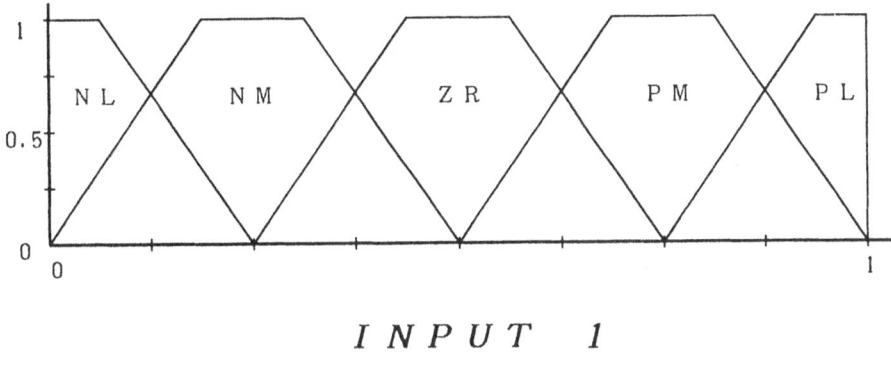

I N P U T   1

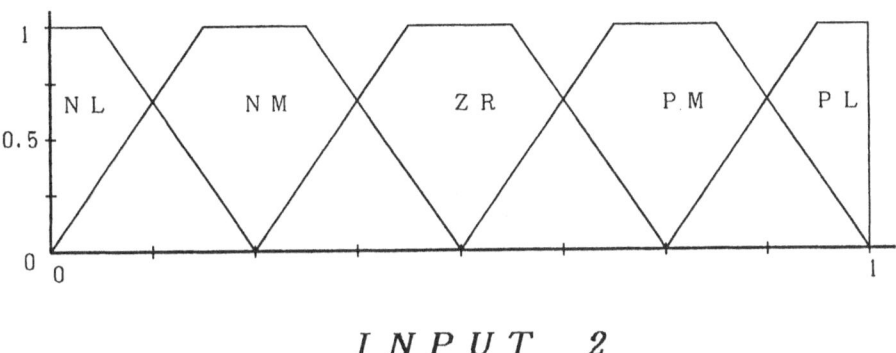

I N P U T   2

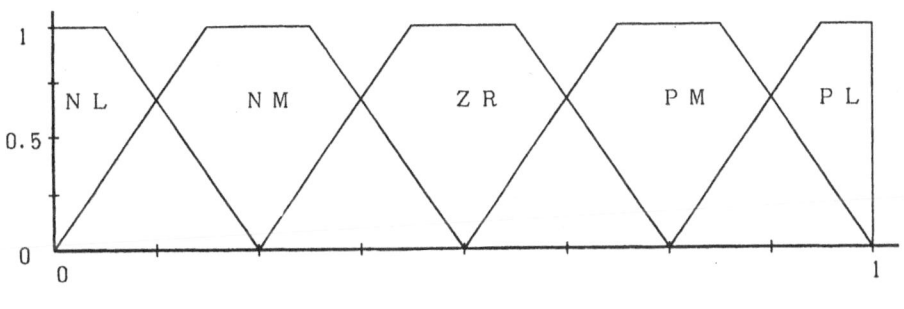

O U T P U T   1

Fig. 4.21 Membership functions.

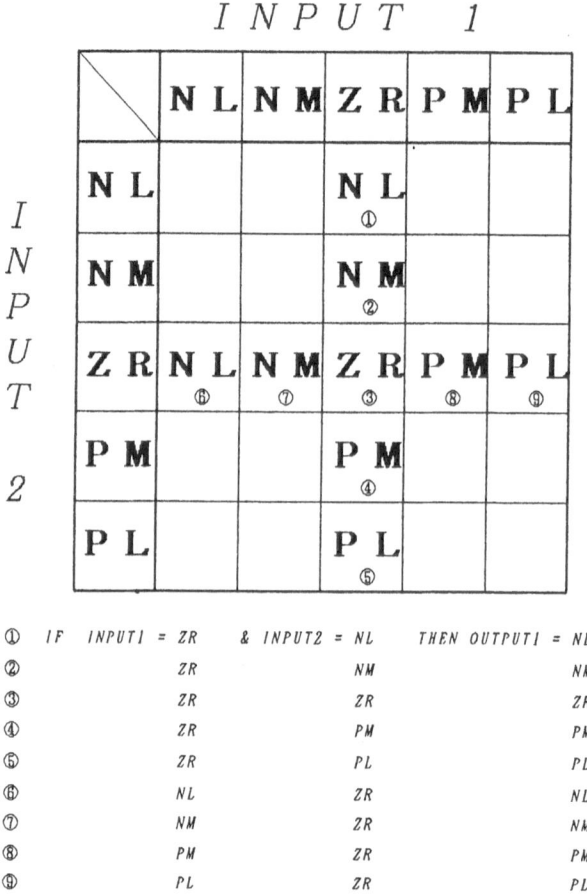

Fig. 4.22 A fuzzy production rule.

4.5.1 Characteristics of the Special Use Fuzzy Inference Chip

High speed inference execution

Standard specifications, with inference speed of 6MFLIPS.

Ability to resist noise

Since this is a digital device, it is less affected by noise than analogue devices. In addition, this processor does not have a command code like a Neumann- type CPU, so there is no runaway sequence.

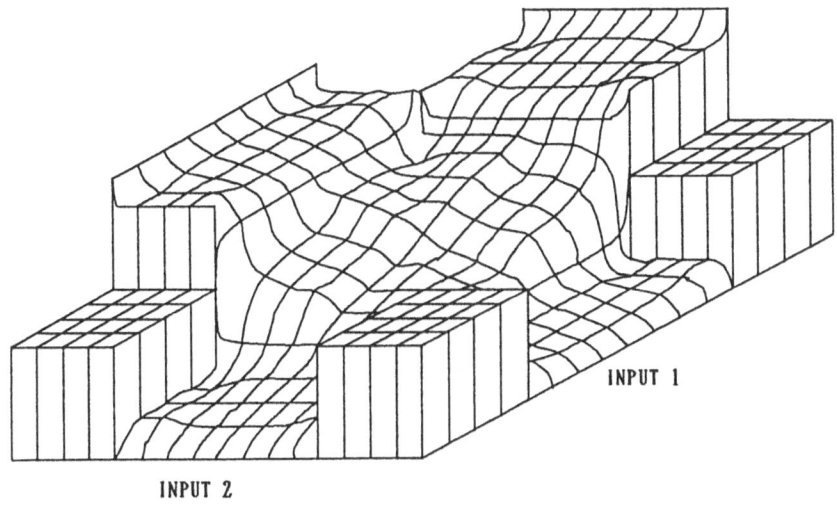

Fig. 4.23 a A fuzzy production rule.

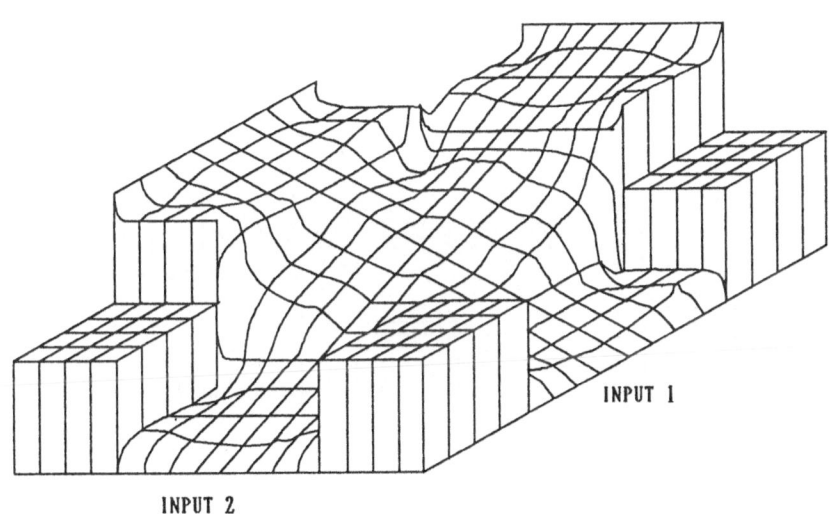

Fig. 4.23 b

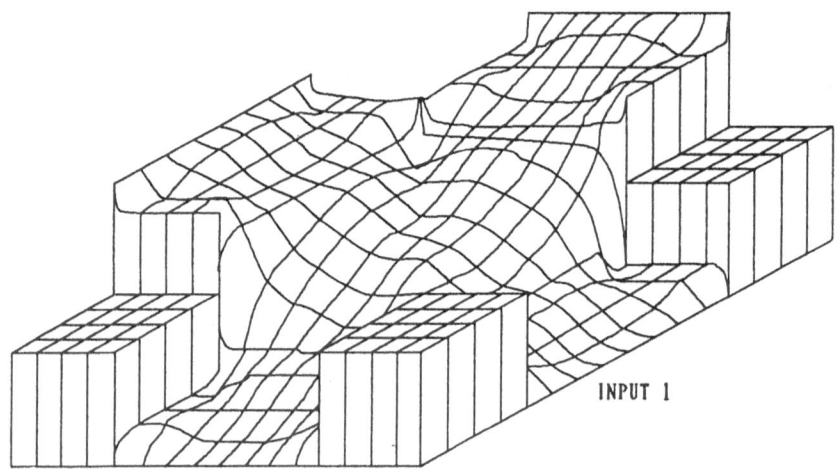

Fig. 4.23 c

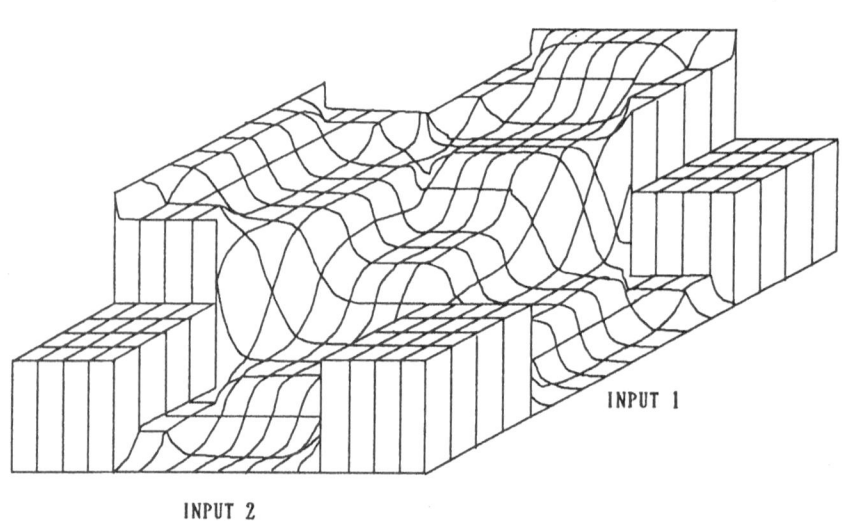

Fig. 4.23 d

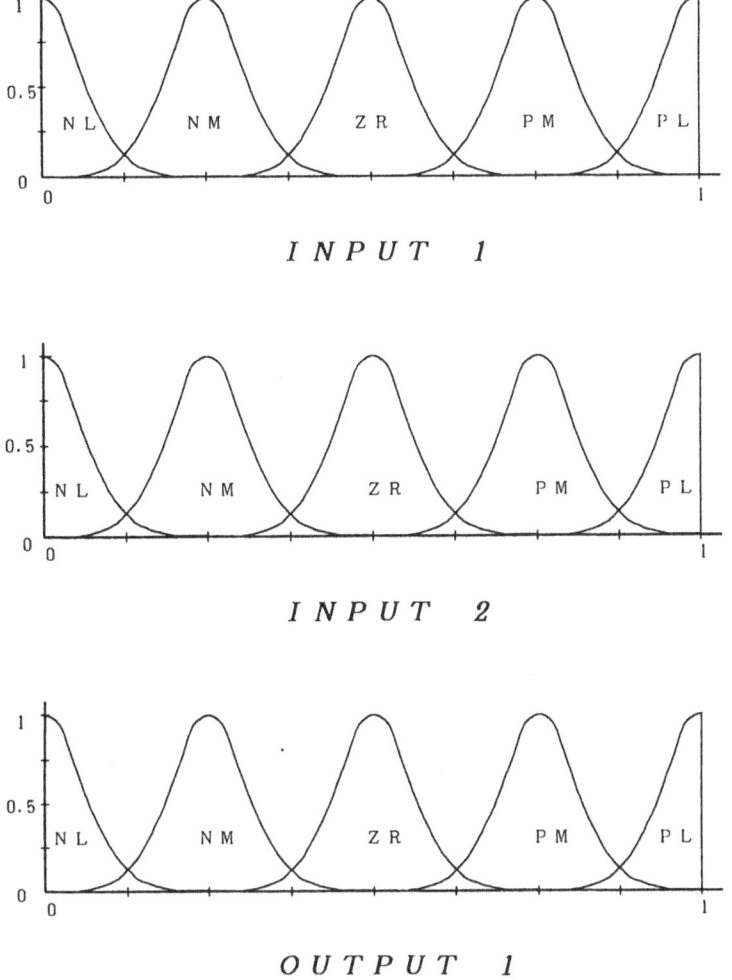

Fig. 4.24 Case of a normal distribution curve.

Linkage is easy

The inference result output signal has bus specifications, so it can be linked by bus to an existing system. Compressor-type action can be performed.

Number of rules, number of labels

The Work Station has constraints on the number of rules and number of labels, but in this processor there is no limit.

Membership function

The membership function shape may be freely selected; the membership values are processed as continuous, not discrete, values.

Operation method

On the Work Station, there are 3 types of operation method, but it has the ability to use other operation methods as well.

Membership function expandability

The number of input terminals is naturally limited. Consequently, it is easy to think that the number of input items would also be limited, but this is not the case.

Suppose, for example, that we have the 2 input items X and Y. This means that there exist membership functions m_x and m_y. However, $m_{f(X,Y)}$ can be thought of as a membership function consisting of X and Y.

For example, one can conceive of m_{X-Y}, $m_{(X_2-Y_2)}$, etc.

Input that is connected as hardware can at least be defined without limit as a membership function. Finite differences, etc. can be treated as variation amounts, so, rather than having a differentiation circuit or performing differentiation before the input section, it is possible to latch past data, then input 2 present data and take the difference internally.

4.5.2 Future fuzzy inference processors

The characteristics of the Special Purpose Fuzzy Inference Chip was described in section 4.5.1, but it is not suitable for a system which requires a great deal of input and output (not input that is created internally, but input that is connected as hardware).

The 8 input 2 output type, which was described in the section on the Fuzzy Work Station with engine specifications, cannot be constructed. At present a processor that will match an analogue input type (an analogue 0 to - digital converter is built in) Fuzzy Work Station with engine specifications is being designed. In industrial fields, high speed and multi - input processors are necessary. Meanwhile, the fuzzy inference processor does not function at all in managing the whole system. It is necessary to perform the management with another CPU. However, if a CPU is equipped with fuzzy operation software or fuzzy operation hardware, it can perform both functions. From this point of view, a fuzzy inference processor that can perform management processing will be constructed in the near future.

4.6 Summary

In this chapter, the explanation has centered on the Mycom Fuzzy Work Station.

When a fuzzy control system is constructed, how knowledge about operation of the control object system is incorporated into the membership function and fuzzy production rule is important. For this reason, tools are needed to perform the incorporation work smoothly.

We believe that improving tool performance will make it easier to introduce fuzzy control.

Another future improvement that we hope to make is to improve the capabilities of tools, for example by incorporating learning functions such as automatic creation of membership functions.

The need for fuzzy inference processors in industrial fields is going to increase, and higher performance processors are going to be required. We intend to develop processors that will meet the need.

Finally, I wish to express my deep thanks to Prof. Hirota of Hosei University, Prof. Sakawa of Hiroshima University and others who guided and assisted us in developing the Fuzzy Work Station.

References

1. Hirota K (1988) Principles and applications of fuzzy control (in Japanese). Automation 33 (6) : 1–66

2. Hirota K, Terano T (1989) Fuzzy control (in Japanese). Computrol 28 : 1–135

3. Hirota K (1990) Fuzzy control (in Japanese). Automation 35 (4) : 9–97

4. (1990) Full volume on fuzzy (in Japanese). Trigger Sha, June Special Volume pp 1–124

5. Tsuchikata M Front line of fuzzy applications. Trigger Sha 9 (14) : 8–27

6. Arikawa H, Hirota K (1988) Fuzzy inference engine by address—look–up and paging method. In: Proc International workshop on fuzzy system application pp 45–46

7. Togai M (1987) A fuzzy logic acceletor for real-time fuzzy control. In: Second IFSA Congress

8. Watanabe H, Dettloff W (1987) Fuzzy logic inference processor
    for real-time control: A second generation full custom design.
    In: Twenty-first Asilomar conference on signals, systems and computers

9. Yamakawa T (1988) Fuzzy microprocessors. In: Proc International
    workshop on fuzzy system applications

10. Fuzzy chips (in Japanese) (Oki Electric Co.) (1990) Nikkan Kogyo
    Shimbun, 12 July 1990

11. Fuzzy chips (in Japanese) (Omron) (1990) Nikkan Kogyo Shimbun, 12
    July 1990

12. Terano T, Asai K, Sugeno M (1989) Introduction to applied fuzzy systems
    (in Japanese). Ohm Sha 1–48.

13. Sakawa M (1989) Principles and applications of fuzzy theory (in Japanese).
    Morikita Shuppan : 2–72

Chapter. 5
FUZZY PROCESSORS

5.1 Introduction

Fuzzy technology applications, which attempt to introduce human inference methods into engineering fields, are rapidly increasing. Fuzzy is expected to have its greatest effect in fields in which conventional PID control is difficult to apply because rigorous models of the control objects are not available and the tendency has been to give up on obtaining improved performance and/or to rely on human control.

Omron developed the FZ-1000 ultra-high speed fuzzy controller, the world's first hardware exclusively for fuzzy use, in 1987, thus starting today's fuzzy application boom. As the fields in which fuzzy is used increase, the need for specialized chips that execute fuzzy inference at high speed is also increasing. The use of specialized chips has resulted in 2 to 3 orders of magnitude higher speed than could be obtained with fuzzy software run on general purpose microprocessors, leading to new fuzzy applications in machine control, image processing, decision making and diagnosis, and other applications requiring processing of large amounts of information.

This chapter introduces the FP-3000 digital fuzzy processor, which Omron has developed and put on the market, and analogue fuzzy hybrid ICs.

The FP-3000 digital fuzzy processor is the heart of Omron's fuzzy processor for controll purposes. It is a processor exclusively for fuzzy use, designed so that it can be connected to all types of microprocessors so that it can be incorporated into various types of controllers and miniature electronic devices. It was awarded a prize as one of the top 10 new products of the year in 1990 by Nikkan Kogyo Shimbun. A variation of this chip, an ultra - low cost version incorporating a built-in ROM, is also announced. Meanwhile, the analogue fuzzy hybrid IC is the first product of Omron's analogue fuzzy processor series. It is being developed based on the architecture of the FZ-1000 ultra-high speed fuzzy controller. Attention has been focused on the continuity and high speed of the analogue circuit; the use of parallel architecture has made ultra-high speed fuzzy operations possible.

Fig. 5.1 Photograph of the FP-3000

5.2 The FP-3000 digital fuzzy processor

5.2.1 Outline of the FP-3000

The FP-3000 is a processor exclusively for fuzzy use; its microprocessing system incorporates fuzzy inference and defuzzification functions in a single chip. Consideration has been given to incorporation into various types of controllers and miniature electronic devices, and high cost performance for the controller series obtained. Its principal characteristics are as follows.

1. High speed inference processing: 650microseconds (5 conditions, 2 conclusions, 20 rules, 24MHz)
2. Has an interface analogous to an SRAM; can be connected to a variety of CPUs.
3. Fuzzy operations (including defuzzification) can be done in a single chip (single mode)
4. Input/output resolution is 12 bits.
5. Number of rules per inference: 128max (extended mode)
6. Weights can be assigned to each rule.
7. Grade data can be monitored for each rule (extended mode)
8. 3 different inference rule groups can be stored in memory.
9. Operations are executed in a fixed processing time corresponding to the rule.
10. The input/output interface is TTL/CMOS compatible.
11. One power supply, +5V

The FP-3000 has been achieved in a CMOS standard cell. An exterior view of it is shown in Fig. 5. 1, and a pin connection diagram for the 64-pin QFP package in Fig. 5. 2.

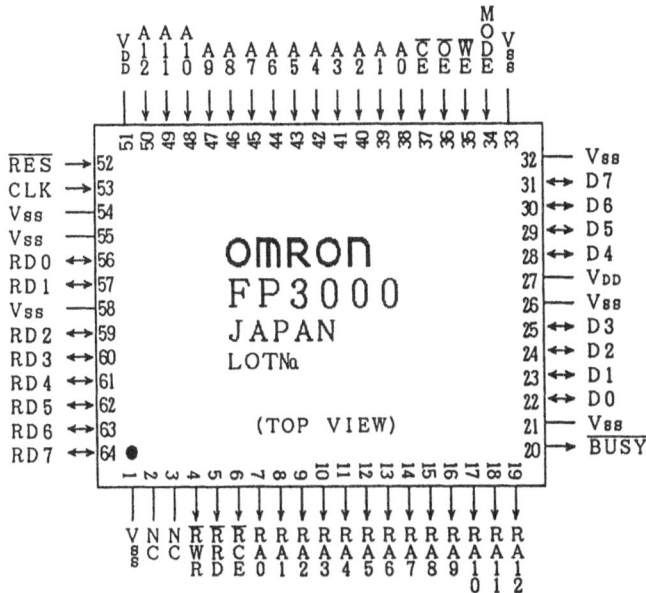

Fig. 5.2 Pin connection diagram

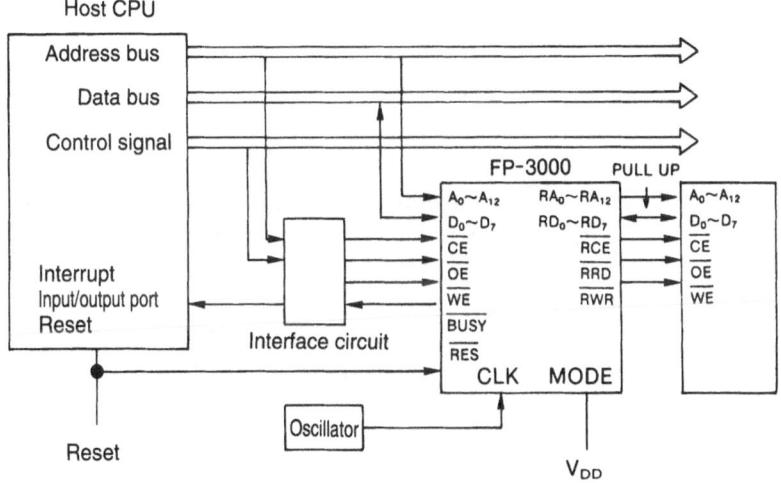

Fig. 5.3 System composition

System composition

The FP-3000 functions as an exclusive-use fuzzy processor through bus interfacing to a host CPU. The system composition is shown in Fig. 5. 3. It can be connected to a variety of CPUs. It has both a single mode and an extended mode; in a system with a large number of rules an SRAM can be connected externally to form an extended mode system. In the case of a 5-antecedent, 2- consequent rule format, up to 29 rules can be set in single mode, up to 128 rules each for 3 different types of rule groups in extended mode. In the case of extended mode, the grade of each rule can be monitored.

The basic system operation is as follows. At the beginning, antecedent membership functions (the membership function is referred to below as the MF), consequent MFs and rules are downloaded to the FP-3000 from the host CPU as input data and knowledge base, and fuzzy operation is activated. The FP-3000 then executes fuzzy inference based on the inference rule in IF ⋯ THEN ⋯ rule format with up to a maximum of 8 conditions and 2 conclusions, with respect to input data in a maximum of 8 inputs, and defuzzification for a maximum of 4 outputs, then informs the host CPU that the fuzzy operations are completed. The host CPU reads the fuzzy operation results, writes the next input data and initiates the next fuzzy operation.

The input/output data resolution is 12 bits; the fuzzy inference method is the MAX-MIN synthesis and logical product method; either the center of gravity method or the maximum height method can be selected as the defuzzification method. In addition, in the maximum height method there is a choise of left or right priority if there are several maximum heights.

Architecture

A block diagram of the FP-3000 is shown in Fig. 5. 4. It is made up of the following 3 blocks, using a microprocessing method.

1. Operation section: Arithmetic and logical operations
2. Control section: Instructions are decoded and microcommands formed.
3. Memory section: Fuzzy rules, MFs input data, and operation results are stored.

1. Operation section: The operation section consists of an ALU, multiplier and registers. The ALU has addition/subtraction, logical operation and bit shift functions, and executes fuzzy inference (MAX-MIN synthesis and logical products) and defuzzification (center of gravity method and maximum height method). The multiplier is formed in combination with a barrel shifter (a shifter that executes multiple bit shifts at high speed) to process antecedent MF decimal data, to increase processing speed.
2. Control section: The control section decodes the 8 types of instructions shown in Table 5.1, and forms micro commands. A micro command consists

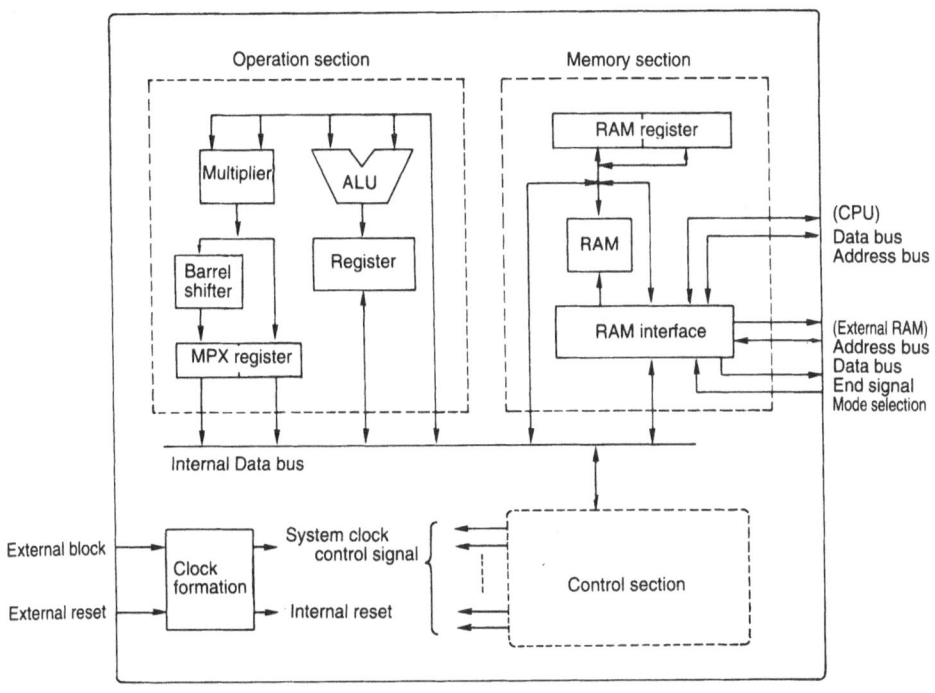

Fig. 5.4 Block diagram

Table 5.1 Instruction

	Instruction	Content
IF	Antecedent processing	MIN operation after grade is extracted from antecedent part.
THEN	Consequent processing	MAX operation by conseqent on grade from antecedent.
MAXH	Maximum height method	Maximum height method (indicate right or left priority)
COG	Center of gravity method	Center of gravity operation
RWIT	Rule weighting processing	Assignment of weights for 1 rule.
NOP	NOP processing	Antecedent weighting processing is skipped, and next instruction is executed.
SKIP	Rule skip	An entire rule is skipped, and the next rule is executed.
END	End of processing	Inference processing is ended.

of 24 bits; there are 3 types: operation commands, register addressing commands and direct addressing commands. Operation commands execute control to the operation section. Register addressing commands transfer data between registers. Direct addressing commands are used to load ROM write-in constants to the register and for memory addressing. The memory can be accessed with any command.

3. Memory section: The memory section has a 1k× 8-bit RAM on-chip. An 8Kbyte SRAM can be connected as extended memory. The interface to the internal data bus is through a 16-bit RAM register. There are the following 3 modes for addressing to the RAM.

   (a) Program counter mode: For rule fetching
   (b) Address register mode: For MF reading
   (c) Internal data bus mode: For addressing operation results

A memory map is shown in Fig. 5. 5. The memory stores antecedent and conseqent MFs, rules, input data, fuzzy inference result output data, defuzzification result output data, etc.

MF and rule settings

Next, let us explain MF and rule settings, which form the FP-3000 knowledge base.

The antecedent MFs have a resolution of 12 bits on the X axis and 12 bits on the Y axis (grade); there are 4 shapes, Λ, Π, S and Z, and the inclination can be set arbitrarily. Up to a maximum of 7 labels (language values) of "large", "small", etc. (for example NL, NM, NS, ZR, PS, PM, PL) can be set. Since MFs having 7 labels can be independently defined for 8 conditions, an antecedent MF setting area that holds 56 settings is provided. In the FP-3000, as shown in Fig. 5. 6, 1 the antecedent MF is expressed by 12 bytes; the memory capacity has been reduced compared to that which was used for the conventional table look-up method. The inflection points X_i are treated as integers ($i = 0, 1, 2, 3$) and the inclination a_j ($j = 0, 1$) as a decimal. As shown in Fig. 5. 7, the inclination coefficient is expressed in 16 bit length. With the upper 4 bits used for the exponent (B) and the lower 12 bits for the mantissa (A), the inclination coefficient is given by the following formula.

$$\text{Inclination coefficient} = A \times 2^{B-12} \qquad (1)$$

By formulating the antecedents fuzzy inference in this manner as a decimal point multiplication, resolution equivalent to that of the table lookup method is obtained.

Since the consequent MF is a singleton, 12 bit locations on the X axis are defined for each of the 4 outputs.

The rule setting method is shown in Fig. 5. 8. The rule format can be matched to a maximum of 8 conditions and 2 conclusions; only the number of

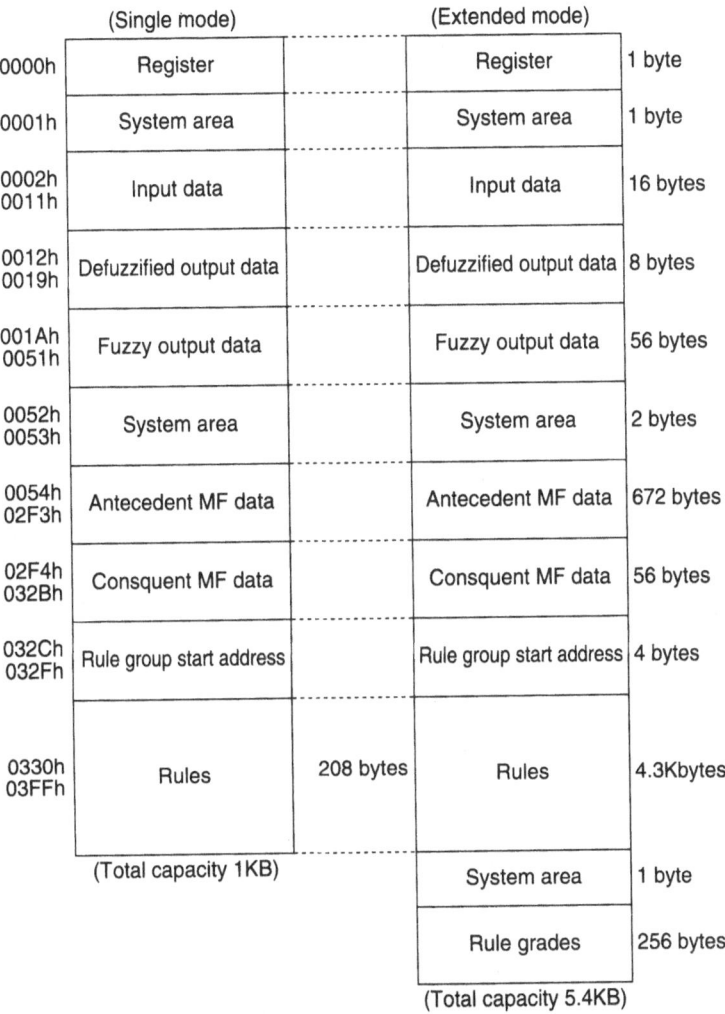

Fig. 5.5 Memory map

rules which are executed are set, with weights assigned to the rules . Then the
output defuzzification method is selected and specified, and 1 rule group is set
in the end code setting. The memory capacity that is needed for rule setting
differs depending on the number of antecedents, the number of conseqents and
whether or not weights are assigned, but in any case the maximum number
of rules that can be executed in 1 rule group is 128. There are 6 weight
values, which are grades found by the MAX operation in the condition part,
that can be assigned to a rule. The rule group setting function is useful in
applications in which it is desired to execute fuzzy operations using different

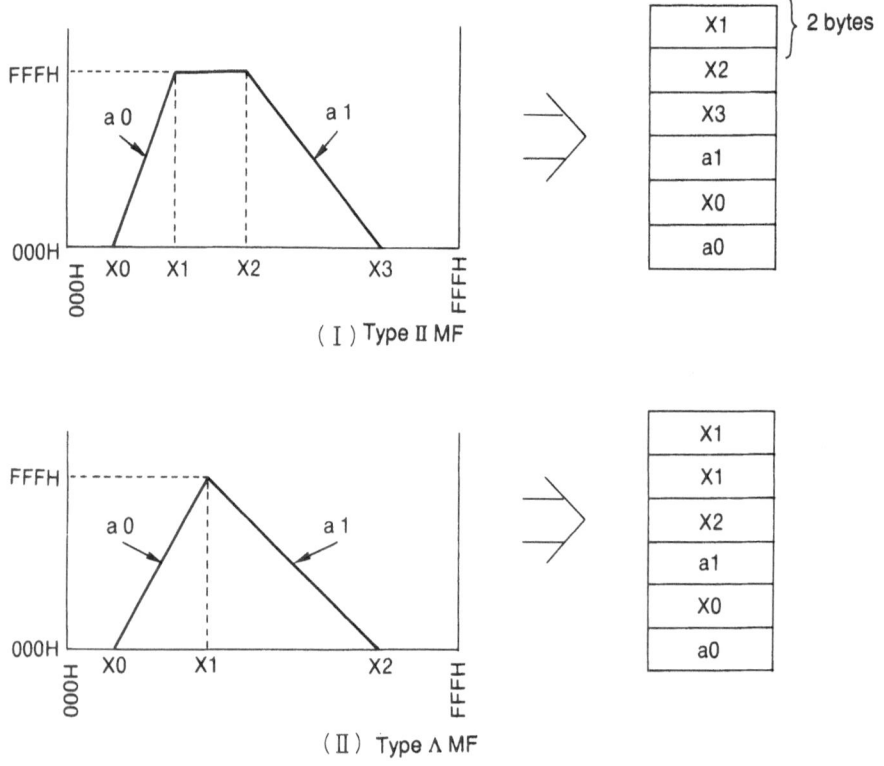

Fig. 5.6 Antecedent MF settings

(LSB)   (MSB)
b 15   b 12 b 11   b 0

Exponent : B (4 bits)	Mantissa : A (12 bits)

A = (0 1 1 0 1 0 0 1 0 0 1 1)₂

B	A
0 0 0 0	. 0 1 1 0 1 0 0 1 0 0 1 1
0 0 0 1	0 . 1 1 0 1 0 0 1 0 0 1 1
0 0 1 0	0 1 . 1 0 1 0 0 1 0 0 1 1
1 0 1 0	0 1 1 0 1 0 0 1 0 0 . 1 1
1 0 1 1	0 1 1 0 1 0 0 1 0 0 1 . 1
1 1 0 0	0 1 1 0 1 0 0 1 0 0 1 1 .

Decimal point

Fig. 5.7 Antecedent inclination coefficient settings

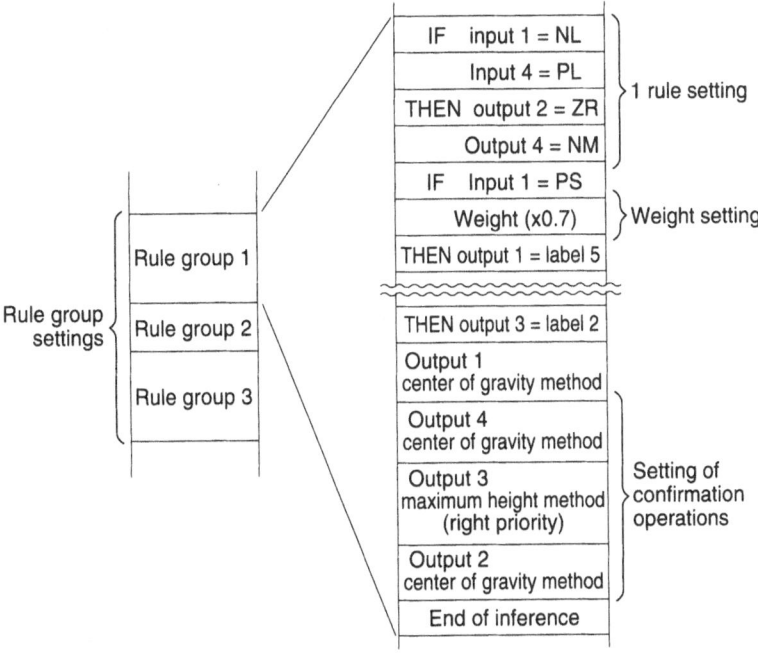

Fig. 5.8 Rule settings

rule groups.

These MFs and rules are created by the Fuzzy Inference Software FS-10AT or FS-30AT.

Fuzzy operation speed

The numbers of machine cycles required for FP-3000 fuzzy operations are as follows.

1. Fuzzy inference speed
   Processing of 1 condition: 23   cycles
   Processing of 1 conclusion: 13   cycles
   Processing of 1 weight assignment: 7   cycles
2. Defuzzification speed (per 1 output)
   Center of gravity method: 169   cycles
   Maximum height method: 89   cycles
3. Overhead time
   Overhead per 1 rule: 2   cycles
   Fixed overhead: 80   cycles
   (Note: 1 cycle = 4 external clock cycles.)

Consequently, the fuzzy operation speed for 5 conditions, 2 conclusions,

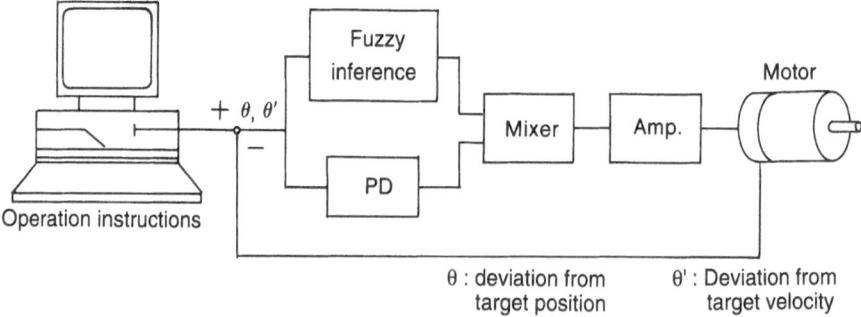

Fig. 5.9   Application of fuzzy hybrid control

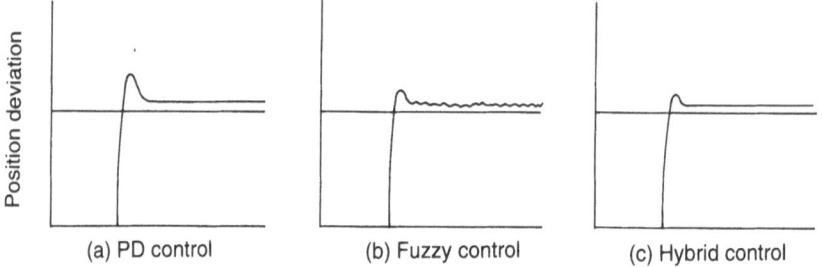

Fig. 5.10   Comparison of control performance

4 outputs (center of gravity method), 20 rules, clock 24MHz is 626 microseconds.

5.2.2   Application examples

The FP-3000 can be used for a wide variety of applications, from high speed machine control to judgment and identification. Here let us explain servo motor positioning control, an application in which the advantage of fuzzy over the conventional control method is easy to see.

A block diagram is shown in Fig. 5. 9. It is required to find the deviation of the present position and speed from the target position and speed in an external operation command. The present position and speed are computed from the output of an encoder attached to the motor. In conventional PD control, the response speed is limited by increase of overshoot that occurs as the response speed is increased. It is stable, however, at the setting time (Fig. 5. 10(a)). In fuzzy control, even in the case of rapid response the overshoot is small. However, even after settlement small oscillations remain (Fig. 5. 10(b)).

PD and fuzzy control can be combined so that fuzzy is weighted more strongly when the deviation is large and PD when it is small; then in either case the overshoot is smaller than it would be with either method used alone, and moreover after the setting time, it is stable (Fig. 5. 10(c)).

The fuzzy inference section has 2 inputs, position deviation and speed deviation, and 1 output. The FP-3000 can use up to 60 rules without need for an external RAM, in single mode, in the case of 2 inputs. The MFs used are standard equidistant triangles crossing adjacent ones at the 50% point.

5.2.3 Development support tools

The FS-10AT Fuzzy Inference Software

This is a software tool that runs on IBM PC/AT computers. Rules and MFs for the FP-3000 are created, and an object code is written. In addition, the FP-3000 inference operation is simulated, so that the inference process can be investigated in detail. If analogue-to- digital and digital-to-analogue boards are added, a control experiment can also be performed. The specifications are given in Table 5.2.

Rules are created by filling tables in (Fig. 5. 11). An MF can be created by specifying X coordinates at 4 points in the antecedent and 1 point in the consequent. 8 sets of MFs that are commonly used in control are provided, and can be used by calling them. Examples of default MF shapes are shown in Fig. 5. 12.

A rule and MF are combined, a defuzzification method (center of gravity method or maximum height method) is specified and an execution format file is created. This file is used as the FS-10AT execution file; or, alternatively, it can be used as the object code for the FB-30AT inference board to be discussed below. There is also a function for converting the execution format file to a format for use with the PROM writer.

An execution format that is created can be evaluated with details of each rule in advance by simulation. The antecedent has capabilities for single inference with details of each rule displayed graphically, and for up to 65,535 continuous inferences using input file data. In continuous inference, input/output trend graphs can be displayed. Consequent MFs and defuzzified values are displayed for both single and continuous inferences. An example of an inference result display is shown in Fig. 5. 13.

The FS-10AT is a tool for use by the FP-3000, and, at the same time, the FS-10AT itself is software that can be used independently for fuzzy learning and experiments. If analogue-to-digital and digital-to-analogue boards are added in the personal computer expansion slots, the personal computer can be used as a fuzzy controller. The software processing control cycles are in units of seconds, which is sufficient for some applications such as temperature

Table 5.2 FS-10AT specifications

Item		Specification/explanation
Form of product		5.25 inch floppy discs 3 discs FS-10AT instruction manual 1
Development languages		C and assembler
Knowledge setting section	Rule format	8 conditions, 2 conclusions per rule. Weights can be assigned for each rule.
	Number of rules	128 rules per group (file); synthesis can be performed for up to 3 groups.
	Numbers of inputs and outputs (signals)	8 inputs, 4 outputs; 12 bits per signal in both cases
	Rule setting method	Table input format: 3 alphanumeric characters or less for both signal name and label name
	Number of labels	7 labels per signal
	Membership functions	Ancedent: types S, Z, Λ and Π (the actual shapes of these are segmented line graphs with 4 points joined by line segments) Consequent: I singleton (1 straight line of grade 1) Antecedent resolution: X axis 4096, Y axis 4096 Consequent resolution: X axis 4096, Y axis 4096 Signal range: either -5.00V to +5.00VC or 0 to 4095 can be selected; this is the maximum range of variation of signal values.
Inference section	Inference mechanism	Forward inference by fuzzy production rules
	Inference method	MAX-MIN logical product
	Defuzzification method	Center of gravity method, left priority height method or right priority height method can be specified for each output signal.
	Simulation	Single inference Continuous inference Number of simulations: maximum 65535 (when HD is used)
	Actual machine control	Input: Can select either direct (use input values as they are) or differential (use difference from previous input value). Can specify sum or difference of values at different input ports. Gain or offset operation can be specified. Calculation results for both input and output must be within the specified upper and lower limits. Control interval: Can be specified as 1 to 999 seconds in 1 second steps; with additional maximum speed mode. Control data filing: data for 512 inferences can be stored for both the fuzzy inference section and the fuzzy output section.
User interfaces		Pull-down menu system; color graphic display; MMI; key board and mouse; hard copy can be printed.

File	Knowledge setting	End										

File name   **[SAMPLE]**   Comment   [Sample rule   ]

Number of rules [7]   Creation date **90/11/15**

Rule	Condition part								Operation part				Weight	
No.	DEG	DEL							SPD				(* = 0)	
1	NM	ZR							NM				1	
2	NS	PS							ZR				1	
3	NS	NS							NS				1	
4	ZR	ZR							ZR				1	
5	PS	PS							PS				1	
6	PS	NS							ZR				1	
7	PM	ZR							PM				1	
8													1	
9													1	
10													1	
11													1	
12													1	
13													1	
14													1	
15													1	
16													1	

M E N U = [ESC]   / Next page =[ROLL UP]   / Previous page = [ROLL DOWN]

Rule creation   FS–10AT

Fig. 5.11   FS-10AT rule setting

control. Functions such as deviation value and derivative computation, scale conversion and switching between position type and velocity type control are provided for convenience in control experiments. Data can be logged and displayed during control experiments.

Thus, the FS-10AT permits rules and MFs to be set interactively, making this software easy for a beginner at fuzzy control to use.

FS-30AT is a higher version tool with C language file output.

The FB-30AT fuzzy inference board

This board, mounted with the FP-3000, is for use with IBM PC/AT series computers (Fig. 5. 14). This board is plugged into a personal computer expansion slot, as discussed above, and is used by calling it from user software. For this reason, fuzzy inference can be easily incorporated into a user system on the personal computer.

This board is equipped with a driver function that is used by being called from user software (large model) created using an MS-C or Turbo C (Table 5.3). In addition, sample programs for use by the board and utilities for confirming operation are stored on the same floppy disc.

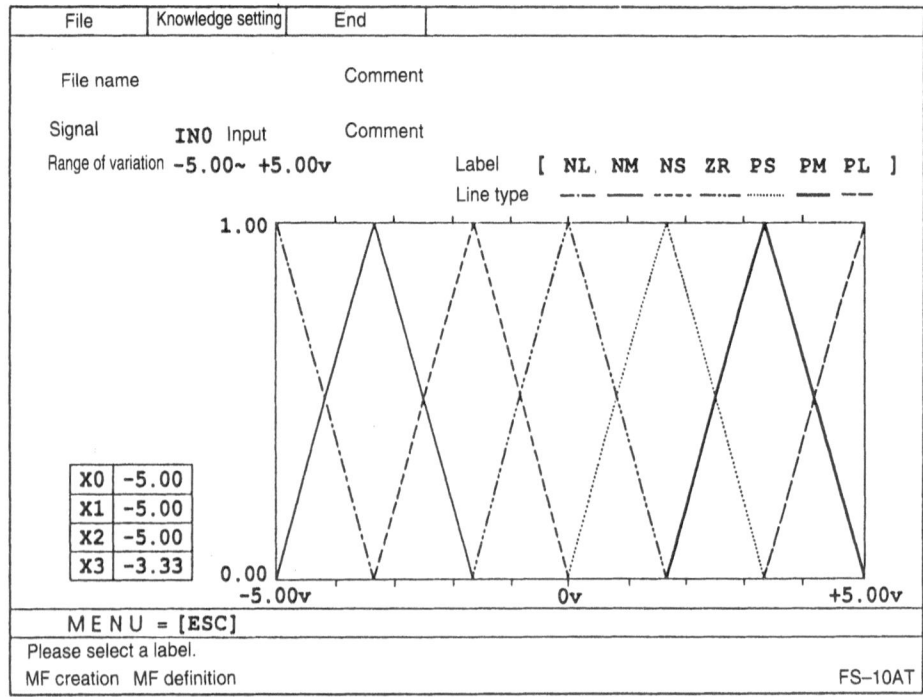

Fig. 5.12   Example of division into equal parts

Table 5.3   List of driver functions

No.	Function name	Designation	Function
1	_fb3_init	Initial processing	Sets address of interface to FB-30AT board.
2	_fb3_int_set	Interrupt processing function setting	Sets interrupt processing function.
3	_fb3_rule_xfer	Rule function transfer	Transfers execution format object to FB-30AT board.
4	_fb3_inp_set	Input value setting	Transfers input value to FB-30AT board.
5	_fb3_run_start	Inference start instruction	Instructs start of inference.
6	_fb3_sense	Status check	Checks status of FB-30AT board.
7	_fb3_out_get	Defuzzified output read-out	Confirmed output is read-out from FB-30AT board.
8	_fb3_fuz_get	Fuzzy output read-out	Fuzzy output is read-out from FB-30AT board.
9	_fb3_rule_get	Rule of grade read-out	Grade is read-out from FB-30AT board.
10	_fb3_end	End processing	Indicates end of use of FB-30AT board.

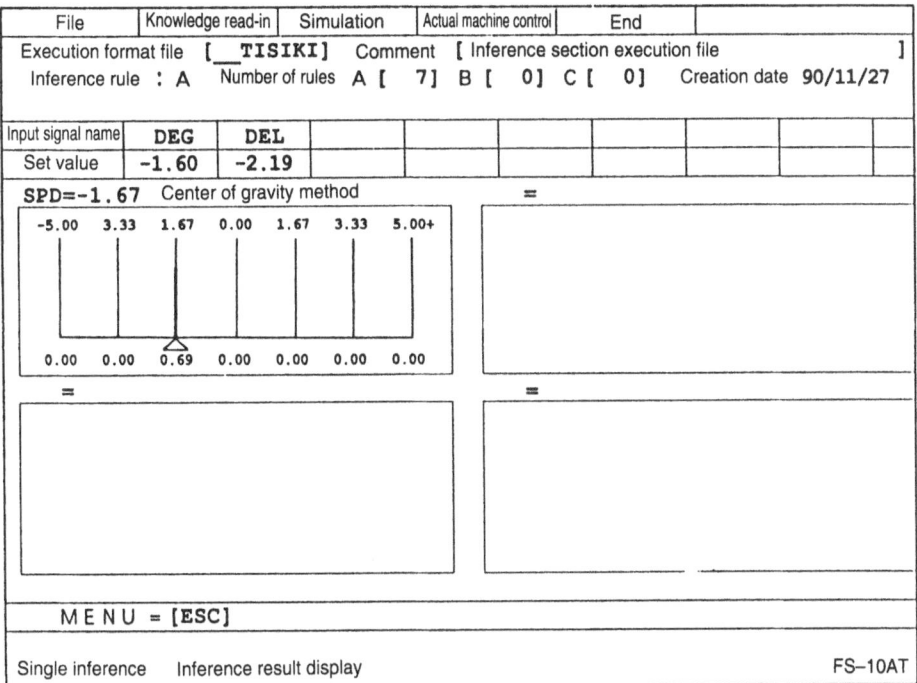

Fig. 5.13 Inference result

Fig. 5.14 Photograph of the FB-30AT

This board permits the capabilities and performance of the FP-3000 to be easily evaluated. The FB-30AT as well as the previously described FS-10AT should be thought of as being mainly for use in development and experimentation. In actual control of machines and production lines, the C500-FZ001 and C200H-F2001 Fuzzy Inference Units for use with the SYSMAC programmable controller are recommended. An in-circuit emulator, the CE-3098, is available for the FP-3000.

The CE-3098 ICE for FP-3000

When the FP-3000 is incorporated into a mass production line, a special circuit board has to be designed. This is the first ICE for fuzzy chips in the world developed to monitor the FP-3000 operational status so that, if trouble occurs, the process leading up to it can be investigated, or, alternatively, to make rule and MF corrections easy during debugging. The principal specifications are given in Table 5.4.

This ICE is connected and used in place of the FP-3000 of the target circuit board (the circuit board is under development). Use of this ICE has the following effects.

1. The operation status of the FP-3000 can be monitored in detail.
2. Rules and MFs can be directly corrected, without going through a higher level CPU.
3. When trouble occurs, inference data for up to 511 previous inferences can be investigated, tracing backward.

It is believed that as a result of these effects, the time required for incorporating an FP-3000 into a system can be shortened by several months.

5.3   Analogue fuzzy processors

5.3.1   Outline of the analogue fuzzy hybrid IC

This IC has been developed based on the architecture of the FZ-1000 ultra-high speed fuzzy controller. There are 2 types, one with the Inference chip that executes fuzzy inference and another with a Defuzzification chip that executes defuzzified operations. The Inference chip executes fuzzy inference with respect to external input based on a rule with a maximum of 3 conditions and 1 result; the inference result is output in the form of an MF. The Defuzzification chip takes the logical sum of 2 or more inference results, finds the center of gravity of the combined MF and outputs it as the defuzzified value. Attention has been focused on the continuity and speed of the analogue circuit, and ultra-high speed fuzzy operations made possible by

Table 5.4 ICE specifications for use with FP-3000

(1) Supervision personal computer (personal computer used to give operation instructions to the ICE and for displays)

• NEC PC-9801 VXZ/RX/RA or Omron FC-985

(2) ICE main unit

Item	Specification
Matching processor	FP-3000 Omron digital fuzzy processor
Clock frequency	Can select either internal 24, 12, 6 or 3MHz, or external up to maximum 6MHz.
Emulation memory	Can select either internal 8KB or external 8KB.
Break conditions	External trigger signal Specified number of inferences (1 to 9,999) completed
Break point	1 point
External trigger signal	Output when a break occurs.
Trace	Trace memory: enough for 512 inferences Trace data: input and output signals, fuzzy output Readout is only possible when trace is stopped.
Host interface	Exclusive-use parallel interface
Exterior dimensions	190(W)×280(D)×80(H)mm
Weight	about 3kg

(3) Management software (software used for ICE operation and monitoring)

Item	Specification
Rule, MF correction	Table entry method
Simulation	Single inference, continuous inference Number of simulations: maximum 65,535
ICE control	Emulation memory operation: Memory clear, fill-in, and verification of execution file and emulation memory are performed. Single inference: Single inference input settings are performed from the keyboard or a file and control 1 cycle of FP-3000 inference operation. Emulation: Continuous inference is executed upon receipt of an input signal given from the target side.
User interface	Pull-down menu type, color graphics display, menu selection, key input, and hard copies can be printed whenever desired.

the adoption of parallel architecture.

Features

1. The world's first chip exclusively for analogue type fuzzy use
2. Parallel architecture is used, and high speed fuzzy operations executed independently of the number of rules.
   Inference speed: 3 microseconds
   Including defuzzification operations: 10 microseconds
3. There is 1 Inference chip which executes 1 rule (3 conditions 1 conclusion)
4. Two or more Inference chips can be combined to increase the numbers of conditions and conclusions.
5. Weights can be assigned to each rule.
6. Analogue input and output (−5V to +5V)

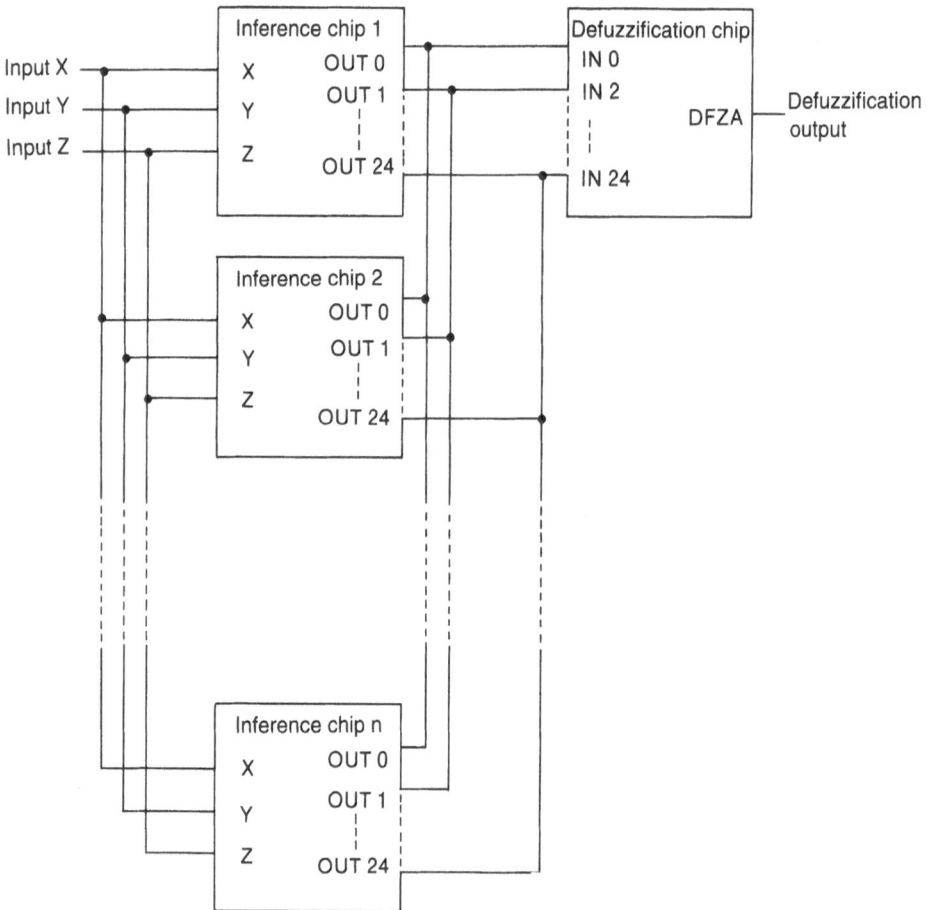

Fig. 5.15 System composition

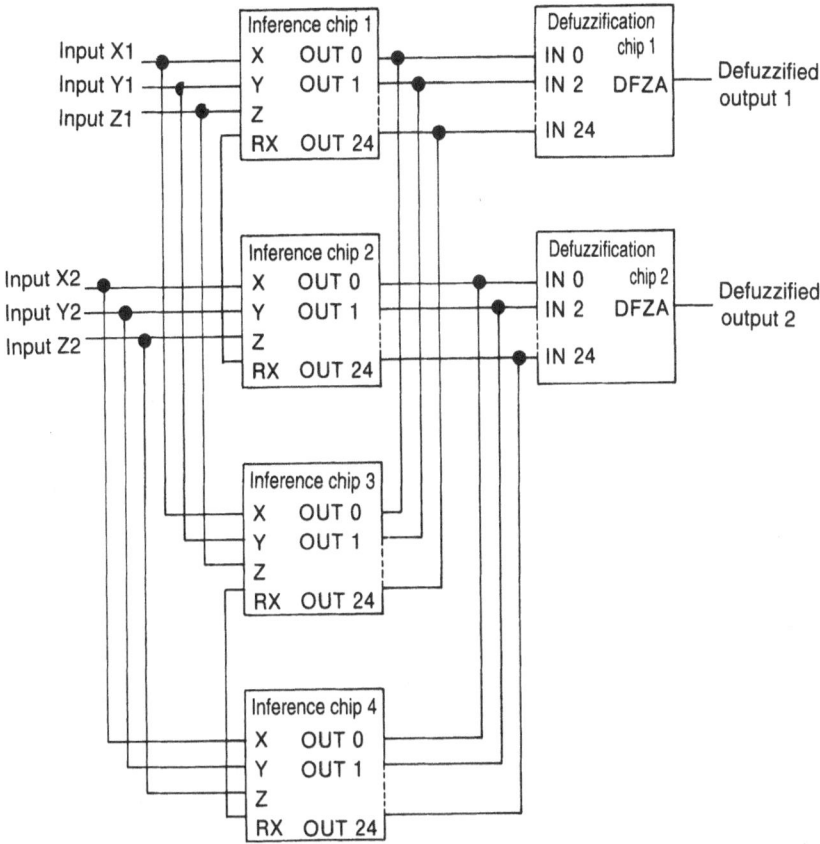

Fig. 5.16   System composition with 6 conditions, 2 conclusions

System composition

The composition of a typical system combining Inference chips and De-fuzzification chips is shown in Fig. 5. 15. A fuzzy controller with n rules consists of n Inference chips and one Defuzzification chip. One Inference chip has a 3-condition 1-conclusion inference function; by combining 2 or more In-ference chips and Defuzzification chips, respectively, the inference capability can be expanded to n conditions and m conclusions. The composition of a 6-condition 2-conclusion system is shown in Fig. 5. 16.

The system basically operates as follows. First MFs and rules are set as the knowledge base. Then the antecedent MFs are set by an externally connected resistor and diode, and antecedent rules are set by 4 terminal bit specifications. The consequent MF is specified by 2 terminal bits, and the consequent rule is set by specifying 3 terminal bits. The inference chip executes fuzzy inference for a maximum of 3 analogue inputs based on an inference rule in IF $\cdots$ THEN $\cdots$ inference format with a maximum of 3

Fig. 5. 17 Photograph of exterior   **Fig. 5. 18** Photograph of the In-
of the Inference chip   ference chip

conditions and 1 conclusion. The inference result is output in the form of
an MF (analogue outputs on 25 buses: bar graph). The logical sum (wired
OR) of multiple Inference chip analogue outputs on 25 bus lines is taken and
input to a Defuzzification chip; then the Defuzzification chip finds the center
of gravity and outputs an analogue voltage as the defuzzified value.

The fuzzy inference method is the MAX-MIN synthesis and logical product method, and the defuzzification method is the center of gravity method.

5.3.2   The TG005MC Inference chip

A photograph of the exterior of this Inference chip is shown in Fig. 5. 17,
and a photograph of the chip itself in Fig. 5. 18.

The Inference chip is a BiCMOS monolithic IC. The Inference chip pin
connection diagram is shown in Fig. 5. 19, a list of Inference chip pins in
Table 5. 5, and the electrical characteristics in Table 5. 6.

The Inference chip consists of the following 4 blocks.

1. Antecedent
2. MIN operation part
3. Operation part
4. Truncation part

A block diagram of the Inference chip is shown in Fig. 5. 20.

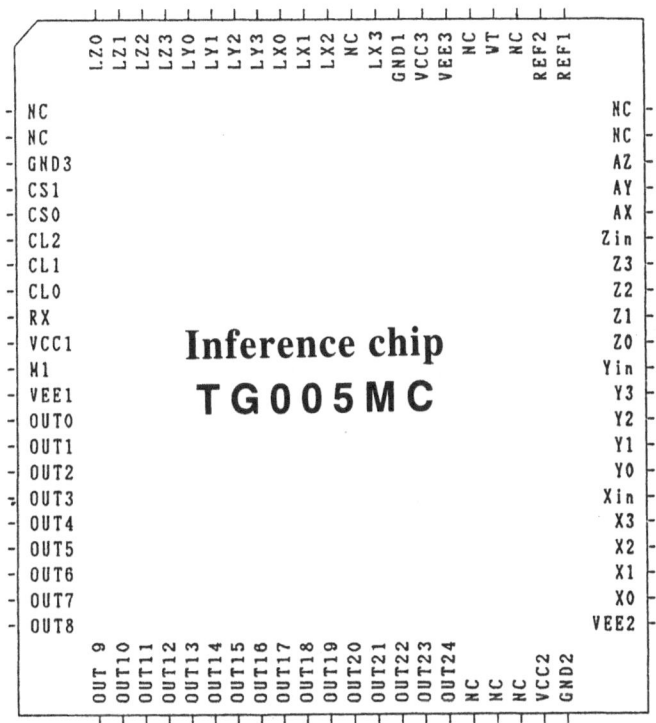

Fig. 5.19 Inference chip pin connections diagram

Table 5.5   List of Inference chip pins

no.	Terminal name	Function
1	LX2	Antecedent (X) label setting terminal
2	LX1	
3	LX0	
4	LY3	Antecedent (Y) label setting terminal
5	LY2	
6	LY1	
7	LY0	
8	LZ3	Antecedent (Z) label setting terminal
9	LZ2	
10	LZ1	
11	LZ0	
12	NC	No connection
13	NC	
14	GND3	Ground terminal (0V)
15	CS1	Consequent MF setting terminal
16	CS0	
17	CL2	Consequent label setting terminal
18	CL1	
19	CL0	
20	RX	Rule expansion terminal
21	VCC1	Power supply terminal (+10V)
22	M1	Antecedent output monitoring terminal
23	VEE1	Power supply terminal (−10V)
24	OUT0	Fuzzy output terminal
25	OUT1	
26	OUT2	
27	OUT3	
28	OUT4	
29	OUT5	
30	OUT6	
31	OUT7	
32	OUT8	
33	OUT9	
34	OUT10	
35	OUT11	
36	OUT12	
37	OUT13	
38	OUT14	
39	OUT15	
40	OUT16	
41	OUT17	
42	OUT18	

Table 5.5 *(continued)*

no.	Terminal name	Function
43	OUT19	Fuzzy output terminal
44	OUT20	
45	OUT21	
46	OUT22	
47	OUT23	
48	OUT24	
49	NC	No connection
50	NC	
51	NC	
52	VCC2	Power supply terminal (+10V)
53	GND2	Ground terminal (0V)
54	VEE2	Power supply terminal (-10V)
55	X0	Antecedent (X) MF setting terminal
56	X1	
57	X2	
58	X3	
59	Xin	Antecedent (X) input terminal
60	Y0	Antecedent (Y) MF setting terminal
61	Y1	
62	Y2	
63	Y3	
64	Yin	Antecedent (Y) input terminal
65	Z0	Antecedent (Y) MF setting terminal
66	Z1	
67	Z2	
68	Z3	
69	Zin	Antecedent (Z) input terminal
70	AX	Antecedent (X) label/analogue setting terminal
71	AY	Antecedent (Y) label/analogue setting terminal
72	AZ	Antecedent (Z) label/analogue setting terminal
73	NC	No connection
74	NC	
75	REF1	Reference voltage setting terminal 1 (+5V)
76	REF2	Reference voltage setting terminal 2 (−5V)
77	NC	No connection
78	WT	Weighting terminal (normally 5V)
79	NC	No connection
80	VEE3	Power supply terminal (−10V)
81	VCC3	Power supply terminal (+10V)
82	GND1	Ground terminal (0V)
83	LX3	Antecedent (X) label setting terminal
84	NC	No connection

Table 5.6 Inference chip electrical characteristics

Item	Symbol	Condition	Min.	Typ.	Max.	Unit
Power supply voltage	VCC		9	10	11	V
	VEE		−11	−10	−9	V
Current consumption	ICC	No output load		16.5	21	mA
	IEE		−19.5	−13		mA
Input voltage	VIN	applied to inputs X, Y and Z	−5		5	V
Output voltage	VO		0		5	V
Weighting input voltage	VWT		2.5		5	V
Input bias current	IIB	applied to inputs X, Y and Z			20	μA
High level input voltage	VIH	applied to digital terminals	3.5		5	V
Low level input voltage	VIL		0		1.5	V

$$V_{CC}=10V, \ V_{EE}=-10V, \ T_a=25°C$$

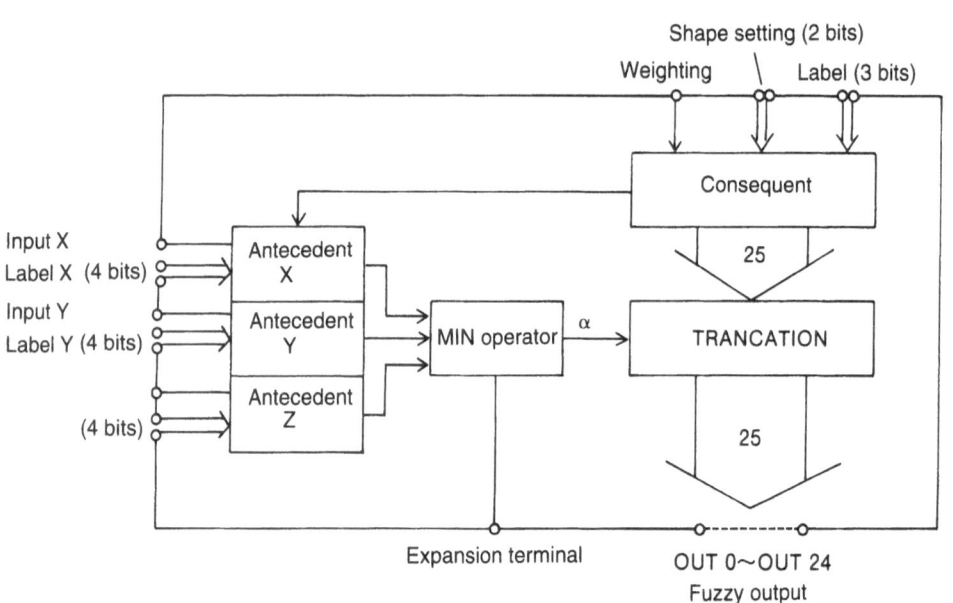

Fig. 5.20 Block diagram of Inference chip

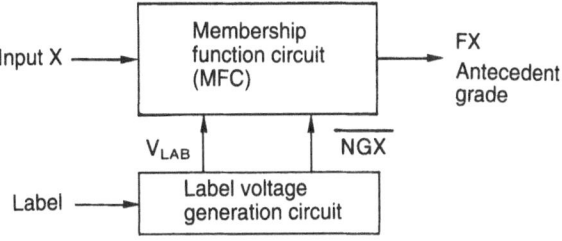

Fig. 5.21 Block diagram of antecedent

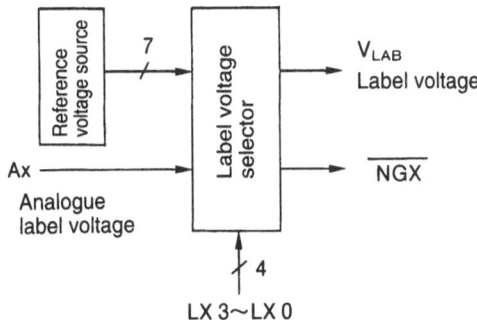

Fig. 5.22 Label voltage generation circuit

Antecedent

The antecedent is a block consisting of a label voltage generation circuit and an MF circuit that compares external input to set conditions and outputs the grade. A block diagram of it is shown in Fig. 5. 21. The Inference chip supports a rule with 3 conditions, so each chip has 3 antecedents.

1. Label voltage generation circuit: The label voltage generation circuit outputs the label voltage V_{LAB} that is supplied to the MF circuit, and and an NGX signal that sets the MF circuit to NG (1 output regardless of the input value) when the antecedent is not used. This circuit consists of a standard voltage source and a label voltage selector. A block diagram of it is shown in Fig. 5. 22.
   The standard voltage source generates 7 equally spaced voltage levels from −5V to +5V. The label voltage selector selects 1 of the 7 voltage levels produced by the standard voltage source and analogue label voltages (−5V to +5V) supplied from outside, as specified by the 4 bits LX3 to LX0, and outputs it as the label voltage V_{LAB}. The settings are listed in Table 5. 7.

   If the antecedent is not used, the setting becomes NG. If the setting is NG, $V_{LAB} = 0V$ and the NG signal $\overline{NGX}$ = "L" is output.

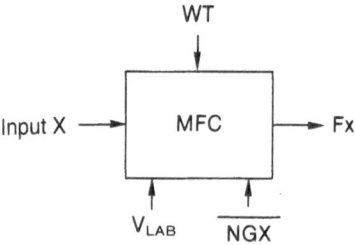

Fig. 5.23 MFC symbols

Table 5.7 Antecedent label settings

	V_{LAB}	LX3	LX2	LX1	LX0
Analogue input specification	V_{AX} (−5V∼ 5V)	0	DON'T CARE		
NL	−5V	1	0	0	0
NM	−3.33V	1	0	0	1
NS	−1.66V	1	0	1	0
ZR	0V	1	0	1	1
PS	1.66V	1	1	0	0
PM	3.33V	1	1	0	1
PL	5V	1	1	1	0
NG	0V	1	1	1	1

Table 5.8 Membership function settings

Shape	Setting method
Λ type	X0 X1 X2 X3 R1 R2
S type	X0 X1 X2 X3 R1 OPEN
Z type	X0 X1 X2 X3 OPEN R2
Π type	X0 X1 X2 X3 D1 R1 D2 R2

(Note) V_{LAB} is the label voltage.

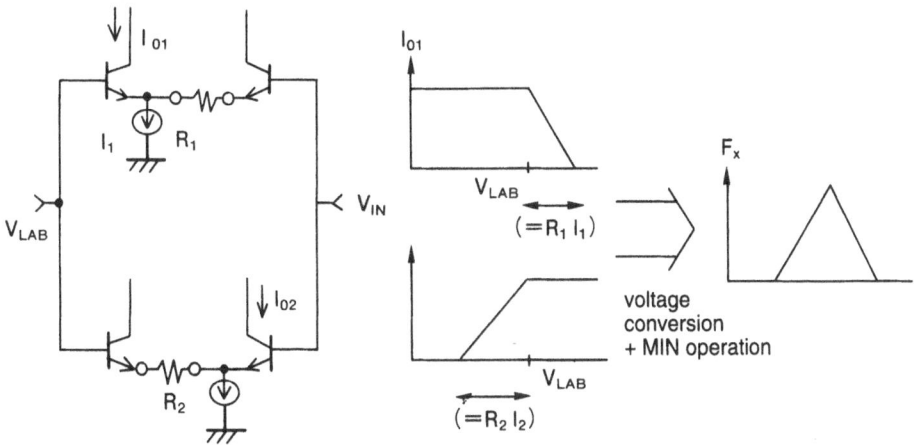

Fig. 5.24 Condition part membership function generation circuit

2. The MF circuit: The MF circuit (MFC) outputs the grade of the antecedent with respect to an external input in accordance with a preset MF. Its symbol is shown in Fig. 5. 23.

The MFC input/output characteristics are controlled by the weighting input WT, the label voltage V_{LAB} and the NG signal $\overline{NGX}$. The weighting input WT sets the MF peak voltage (normally $V_{WT} = 5V$). The label voltage V_{LAB} sets the position at which the MF shows the peak voltage. The NG signal $\overline{NGX}$ is $\overline{NGX}$ = "L"; the MFC output is fixed at V_{WT} regardless of the input. Also, the MFC MF setting is performed by an externally connected resistor and diode. Its settings are listed in Table 5. 8.

The MF is formed by varying the form of input and output characteristics of a differential amplifier. Specifically, the slope of the function is expressed by connecting a resistance in series with the differential transistor emitter and then MIN synthesizing those values. The MFC is shown in Fig. 5. 24.

The MIN operation part

The MIN operation part is a 3-input MIN circuit that selects the minimum voltage having the minimum grade, as the rule grade α. This circuit is shown in Fig. 5. 25.

The 3-input MIN circuit has 3 pnp transistors with emitter follower cascade connection, and executes comparison operations in parallel.

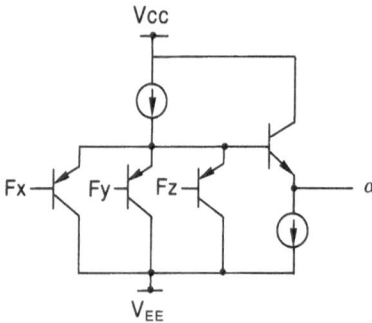

Fig. 5.25 A 3-input MIN circuit

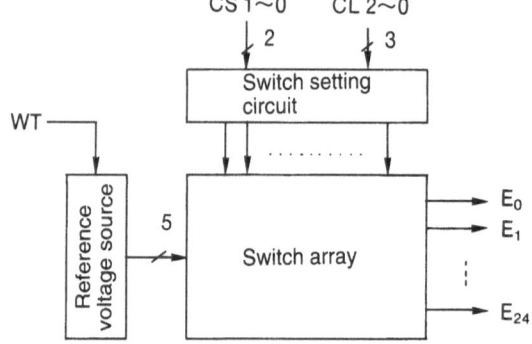

Fig. 5.26 Block diagram of consequent

Table 5.9 Conseqens MF shape settings

Shape	CS1	CS0
Λ type	0	0
S type	0	1
Z type	1	0
Π type	1	1

Note) "1" is 5V; "0" is 0V.

Consequent

The consequent is a circuit consisting of a standard voltage source, a switch setting circuit and a switch array that outputs the consequent MF as analogue voltages on 25 lines. A block diagram of it is shown in Fig. 5. 26.

The standard voltage source generates 4 equally spaced voltage levels between the input voltages VWT and 0V. The switch setting circuit turns one of the switch array analogue switches ON so that the MF set by the 5 bits CS1, CS0 and CL2 to CL0 is output. The switch array in turn consists of analogue switches arranged so that the 5 levels of voltage produced by the standard voltage source are output to the 25 output lines.

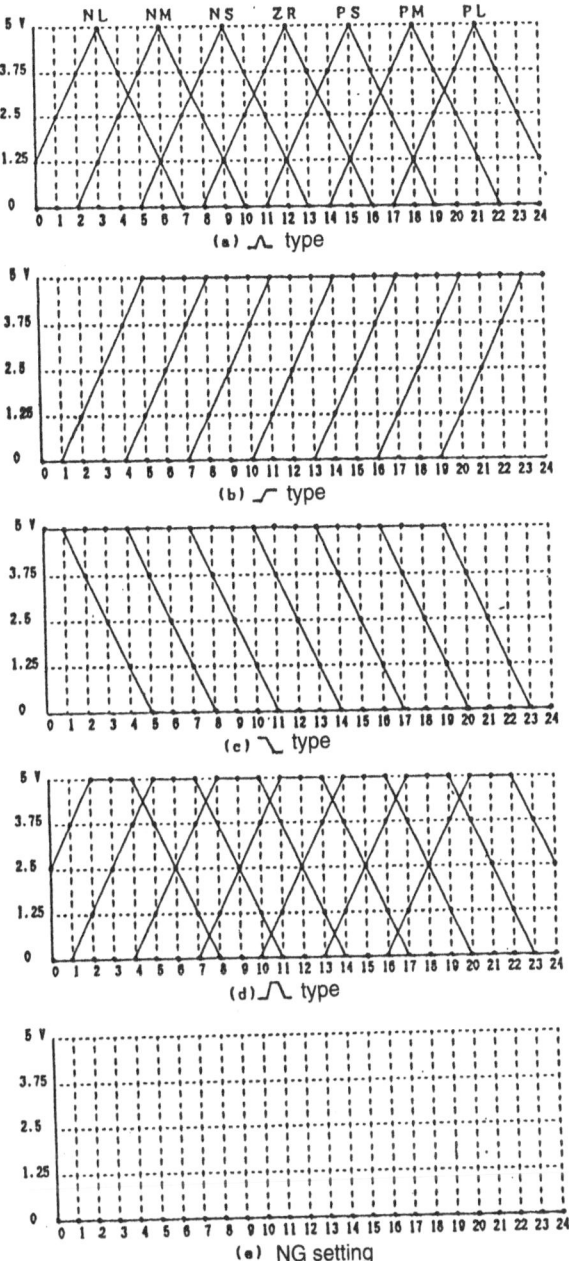

Fig. 5.27 Consequent output characteristics

The MF shape is set by CS1 and CS0; the label is set by CL2 to CL0. These settings are listed in Tables 5. 9 and 5. 10; the output characteristics are shown in Fig. 5. 27.

Table 5.10 Consequent label settings

Label	CL2	CL1	CL0
NL	0	0	0
NM	0	0	1
NS	0	1	0
ZR	0	1	1
PS	1	0	0
PM	1	0	1
PL	1	1	0
NG	1	1	1

Note) "1" is 5V; "0" is 0V.

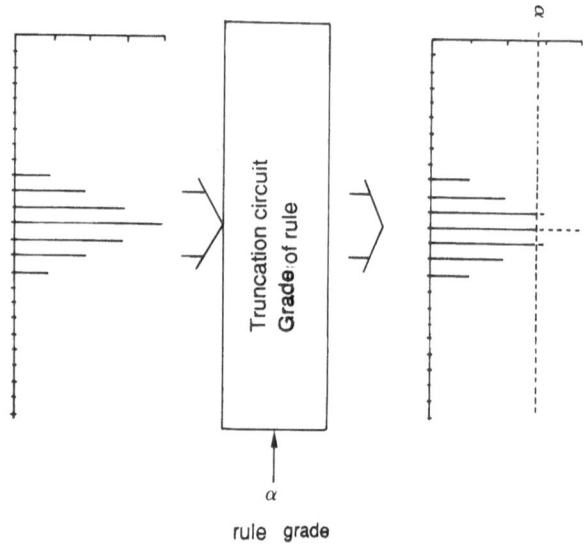

rule grade

Fig. 5.28 Truncation circuit action

Here the NG setting is used when it is desired to ignore the inference result; then 0V is output on all 25 output lines.

Truncation part

The truncation part executes the rule grade α and consequent MF operations; its operation is shown in Fig. 5. 28.

The consequent MF is truncated according to the rule grade α; the truncated MF becomes the inference result.

This truncation operation involves taking the MIN of the values of each element and the rule grade α. Therefore, the truncation circuit consists of 25 2-input MIN circuits.

Fig. 5.29 Photograph of exterior of the Defuzzification chip

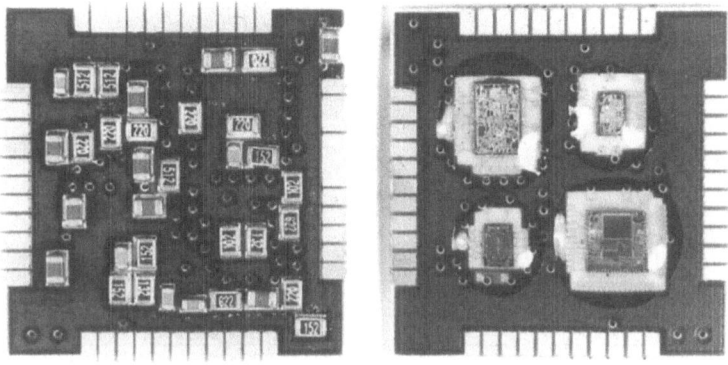

Fig. 5.30 Photograph of Defuzzification chip

5.3.3   The TB010PL Defuzzification chip

A photograph of the exterior of the Defuzzification chip is shown in Fig. 5. 29, and one of the chip itself in Fig. 5. 30.   The Defuzzification chip is a hybrid IC that consists of 4 bare chips, resistors and 44 capacitors (44 resistors and capacitors) on a 24mm×24mm ceramic circuit board. The Defuzzification chip pin diagram is shown in Fig. 5. 31, a list of Defuzzification chip pins in Table 5. 11, and the electrical characteristics in Table 5. 12.

The Defuzzification chip consists of the following 2 blocks.

1. Input section
2. Center of gravity operation section

A block diagram of the Defuzification chip is shown in Fig. 5. 32.

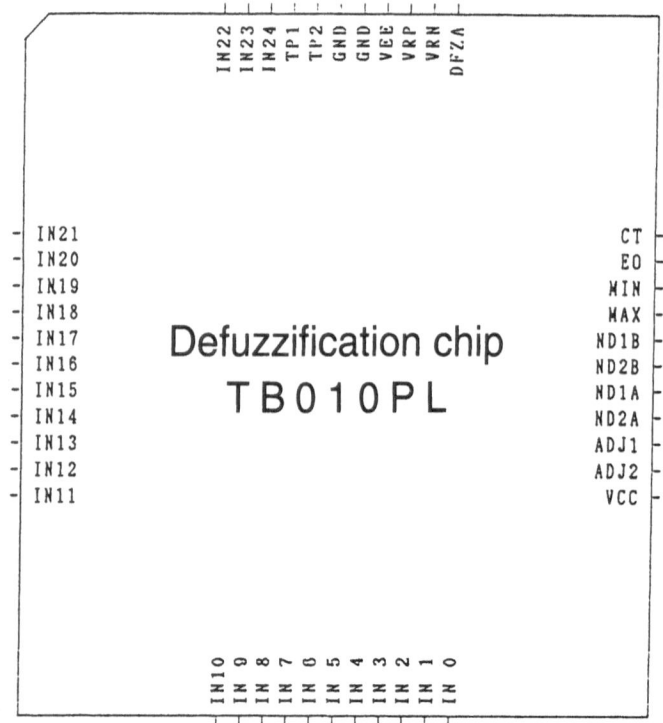

Fig. 5.31 Defuzzification chip pin connection diagram

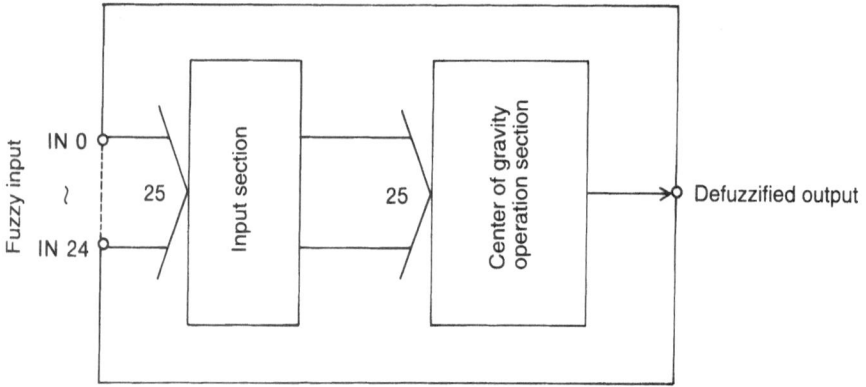

Fig. 5.32 Block diagram of the Defuzzification chip

Table 5.11 List of Defuzzification chip pins

Pin no.	Terminal name	Function
1	GND1	Ground terminal (0V)
2	TP2	Monitor terminal 2
3	TP1	Monitor terminal 1
4	IN24	Fuzzy input terminal
5	IN23	
6	IN22	
7	IN21	
8	IN20	
9	IN19	
10	IN18	
11	IN17	
12	IN16	
13	IN15	
14	IN14	
15	IN13	
16	IN12	
17	IN11	
18	IN10	
19	IN9	
20	IN8	
21	IN7	
22	IN6	
23	IN5	Fuzzy input terminal
24	IN4	
25	IN3	
26	IN2	
27	IN1	
28	IN0	
29	VCC	Power supply terminal (+10V)
30	ADJ2	Output dynamic range adjustment terminal 2
31	ADJ1	Output dynamic range adjustment terminal 1
32	HD2A	Adjustment terminal 2A
33	HD1A	Adjustment terminal 1A
34	ND2B	Adjustment terminal 2B
35	ND1B	Adjustment terminal 1B
36	MAX	Maximum input setting terminal
37	MIN	Minimum input setting terminal
38	E0	Abnormal signal output terminal
39	CT	Output zero adjustment terminal
40	DFZA	Defuzzifier output terminal
41	VRN	Reference voltage output terminal ($-5V$)
42	VRP	Reference voltage output terminal ($+5V$)
43	VEE	Power supply terminal (-10V)
44	GND2	Ground terminal (0V)

Table 5.12 Defuzzification chip electrical characteristics

Item	Symbol	Condition	Min.	Typ.	Max.	Unit
Power supply voltage	VCC		9	10	11	V
	VEE		−11	−10	−9	V
Current consumption	ICC	no output load		42.5	55	mA
	IEE		−66	−49.5		mA
Input voltage	VIN	applied to fuzzy input	0		5	V
Output voltage	V0		−5		5	V
Input current	IIN	VIN=0V	100	200	320	μ A
Defuzzified output accuracy	A0	accuracy for output full-scale $\pm$ 5V	−2		2	%

$$V_{CC}=10V,\ V_{EE}=-10V,\ T_a=25°C$$

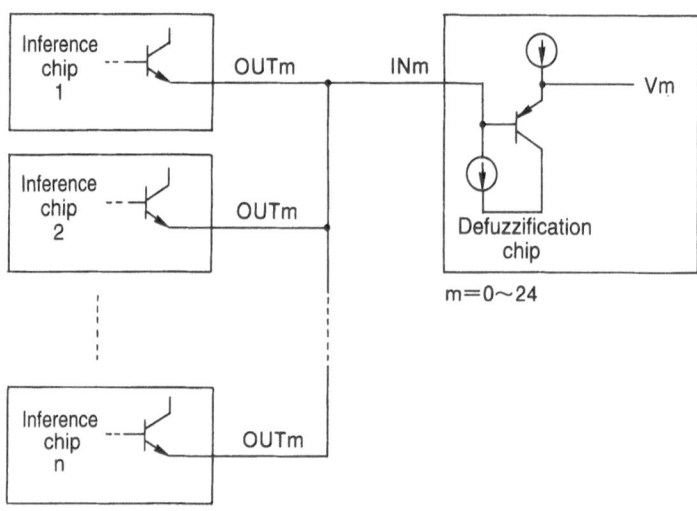

Fig. 5.33 Composition of MAX circuit consisting of the Inference chip and the Defuzzification chip

Input section

The input section MAX synthesizes the outputs from 2 or more Inference chips on each line. The MAX circuit is formed by connecting the Inference chip to the Defuzzufication chip, as shown in Fig. 5. 33.

Center of gravity operation section

The center of gravity operation section is a block that finds the center of gravity by taking the weighted average of the MAX-synthesized inference result MFs, and outputs it as the defuzzified value. It consists of an addition circuit, a weighted addition circuit, a division circuit and an amplification circuit. A block diagram of it is shown in Fig. 5. 34.

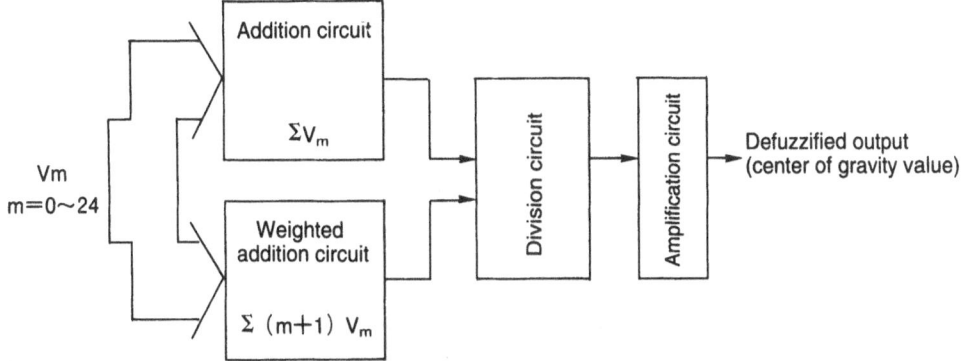

Fig. 5.34   Center of gravity operation section

The center of gravity value V_G is found from the following formula:

$$V_G = K \cdot \frac{\sum_{m=0}^{24}(m+1)V_m}{\sum_{m=0}^{24}V_m} \tag{2}$$

Here K is a coefficient to make the devision circuit output $\pm 5V$, and indicates the degree of amplification of the amplification circuit.

5.3.4   Development support tools

In this processor, which consists of the Inference chips that process one rule per chip and the Defuzzification chip, the combination of chips differs for each application. In addition, many switches and potentiometers are used for the Inference chip settings. The processing speed of each chip is very fast. A high density, multi-layered circuit board is indispensable for mounting these chips. There are also some characteristics peculiar to analogue operation; complete software simulation is difficult.

To cope with these conditions, the FZ-5000 controller and monitor software are provided as development tools.

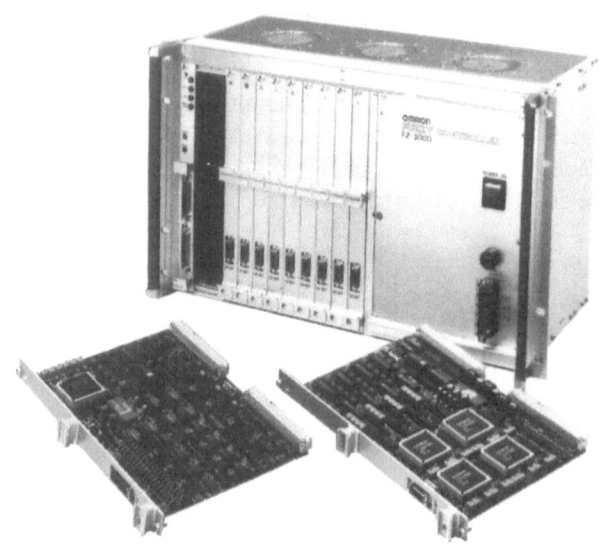

Fig. 5.35 Photograph of exterior of the FZ5000

The FZ-5000 analogue fuzzy controller

This is an ultra-high speed controller containing 2 or more inference boards on which 4 Inference chips are mounted and a defuzzification board on which a defuzzification chip and an input/output interface are mounted, in a VME rack. Each of the boards is double height (Fig. 5. 35). The upper connector is connected to the CPU through a VME bus. The lower connector is connected to the analogue fuzzy bus that transmits the fuzzy output. The principal specifications are given in Table 5. 13. The composition of the inference board is shown in Fig. 5. 36, and that of the defuzzification board in Fig. 5. 37.

Programming is done using the dip switches and jumper pins on the inference board. Programming can also be done by FT-6000 (described below) using a personal computer.

It is also possible to give an MF center value directly from outside through the inference board connector. There are a number of capabilities that are useful in research and development, including using 2 Inference chips in parallel to handle 6 antecedent, assigning a weight to each rule, and forming 2 or more inference groups.

The FT-6000 software tool

This is a programming and monitoring tool for use with the FT-5000. It runs on the PC-9801VX, OMRON FC-985 factory computer, etc. In order

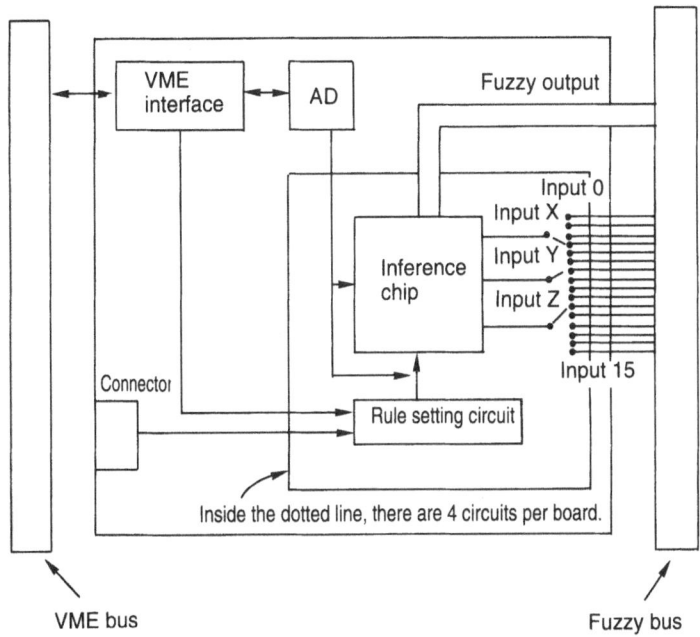

Fig. 5.36 Inference board composition

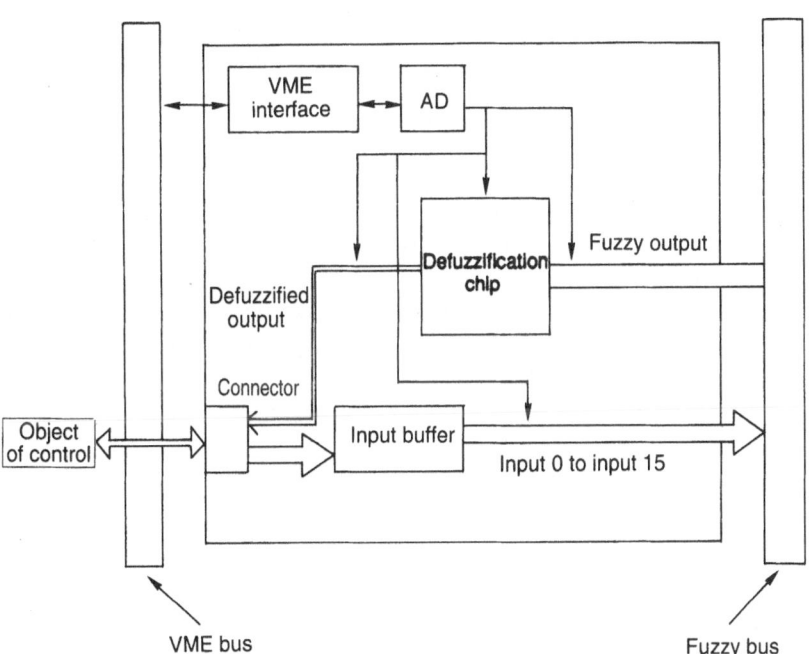

Fig. 5.37 Determination board composition

Table 5.13 FZ-5000 specifications

Item		Content
		12 slots
Fuzzy inference section	Rule format	3 conditions 1 conclusion (standard)
	Number of rules	Maximum 44 rules
	Inference direction	Forward-directed inference
Fuzzy inference section	Inference format	MAX-MIN logical product
	Linguistic value	7 labels
	Inference speed	15 μs(including MAX value and defuzzification)
	Defuzzification method	Center of gravity method
	Inference accuracy	$\pm 10\%$ full scale or less
Membership function	Antecedent	Continuous function Supports Λ function, S function, Z function, Π function.
	Conseqent	Discrete function divided into 25 parts Supports Λ function, S function, Z function, Π function.
	Settings section	Labels can be set for each input and output.
Analogue input section	Number of input points	Maximum 16 points
	Full-scale range	-5.0V$\sim +5.0$V
	Maximum input range	-10.0V$\sim +10.0$V
	Input impedance	200kΩ
Analogue output section	Number of output points	1 output per Defuzzification board
	Full-scale range	-5.0V$\sim +5.0$V
	Output current	5mA or less
	Output temperature drift	1.5% (full-scale)

to use this software tool, a CPU board (Omron 3G8B3-M0023) is set in the FZ-5000, and connected to the personal computer through an RS-232C.

This software tool has the following functions.

1. Antecedent part MF adjustment: The potentiometer on the board is adjusted while watching the MF shape on the screen.
2. Rule setting and downloading: Rules are created by filling tables in, then downloaded to the FZ-5000.
3. Monitoring (Fig. 5. 38): Input/output signals and fuzzy output signals are monitored. Monitoring does not affect inference speed.

FT-6000 software is supplied free of charge to all purchasers of the FZ-5000 who request it (the CPU board, cables, etc. must be paid for).

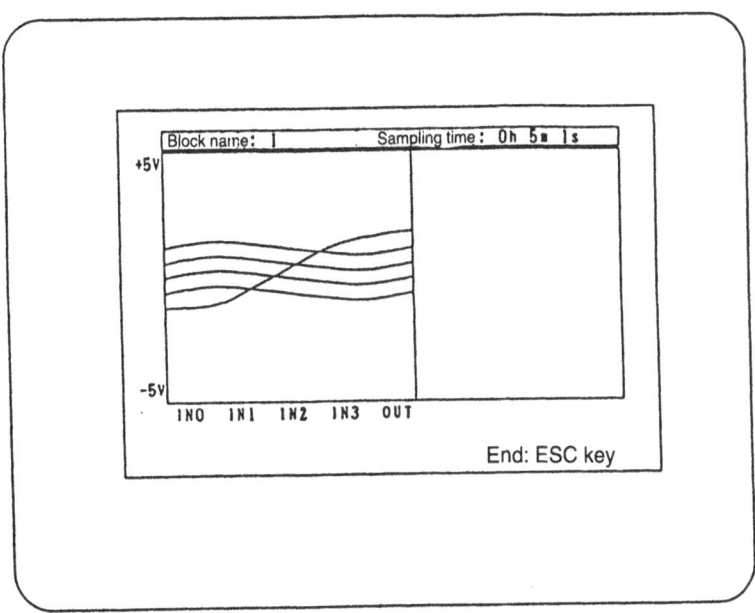

Fig. 5.38 Monitor screen

5.4 In conclusion

This chapter has introduced 2 different types of fuzzy processor developed by Omron, the FP-3000 digital fuzzy processor which is suitable for control applications and the analogue fuzzy hybrid ICs (Inference chip and Defuzzification chip), and their respective fuzzy development support tools.

Digital processing provides good interfacing with existing microprocessor systems. As fuzzy application fields expand in the future, it will become necessary to promote specialized chips for specific applications, such as low cost chips for incorporation into systems, large capacity chips for large volume fuzzy information processing, and fuzzy decision making support chips. In addition, it will not be long before fuzzy capabilities are provided on-chip to various types of existing microprocessors. Meanwhile, analogue processing makes it possible to directly perform fuzzy processing of analogue signals such as sensor output. It offers the advantage of ultra-high speed inference independent of the number of rules; it is believed that it will be possible to make use of this feature to incorporate fuzzy capabilities into various types of sensors. To bring this about it will be important to increase the degree of integration of the Inference chips (meaning to increase the number of rules for which 1 chip can perform inference) and to make the Defuzzification chip monolithic.

In any case, the increased use of fuzzy processing in electronic devices, using human-like inference to improve capabilities and performance, is expected to penetrate into all fields of electronics; we believe firmly that fuzzy chips and their development support tools will play an important role in these developments. The analogue fuzzy hybrid ICs discussed in the latter half of this chapter were developed by OMRON under the guidance of Prof. Takeshi Yamakawa of Kyushu Institute of Technology; I express my thanks to him for extending his assistance.

Chapter. 6

FUZZY CONTROLLERS AND THEIR APPLICATION TO WATER TREATMENT

6.1 Introduction

In the last several years, applications of fuzzy control have been rapidly increasing; recently such control has become widely used in household electrical appliances. Fuji Electric Co., Ltd. moved quickly to establish expertise in the area of fuzzy control, and has demonstrated its usefulness in controlling the injection of chemicals in a water treatment plant. In 1985 Fuji began to market FRUITAX, the first general-purpose fuzzy control system in Japan[1,2]. Since then, with the increase in demand for fuzzy control technology, Fuji has been steadily increasing sales, mainly in the process control field. In addition, whereas at first fuzzy controllers had to run on personal computers and minicomputers, they can now be incorporated into 1-chip microprocessors and programmable controllers, making it possible to use them in many fields.

Here we introduce the design procedure for fuzzy controllers as carried out by Fuji Electric Co., Ltd. and the FRUITAX series of fuzzy controllers with their user-friendly fuzzy development environment, and explain an example of application to a water treatment process, which has contributed greatly to making fuzzy control practically useful.

6.2 The general fuzzy controller design procedure

6.2.1 What fuzzy control is

Fuzzy control permits flexible, highly responsive control, which in the past had to be performed by humans, to be done on a computer. In an industrial plant, when a system is run by a skilled operator the quantities which the operator inputs are often determined through an overall judgment based on the operator's past experience and consideration of a number of factors. In such a situation, fuzzy control permits the control to be optimized

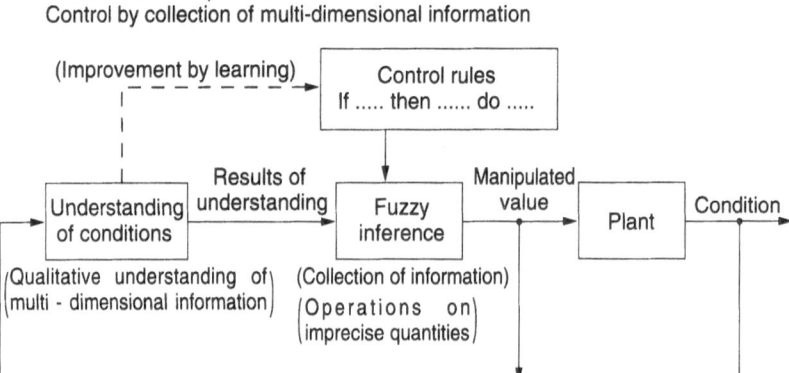

Fig. 6.1 Fuzzy control flow diagram.

to fulfill its purpose. One of the advantages of this kind of control is that the operator's experience can be reflected in the if··· then··· production rules. In general, the flow of fuzzy control is as shown in Fig. 1. The control rule (fuzzy production rule) used here can be considered to perform fuzzy partition on the input values determined by many input values and then put each subspace into 1-to-1 correspondence with an output value (manimulated value). Consequently, as the number of inputs increases the control rules become complicated and construction and maintenance of the fuzzy control system becomes more difficult.

6.2.2   Fuzzy inference methods

A number of fuzzy inference methods have been proposed. Here we introduce the method that is used as standard in fuzzy control systems by Fuji Electric Co., Ltd. and FUJIFACOM CORPORATION. This inference method is based on the implication used in steam engine control by E.H. Mamdani; This is the fuzzy inference method that is generally used in fuzzy controllers that run on software. This method has the following characteristics.

1. The operations themselves are simple and clear, and processing speed is fast.
2. The proportions of firings of each control rule are easy to see visually.
3. It is easy for the operator's operation judgments to be reflected.

The specific method is discussed below. For simplicity only the case of 2 input values is considered.

Control rules are described in the format:

$$\text{IF } x_1 = A_i, x_2 = B_i \text{ THEN } y = C_i \tag{1}$$

Here, if we let x_A and x_B be the input values for x_1 and x_2, respectively, the grade of proposition of antecedent of the control rule is calculated from the implication used by E.H. Mamdani to be:

$$\omega_i = h_{A_i}(x_A) \wedge h_{B_i}(x_B) \tag{2}$$

Here h_{A_i} and h_{B_i} are the membership functions (fuzzy labels) with respect to the fuzzy variables A_i and B_i; $\wedge$ indicates the min operation. When the grade of the ith control rule is found, the membership function that describes the fuzzy variable C_i described in that proposition of consequent is multiplied by ω_i to give the following function $h_{C_i}^*(z)$.

$$h_{C_i}^*(z) = \omega_i \cdot h_{C_i}(z) \tag{3}$$

The membership function $h(z)$ relating to the output value is obtained by performing a max operation for each i in equation (3). Finally, the manipulated value u_0 is computed as the center of gravity of the area generated by the function $h(z)$.

$$h(z) = \bigvee_{i=1}^{n} h_{C_i}^*(z) \tag{4}$$

$$u_0 = \frac{\int h(z)z\,dz}{\int h(z)\,dz} \tag{5}$$

An example of fuzzy inference for the case of 2 control rules is shown in Fig. 6.2. Thus, a feature of this method is that the inference itself is easy to understand visually. For this reason, it is suitable for checking inference conditions and reinforcing and correcting rules.

6.2.3   Design of control rules

1. Design is carried out so that process characteristics are understood, and specialized knowledge about control engineering is put to good use.
2. A skilled worker's intuition, experience and knowledge are expressed linguistically, and design is based on them.
3. Design is based on data from actual operations performed by a skilled operator.
4. Process characteristics are modeled in if...then... format, and design is performed by using an evaluation function to determine the optimum formula.

Design method 1 involves obtaining a grasp of the plant's characteristics, then designing control rules such that control operations that are desirable from the point of view of control engineering are performed. This method is often used in adapting to a plant that has not yet operated or has only operated for a short time. In method 2, a skilled operator's operation experience

IF $x_1 = A_1$ THEN $u = C_1$

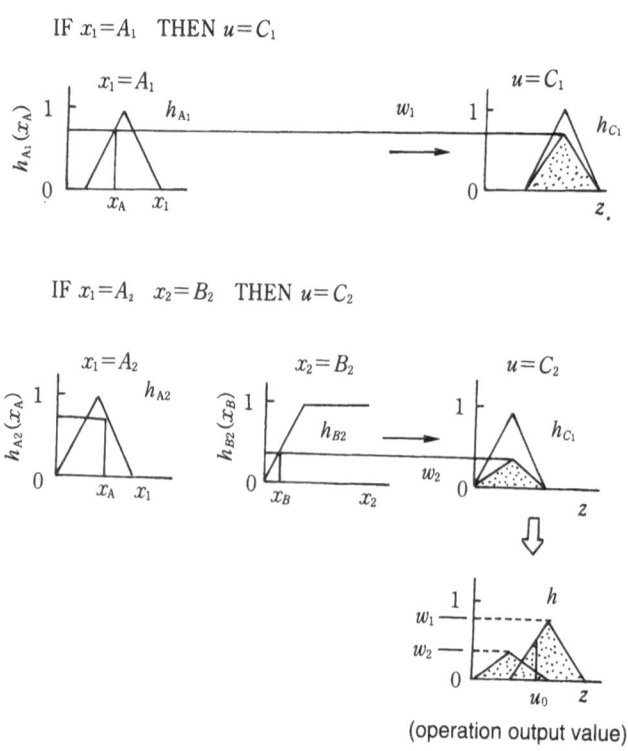

Fig. 6.2 Fuzzy inference rules.

and knowledge are expressed linguistically and then made into control rules. Basically, control rules can be created with either method 1 or 2, but a problem that must be noted here is that the skilled operator's operation method is not necessarily stated clearly. For this reason it is difficult for knowledge that is used subconsciously to be openly articulated, so that later verification and improvement become necessary. In method 3 the operation data include operational know-how that is difficult to obtain in an interview, so there is the advantage that when a control rule is designed its latent operational know-how can be extracted and organized. Method 4 makes it possible to design control rules based on plant characteristics, making it possible to exceed a skilled operator's performance. In addition, fuzzy model research and development are proceeding in several directions, based on learning control, etc.

6.2.4 Fuzzy controllers

The greatest problem in making fuzzy control a reality is obtaining the necessary knowledge. Specifically, this means the design of the control rule and identification of the membership functions. Consequently, in constructing a fuzzy controller, it is necessary to develop a fuzzy environment in which

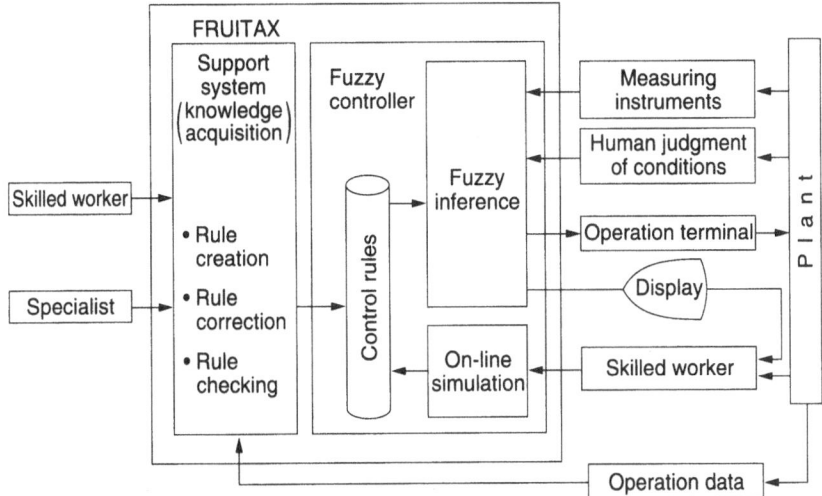

Fig. 6.3 Example of a fuzzy-controlled system.

it is possible for a user skilled in actual plant operation to design and verify the rules himself. As an example, a "fuzzy control system with development environment" designed and developed from this point of view is shown in Fig. 6.3. This fuzzy controller runs on a personal computer or control work-station. Values from sensors, instruments and/or the operator's judgment of conditions are input; then fuzzy inference is applied and the result is output to the operation terminal. At this time, the kind of control rule on which fuzzy inference is based is indicated visually on the status indication screen (Fig. 6.4). On the status indication screen in Fig. 6.4, the inference result is shown at the upper left and up to 3 rules which fired strongly are shown at the bottom.

The fuzzy inference status can be read at a glance by watching this monitor, providing a sense of reassurance while using the controller. In addition, when the inferences themselves are unsatisfactory, the rules can be altered and corrected to improve the controllability. For this purpose, there is an on-line simulation function.

This function computes what the inference result would be if a control rule were changed, something that can be confirmed on the screen. An example is shown in Fig. 6.5. Here an example of simulation result

IF FLOC=SA, TUSE=IS THEN DDOS=NS

is shown. When this function is used, the control operation can be improved although some trial and error is involved.

In contrast, the support system is a program package for designing, correcting, strengthening and verifying the control rules from data from operations by skilled operators.

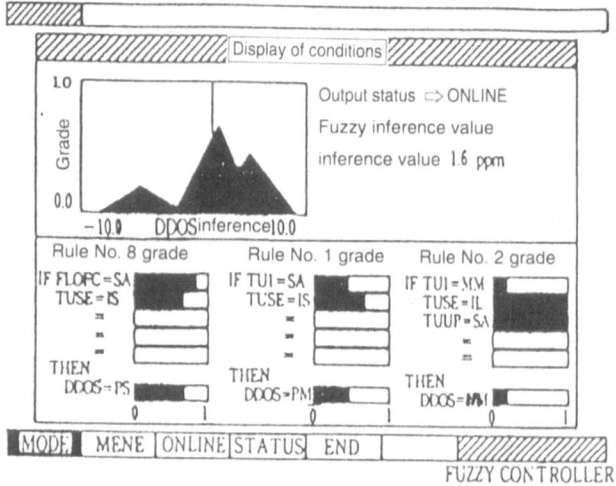

Fig. 6.4 Example of screen on which conditions are displayed.

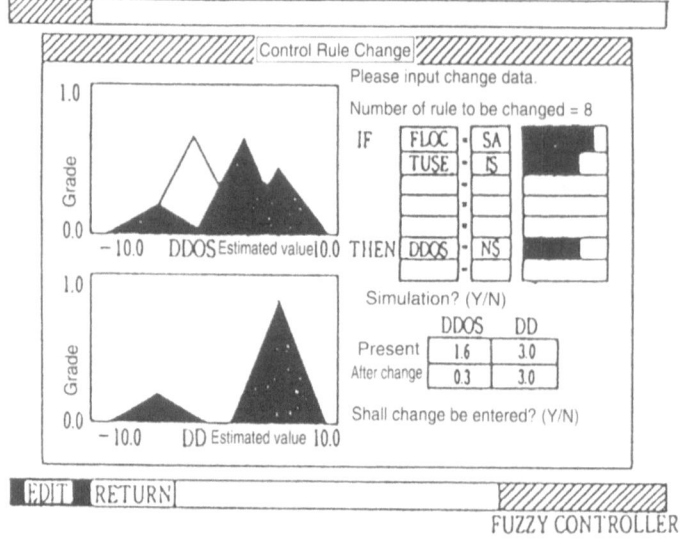

Fig. 6.5 Example of on-line simulation screen.

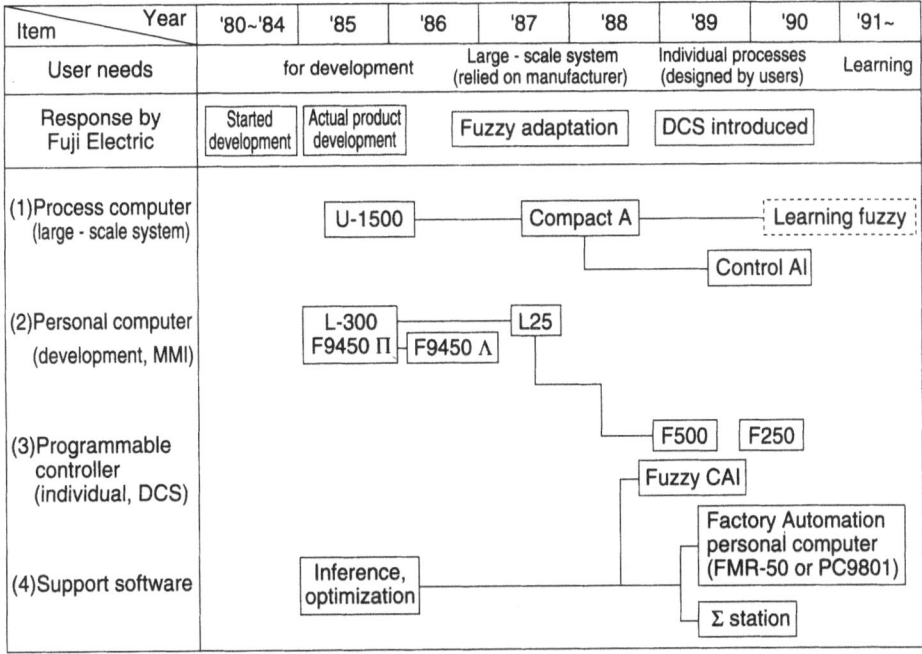

Fig. 6.6   History of efforts to apply fuzzy control to Fuji Electric machinery

6.3   The FRUITAX general purpose fuzzy control system

6.3.1   Development of FRUITAX

The work done by Fuji Electric Co., Ltd. and FUJIFACOM CORPO-
RATION is depicted in Fig. 6.6. The world's first general purpose fuzzy
controller, FRUITAX, was put on the market in 1985. Since then, there have
been a number of improvements, such as line-up, reduction of size and increase of speed, to create a fuzzy controller that is more user friendly. As of
this writing, development of a one-chip microprocessor and a fuzzy controller
that can be mounted on a general purpose programmable controller has been
completed, and efforts are now directed at developing applications in many
different fields.

6.3.2   The FRUITAX series

The FRUITAX series lineup is shown in Fig. 6.7. FRUITAX controllers
can be functionally classified into those that are for in-line control and those
that are for off-line development support. Models for on-line control use include the Compact A (minicomputer) edition FRUITAX-RX and FRUITAX-
L for incorporation into large-scale systems, and the FRUITAX-MF for use

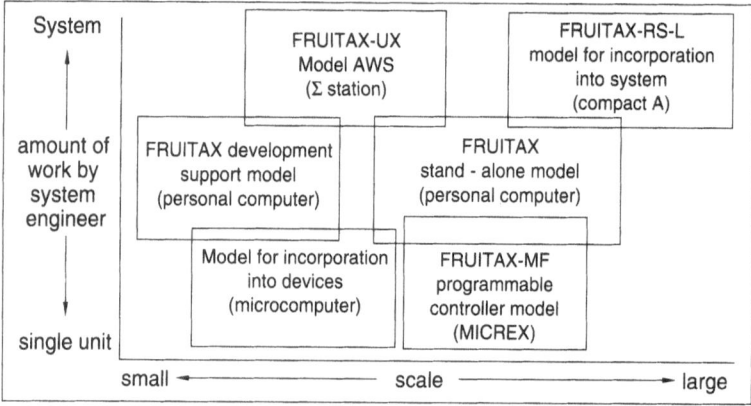

Fig. 6.7 FRUITAX composition

with the MICREX general purpose programmable controller. Models for off-line development support use include the FRUITAX-UX for use with the Model AWS (AI Advanced Work Station) (SIGMA) station, and the FRUITAX-FS, FRUITAX-FR and FRUITAX-3D development support tools for use on personal computers (Fujitsu Factory Automation Personal Computer, FMR-50, etc.). The features of these models of the FRUITAX series are given below.

1. FRUITAX-RX:

Used to incorporate fuzzy control into computers used to control large-scale monitoring control systems.

2. FRUITAX-L:

Used in large-scale control systems, as is the FRUITAX-RX. Runs under management of the AIMAX-C expert control system (Fuji Electric Co., Ltd.), making it possible to incorporate fuzzy operations into expert control systems for plant applications, optimum planning, etc.

3. FRUITAX-MF:

This fuzzy controller and development support tool can be used on the MICREX series F500 and F250 general purpose programmable controllers. FRUITAX can be used with function images from a sequence program; construction of a system that is coordinated with existing sequence control is possible.

4. FRUITAX-UX:

Since this is mounted on an AWS that can adapt to a wide variety of types of control, from conventional type control to AI control, it makes possible not only development support related to fuzzy control but also design support for unified control systems.

5. FRUITAX-FS, FRUITAX-FR, FRUITAX-3D:

These are provided as tools that will run on personal computers but are based on fuzzy development support tools developed on main frame computers. They are convenient because they are low in cost while preserving the basic FRUITAX capabilities.

6.3.3 Functions

FRUITAX is a system that is capable of designing and adjusting fuzzy control systems. It provides a fuzzy development environment which emphasizes the ability to adjust control rules on-line, compactness and user friendliness. In general, the main problem in implementing fuzzy control is efficiency of designing control rules. FRUITAX is provided with the following functions for the purpose of design from operation data.

1. Off-line simulation: Output status is checked, using collected operation data as input. This simulation function can be used to verify, strengthen, and evaluate the suitability of control rules. In addition, the following functions are provided:
   (a) Design of control rules and membership functions, and checking of changes in their inference values on a 3-dimensional display screen.
   (b) Tuning of condition part membership functions using operation data. This is called the optimum fuzzy variable search function.
2. Editing of control rules: This function edits membership functions, control rules, etc. A sample screen display is shown in Fig. 6.8.
3. Status indication: This function permits the fuzzy inference result to be visually checked. A sample screen display is shown in Fig. 6.4.
4. On-line simulation: This function alters rules on-line, and also simulates such alterations. This function is mainly used for altering rules on-site.
5. Time sequence data display: This function displays process data as trend graphs.

FRUITAX basically performs max-algebraic product-center of gravity calculation; FRUITAX-L also permits algebraic product is addition-center of gravity calculation to be selected. The membership function shape is basically triangular. The specifications of these methods are given in Table 6.1.

FUZZY CONTROLLER ──────── <list of control rules> ────────
Injection of chemicals into water purification system   page   1/16

NO	Condition part statement					Operation part statement		
1	IF SUNA	= SA DEN = SA	=	=	=	THEN PRI = LA	=
2	IF SUNA	= SM DEN = SA	=	=	=	THEN PRI = ML	=
3	IF SUNA	= MM DEN = SA	=	=	=	THEN PRI = MM	=
4	IF SUNA	= ML DEN = SA	=	=	=	THEN PRI = SA	=
5	IF SUNA	= LA DEN = SA	=	=	=	THEN PRI = SS	=
6	IF DEN	= MM	=	=	=	THEN PRI = ML	=
7	IF DEN	= ML	=	=	=	THEN PRI = LA	=
8	IF DEN	= LA	=	=	=	THEN PRI = LL	=
9	IF	=	=	=	=	THEN	=	=
10	IF	=	=	=	=	THEN	=	=
11	IF	=	=	=	=	THEN	=	=
12	IF	=	=	=	=	THEN	=	=
13	IF	=	=	=	=	THEN	=	=
14	IF	=	=	=	=	THEN	=	=
15	IF	=	=	=	=	THEN	=	=
16	IF	=	=	=	=	THEN	=	=

PF3: previous screen   PF7, 8: return, forward feed   ENTER: enter

Fig. 6.8   example of control rule input.

Table 6.1   FRUITAX performance comparison table

Item	FRUITAX -RX	FRUITAX -UX	FRUITAX -L	FRUITAX -MF	FRUITAX-FS FRUITAX-FR FRUITAX-3D
Machine to which applied	Compact A series	Compact A series Σ station series	Compact A series	MICREX -F500 F250	FMR-50 PC-9801 PC-286
Number of controllers	8	—	8	4	—(—)*3
Input Process inputs	96	20	96	16	16(32)
Input Keyboard inputs	32	32	32	16	—(—)
Operation outputs	8	8	8	4	4(8)
Control rules Number of rules	256	256	256	128	96(256)
Number of proposition of antecedent	5	5	5	5	5(5)
Number of proposition of consequent	2	2	2	4	2(2)
Control interval (minimum)*1	10 seconds	—	10 seconds	20ms*2	—(—)
Type	incorporated into system	AWS type	incorporated into system	Programm-able con-troller type	Development support type
On-line output	○	×	○	○	×

*1: The control period given is the minimum time when 64 rules are used and PIO is included.

2: Inferred time when 32 rules are used (not including PIO)

3: Symbols in parentheses are for FRUITAX-3D.

6.3.4 Application fields

FRUITAX series controllers have been used successfully in many applications, principally process control fields. They can be selected according to the scale and characteristics of the object of control.

(1) FRUITAX for large-scale systems:

In applications such as water treatment, cement, iron and steel, and electric power, it is necessary to handle large quantities of input information. FRUITAX-RX is used when fuzzy control is applied to a large-scale system having a relatively long control cycle. FRUITAX-L is used for fuzzy control applications including operation and planning in which higher level monitoring and coordination are performed. For public services such as water treatment, Fuji Electric Co., Ltd. designs the control rules; in the private sector this is often done by the customer. In such a case, FRUITAX-UX is used.

(2) FRUITAX for small and medium-scale systems:

For processes such as chemicals, foods, construction and machinery, in which the amount of input information is small and mainly loop control is applied, the high-speed, low-cost FRUITAX-MF is often used. The user designs the control rules.

The following benefits can be expected from using these tools. *Construction, maintenance and adjustment of the fuzzy control system by the user *On-site status monitoring, checking and adjustment *Coordination with existing control systems using conventional control, AI control, etc.

6.3.5 The development of FRUITAX-L

Characteristics of FRUITAX-L

FRUITAX-L is provided as a library for fuzzy operation use, having an AI tool-AIMAX-C man-machine interface for control use. This FRUITAX-L was developed for the purpose of applying fuzzy operation to a variety of fields from conventional control to planning and operations.

(1) What AIMAX-C is:

AIMAX-C is an expert control system construction tool to enable the user himself to perform process control that is unified with the control system. In addition, the processing speed can be increased and the system made more compact by using FORTRAN/KR, which is a construction support tool incorporating the object oriented concept.

IF part

No.	Content	Weight
1	Platform holding load is oscillating with large amplitude in direction of motion.	1.00
2	The distance to the target position is large.	1.00

THEN part

No.	Content
1.	Increase the speed of the platform holding the load by a large amount.

Fig. 6.9   Example of AIMAX-C rule editing. (Japanese)

(2) How to use FRUITAX-L:

FRUITAX-L was developed for the purpose of cooperation between mathematical models and fuzzy models that cannot be described by numerical models and formulas by using it in conjunction with AIMAX-C. Consequently, it is possible to construct a system in which it is possible to call fuzzy inference from a production rule that has binary truth values which are often used in AI-type control, and to execute fuzzy inference using numerical model values as prediction values.

(3) How to start a fuzzy inference engine from AIMAX-C: (a) A fuzzy inference engine can be started by calling a subroutine within a program. (b) A production rule described by AIMAX-C is described in the Japanese language by a linguistic expression (Fig. 6.9), and then converted to the FRUITAX-L knowledge base by FORTRAN/KR.

FRUITAX-L functions

FRUITAX-L basically inherits the functions of FRUITAX-RX, but the functions are expanded as follows.

(1) There is a choice of inference methods (the max-algebraic product-center of gravity calculation method or the algebraic product-addition-center of gravity calculation method).

(2) A trapezoidal shape can be used for the condition part membership function.

(3) A bit map display can be used (previously only industrial display could be used).

Application fields

FRUITAX-L application fields include plant operation, decision making, process status judgment, etc. Potential application fields are much broader than for methods using the existing loop control concept.

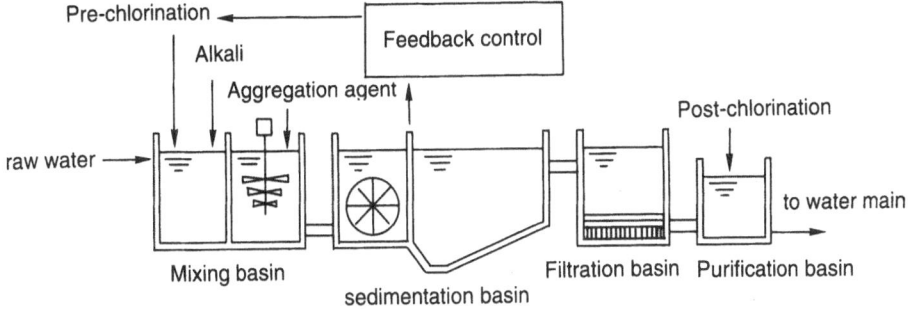

Fig. 6.10 Water purification plant process.

6.4 An example of fuzzy control in the water treatment field

A water treatment facility is a public facility, and therefore its operation must be reliable. However, since control of chemical injection in a water treatment plant directly affects water quality, there is heavy reliance on the experience of skilled operators, and the burden on them is heavy. In this section we introduce an example in which fuzzy control has been applied to control chemical injection at a water treatment plant. An outline of the water treatment process in a treatment plant is shown in Fig. 6.10. The coagulant is a chemical that is injected to remove impurities in the raw water that cause it to become turbid. After the coagulant is injected, the suspended matter in the raw water is removed in a sedimentation basin or a filtration basin. This treatment process includes nonlinearities such as dead time, and in addition the coagulation process cannot be described by a numerical model. For these reasons predictive control based on many measurements and an operator's judgment of conditions is applied. In this treatment process, fuzzy control is applied to the parts of the process involving an operator's judgment of conditions and predictive control. Specifically, average values for the amount of chemical to be injected when there is a certain change in water quality (from a regression model) are supplemented by fuzzy inference. An example of a field test is shown in Fig. 6.11. It can be seen that the fuzzy inference results agree well with the actual injected amounts based on an operator's judgments and adjustments. In fact, the residual mean and its standard deviations between the actual amounts that the operator injected and the fuzzy inference results were 2.28ppm and 1.36ppm, respectively; fuzzy inference gave the better result.

Let us introduce another example of the introduction of fuzzy inference into this process. This is control of the injection of chlorine to remove heavy metals and organic matter from, and to sterilize, the raw water. This prechlorine injection process has the following problems, making automation of it difficult. (1) The amount of chlorine that is dispersed after injection varies, depending on environmental influences such as weather, season and amount of sunlight, and cannot be described by a numerical model. (2) The water

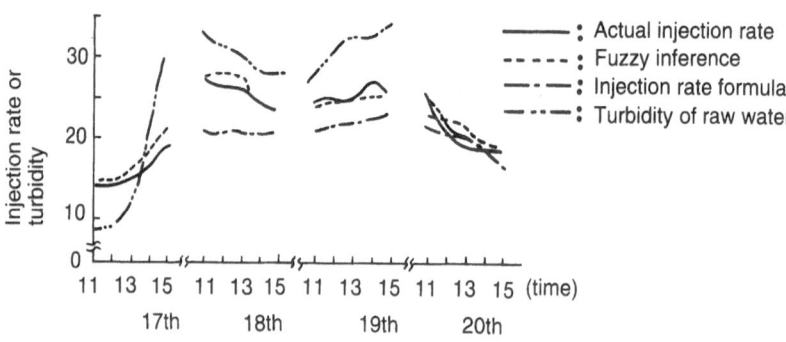

Fig. 6.11 Example of field test results.

has to stand for 6 to 8 hours, and fluctuations in the amount of water drawn from the system affect the amount of chlorine consumed. (3) There are many external disturbances, and input factors that cannot be measured.

Fuzzy control was introduced to improve the situation; the following improvements resulted. (1) The load on the operator was lightened. (2) Even a beginning operator could perform control equal to that of a skilled operator; in addition, the control became more reliable. (3) The control became smoother.

It is believed that these results are due to the following characteristics of fuzzy control. (1) Since the operation is at the level of a highly skilled operator, it is reliable. (2) Control algorithms based on operational know-how can be constructed. (3) Overall judgments incorporating the effects of a number of factors can be made. (4) Whereas in conventional control the response to a change in a process is to change one or more parameters, in fuzzy control it is possible to deal with the trends in a large quantity of information; by having this point reflected in the control rules, predictive control can be performed to a certain extent. (5) By modeling operational know-how in the form of control rules, latent technology can be put into usable form and shared.

6.5   Cooperative control of rain water pumps by an adaptive type controller

6.5.1   Outline of a rain water pumping station

A rain water pump performs the functions of rapidly removing rain water that flows into sewer pipes to rivers, and preventing flooding within the water removal district. In recent years green areas and vacant land have decreased in large cities, the ground surface has become more completely

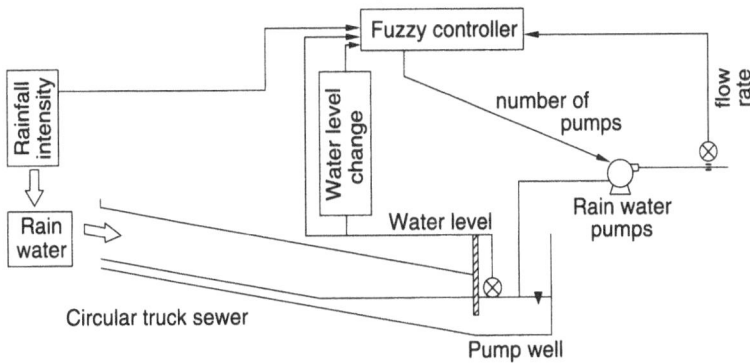

Fig. 6.12   Outline of process.

covered with concrete, and the amount of land that can serve as a buffer into which rain water can permeate has greatly decreased. Due to these environmental changes the amount of rain water runoff has greatly increased, and there is rapid rain water runoff even in light rain. In heavy rain, such as during a thunderstorm, toward the end of the monsoon season or during a typhoon, damage is heavy, with what are called urban type floods occurring. The number and capacity of pumps have been increased in an attempt to deal with this problem, leading to a need to control the increased frequency of pump starting and stopping due to the difference between normal rain water inflow and pump removal capacity and the inflow of rain water at inflow gates to optimize pump operation when inflow exceeds pumping capacity.

In many cases, rain water pumps are controlled by number of pumps control according to pump well level. The rain water pump control itself is very simple, but the following problems occur because of failure to grasp the dynamic change of the process.

1. It takes time to start and stop the pumps, and, since the pump well storage capacity is small, it is difficult to respond to rapid increase of inflow.
2. Flexible response to the amount of inflowing rain water is impossible, and the pump starting and stopping frequency increases, putting a heavy load on the pump.
3. It is difficult to couple to or cooperate with the pumps inflow gates.

For these reasons, attempts have been made to apply predictive control using rain water inflow predictions obtained from both physical and regression models. However, there are few reliable measured values, such as rainfall and water level, and in addition locations at which measurements are made are limited. Moreover, rainfall is hard to predict, and the process is strongly nonlinear so the inflow amount cannot be directly measured and it is difficult to identify a mathematical model. For such reasons, it is presently difficult

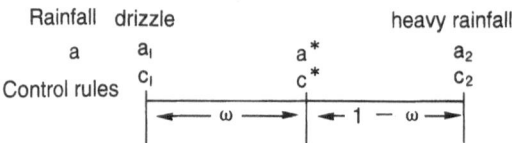

Fig. 6.13   The concept of applying fuzzy control.

to put this system into actual use. So at present it is necessary to rely on skilled pump operators at many pumping stations. Skilled operators start and stop rain water pumps and open and close inflow gates in response to rainfall and rain water inflow conditions. In light rain they accumulate water in the sewers to reduce the frequency of pump starting and stopping; while in heavy rain safety has priority. Such complicated control is very difficult; to automate it, it will be necessary to have fuzzy control that reflects the knowledge of skilled operators.

6.5.2   Fuzzy control of pumping stations

Fuzzy control changes a control algorithm in response to variations of plant parameters. This method simplifies response to changing conditions such as process gain and factors that cause dead time, as well as coordination among multiple control objectives.

In fuzzy control that focuses attention on the parameters that indicate plant fluctuations, multiple plant models that correspond to those parameters are created. Then control rules are created for those plant models. At the pumping stations skilled operators perform control in response to their overall assessment of rain water inflow conditions. The rain water inflow condition is expressed by a plant parameter a. In analyzing the model plant, the extremes of drizzle (condition a_1) and a downpour (condition a_2) are considered, and control rules c_1 and c_2 for those respective conditions are prepared. Next, from judgment of the present rain water inflow condition (called a^*), the degrees of nearness w to the conditions a_1 and a_2 are found. Here we assume that w takes the values [0,1], which means that w is 0 when condition a_1 exists and 1 when condition a_2 exists. From control rule c_1 which applies when condition a_1 exists and control rule c_2 which applies when condition a_2 exists, the control rule c^* for condition a^* is computed from formula (6).

$$c^* = c_1(1 - w) + c_2 w \tag{6}$$

As is shown in Fig. 6.14, the value of the weighting coefficient w is computed from the total pump removal capacity and the rainfall intensity. Here only the total pump removal capacity is used because the rain water inflow is hard to measure directly. Next, the pump well level, which is used to infer the changes of pump rpm and pump removal capacity, and the parameters of the membership function for the slope of that change, become the parameters of the control rule that is adjusted by the parameter w. The reason for this is

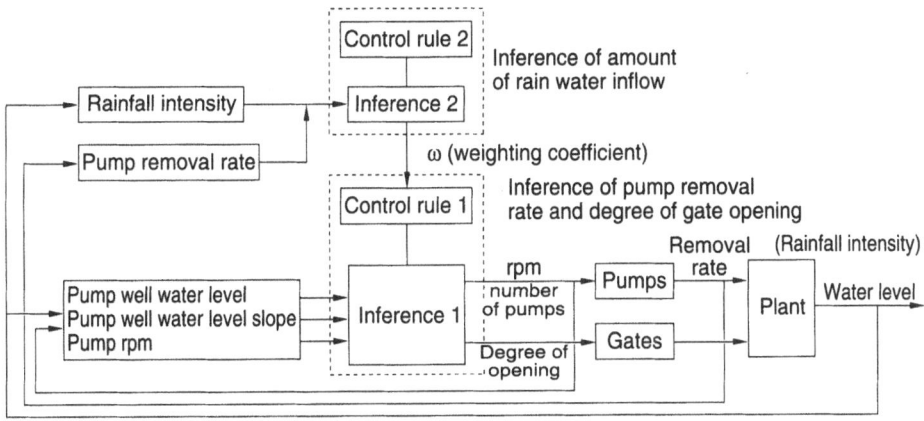

Fig. 6.14 Flow diagram of fuzzy control.

that the judgment values that indicate danger due to the inflow conditions differ. Condition part membership functions for the purpose of restricting the frequency of rain water pump starting and stopping are prepared for the two extremes of drizzle and downpour, and the actual membership function used is adjusted to the actual rainfall by varying ω. However, the pump rpm used in gate operation judgments is not varied; there is no change in the form of control rule 1.

The membership functions used here are triangular in shape and are described by the 3 parameters P_1, P_2 and P_3. The condition part membership function adjustment calculation is done as follows, using the parameters, ω.

$$P_1 = P_{10}(1 - \omega) + P_{11}\omega \tag{7}$$

$$P_2 = P_{20}(1 - \omega) + P_{21}\omega \tag{8}$$

$$P_3 = P_{30}(1 - \omega) + P_{31}\omega \tag{9}$$

In Fig. 6.15 the membership funct6ions for the target water levels during drizzle and downpour are shown by solid lines, and the membership function for the target water level when $\omega = 0.5$ by a dashed line.

In this fuzzy control, the general fuzzy inference rule given in section 6.2.2 is used. Here, the input space is fuzzy - partitioned into 5 parts; the number of control rules used is 60 for control rule 1, 25 for control rule 2. In contrast, if adaptive control is not performed then 625 (5^4) control rules are needed for the rain water pump. Using adaptive control greatly reduces the number of rules.

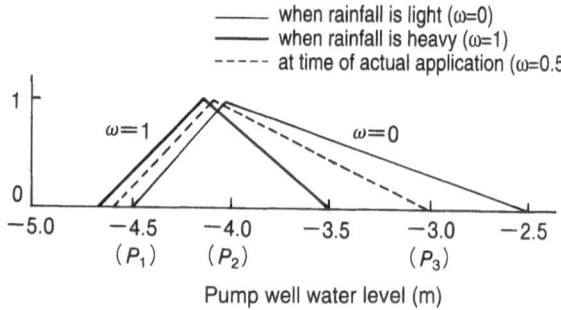

Fig. 6.15 Adaptation of membership function

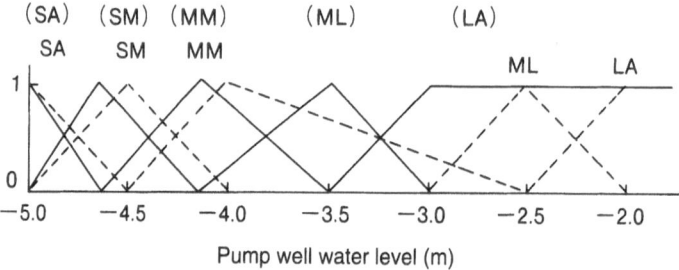

Fig. 6.16 Example of pump well water level membership function

6.5.3 Pumping station coordination simulation

Outline of the model

Rain water flows down to the pumping station through a sewerline 2km long, 3.5m in diameter and having a slope of 0.1%. Rain water flowing into the pumping station passes from the sewerline through the inflow gate to pump well 1 and pump well 2. Rain water that flows into well 1 is removed by a speed - controlled pump, and rain water that flows into well 2 by 4 pumps subject to control of the number of pumps operating. All of the pumps have the same rated water removal capacity, $1.8\text{m}^3/\text{s}$. In this simulation, a downpour of 30mm/h falls, and all of the water enters at the upstream end of the inflow sewerline. An outline of this process is shown in Fig 6.17.

A numerical model of rain water outflow

We assume that the amount of rain water entering a sewer line is found from a 1-stage tank model as shown in Fig. 6.18. In this model the amount of water flowing in is balanced by water flowing out through the side surfaces (penetration).

$$\frac{dX}{dt} = -(y + z) + x \tag{10}$$

$$y = \alpha X \tag{11}$$

$$z = \beta X \tag{12}$$

$$Q = yS/360 \tag{13}$$

Table 6.2 Examples of control rules

NO	Condition part statement		Operation part statement
1	IF $H_1 = SA$	$HD_1 = SA$	THEN $QKAI = NB$
2	IF $H_1 = SA$	$HD_1 = SM$	THEN $QKAI = NM$
3	IF $H_1 = SA$	$HD_1 = MM$	THEN $QKAI = ZE$
4	IF $H_1 = SA$	$HD_1 = ML$	THEN $QKAI = ZE$
5	IF $H_1 = SA$	$HD_1 = LA$	THEN $QKAI = PS$
6	IF $H_1 = SM$	$HD_1 = SA$	THEN $QKAI = NM$
7	IF $H_1 = SM$	$HD_1 = SM$	THEN $QKAI = NS$
8	IF $H_1 = SM$	$HD_1 = MM$	THEN $QKAI = ZE$
9	IF $H_1 = SM$	$HD_1 = ML$	THEN $QKAI = PS$
10	IF $H_1 = SM$	$HD_1 = LA$	THEN $QKAI = PM$
11	IF $H_1 = SA$	$R = LA$	THEN $GATE = PM$
12	IF $H_1 = SM$	$R = LA$	THEN $GATE = PS$
13	IF $H_1 = MM$	$R = LA$	THEN $GATE = ZE$
14	IF $H_1 = ML$	$R = LA$	THEN $GATE = NS$
15	IF $H_1 = LA$	$R = LA$	THEN $GATE = NB$
16	IF $H_1 = SA$	$R = SA$	THEN $GATE = PB$
17	IF $H_1 = SM$	$R = SA$	THEN $GATE = PM$
18	IF $H_1 = MM$	$R = SA$	THEN $GATE = PS$
19	IF $H_1 = ML$	$R = SA$	THEN $GATE = ZE$
20	IF $H_1 = LA$	$R = SA$	THEN $GATE = NS$

H_1 : Pump well water level 1, HD_1 : Pump well water level 1 slope
R : Pump rpm
$QKAI$: Change of pump rpm
$GATE$: Change of degree of gate opening
$NB, NM, NS, ZE, PS, PM, PB$: Reduce greatly, reduce, reduce slightly, leave unchanged, increase slightly, increase, increase greatly.

$\cdots\cdots\cdots$ when rainfall is light $(\omega = 0)$		
———— when rainfall is heavy $(\omega = 1)$		
when lainfall is light	when lainfall is heavy	Judgment of water level
SA	(SA)	low
SM	(SM)	a bit low
MM	(MM)	target level
ML	(ML)	a bit high
LA	(LA)	high

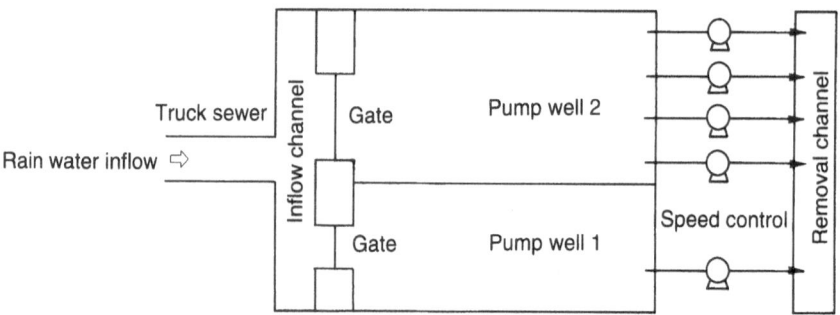

Fig. 6.17　Pump station model.

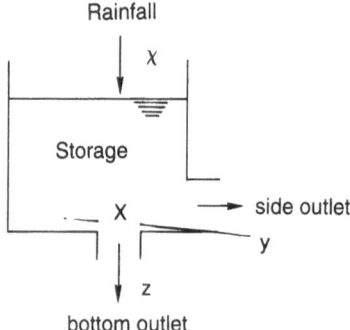

Fig. 6.18　Tank model.

Here, x is the rainfall intensity (mm/h), X is the storage height (mm), y is the side outlet (mm/h), z is the bottom outlet (mm/h), t is time (h), Q is the outflow rate (m^3/s) and S is the area (ha) of the district from which water is being removed; α and β are constants.

A sewerline model

The flow of water inside the sewerline is described using the equations of motion of unsteady flow in an open channel (14) and the continuity equation (15). Here v is flow velocity (m/s), x is distance (m), i is the water channel slope, g is the acceleration of gravity (m/s^2), A is the cross-sectional area through which water flows (m^2), h is water depth (m), n is the Manning roughness coefficient, R is hydraulic radious (m) and t is time (s). The method of solving equations (14) and (15) that is used here is the Leapfrog Method.

$$\frac{1}{g}\frac{\partial v}{\partial t} + \frac{v\partial v}{g\partial x} - i + \frac{\partial h}{\partial x} + \frac{n^2|v|v}{R^{3/4}} = 0 \qquad (14)$$

$$\frac{\partial A}{\partial t} + \frac{\partial vA}{\partial x} = 0 \qquad (15)$$

Pumping station model

As indicated by equation (16), in the pump well tank model that is used here the water level variations are found from the difference between inflow and outflow. In the pump model, the pump stroke curve and the pipe resistance curve are approximated by the quadratic expressions in equations (17) and (18), and the amount of water removed by the pump is found from their intersection. The flow rate through the input gate is proportional to the square root of the difference in water levels before and after the gate.

$$S\frac{dH}{dt} = (Q_1 - Q_2) \tag{16}$$
$$H = aQ^2 + bQ + c \tag{17}$$
$$H = \Delta H + KQ^2 \tag{18}$$
$$Q_g = CA\sqrt{2g(H_1 - H_2)} \tag{19}$$

Here S is the pump well surface area (m^2), H is the pump well water level (m), Q_1 is the inflow (m^3/s), Q_2 is the outflow (m^3/s), a, b and c are constants, $(\Delta)H$ is the actual pump stroke (m), K is the pipe resistance coefficient, Q_g is the flow through the gate (m^3/s), C is the flow rate coefficient, g is the acceleration of gravity (m/s^2), H_1 is the inflow side water level (m) and H_2 is the outflow side water level (m).

Simulation results

In this section we present the results of simulation for the case in which fuzzy adaptive control is used to apply cooperative control of the pumps and gates corresponding to rain water inflow conditions.

The results of a simulation of 90 minutes after the start of rainfall are shown in Figs. 6.19 and 6.20. The solid lines in Fig. 6.19 show the changes in the amounts of water removed by each pump and the pump well water levels when fuzzy adaptive control is used. The dashed lines show the results when conventional control methods are used: PID control with the target water level in pump well 1 set to -4m, and pumps 2 to 5 started and stopped to control the number in operation to keep the pump well 2 water level in the range -5m to -2m. The control parameters used with these conventional control methods were determined from the simulation response waveform.

The median values (P_2) of the membership functions for pump well water level "low" and "high" in drizzle correspond to -5m and -2m, respectively. In this simulation, the inflow gates that control flow into each pump well are fixed at wide open.

As is clear from the simulation results, in fuzzy adaptive control for about 50 minutes from the start of the rainfall speed control operates to limit the

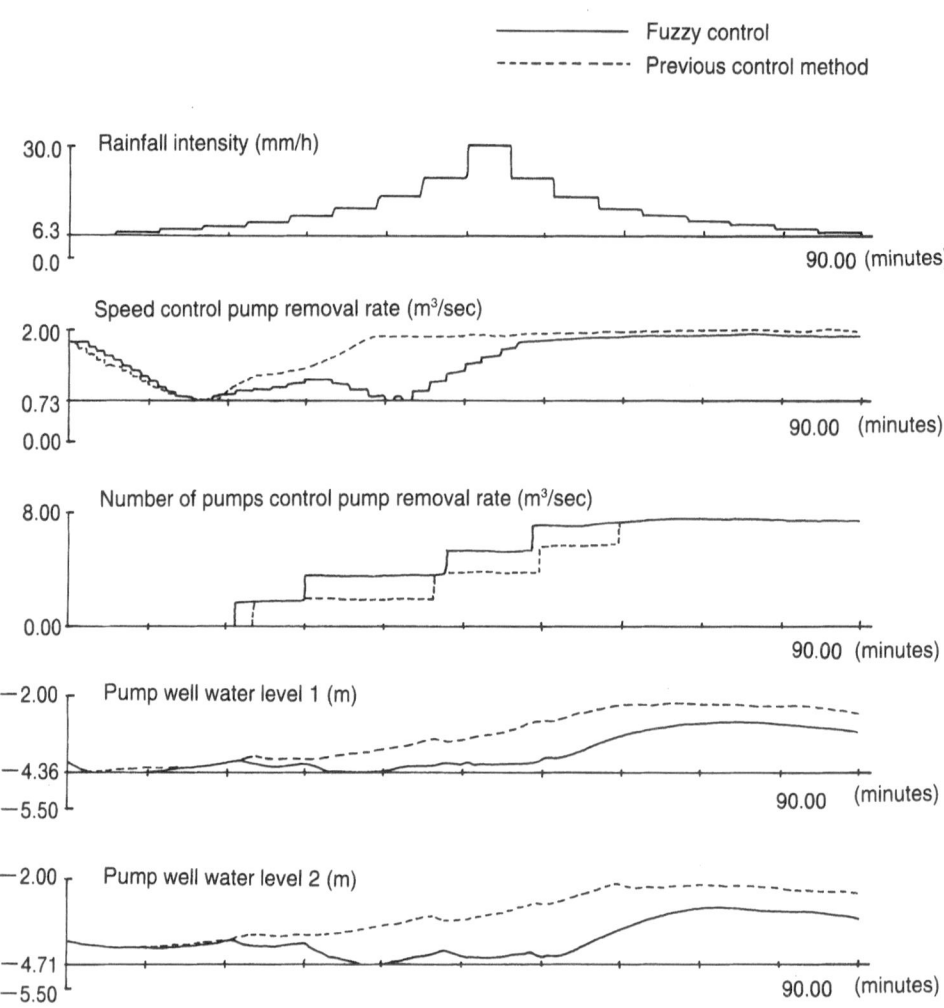

Fig. 6.19 Results of pump flow simulation.

amount of water removed by the speed-controlled pump; the rise of water level is suppressed by quickly starting the pumps in pump well 2 that are subjected to number-of-pumps control. In the case of the conventional control method, after about 35 minutes from the start of rainfall, pump speed becomes full power, then, speed control is not applied. In addition, compared to the case of fuzzy adaptive control, the maximum pump well water level is raised by 70cm. This is because when fuzzy adaptive control is used, when the rain water inflow increases to a level judged to be dangerous based on the pumping capacity, the membership functions for the water level and water level gradient are adjusted to the membership functions for downpour, so that even a slight increase in water level automatically triggers the early starting of a pump to keep the water level down. That is, by using fuzzy adaptive control,

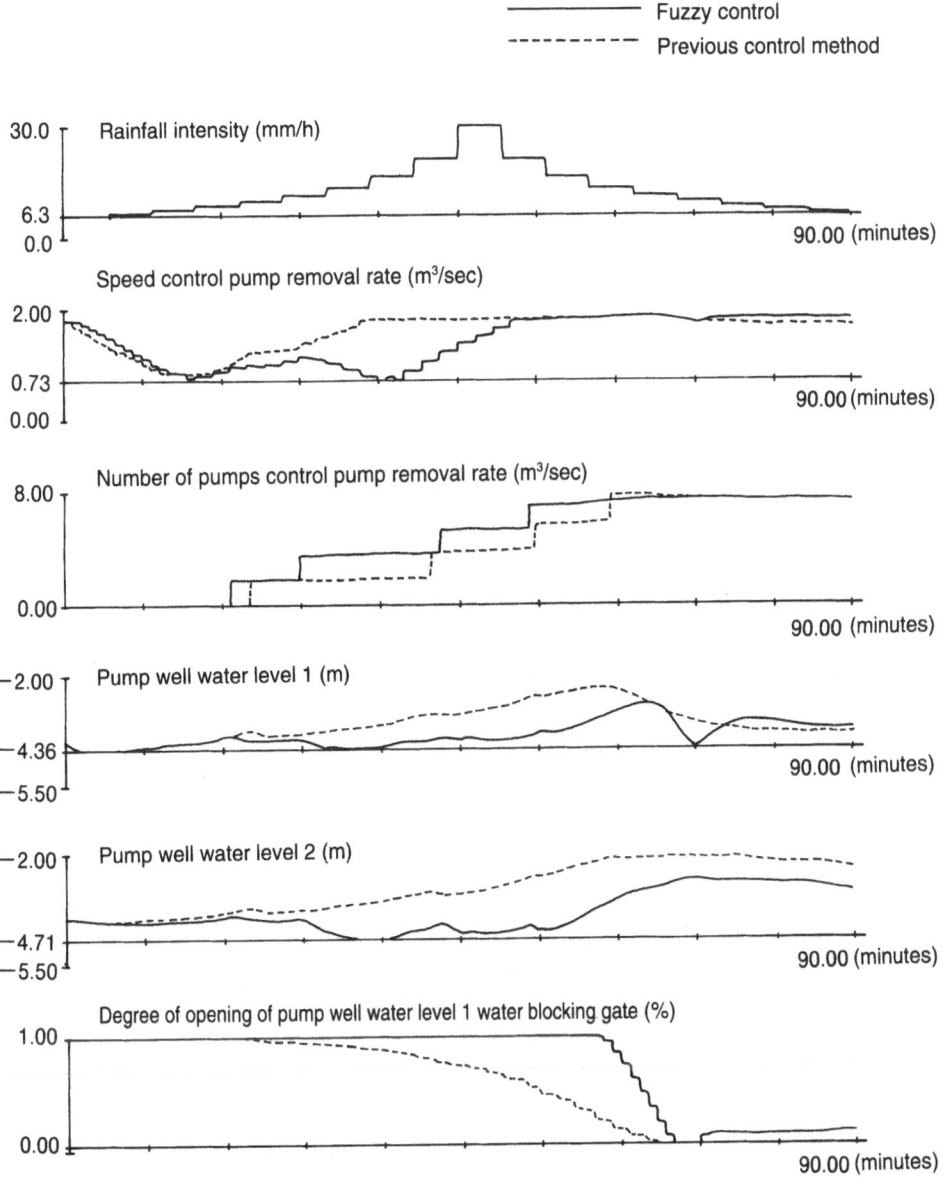

Fig. 6.20 Pump gate coordination control simulation.

the pump well water level and allowable width target values are changed in response to changes in rain water inflow, and rain water pump operation is coordinated with these. In contrast, in the conventional control method the control is very simple, with pump operation not being started until the water level rises to the set water level. In addition, in the conventional control method if the water level that triggers pump starting is set low to keep the water level from rising very high, even slight changes in water level cause the pump to start and stop frequently, resulting in inefficient operation.

The results of simulations with inflow gate control to control the inflow of water to pump well 1 are shown in Fig. 6.20. The dashed line shows the case in which PI control is used for gate control and the pump well 1 target water level is set to -4m. Consequently, gate control that accompanies rising water level starts to be applied before all of the pumps are operating. Conversely, reduction of the inflow amount from the inflow gate lowers the pump well water level, and from the change in amount of water removed by pumping control is applied to suppress the pump rpm. The cause of this is that gate control is applied only on the basis of water level input information; the pump operation status is ignored.

In fuzzy adaptive control, all that is necessary is to prepare control rules, then control that considers both pump well water level and pump operation status can easily be applied. This effect is shown by the solid line in Fig. 6.20. Here, gate control is applied after the pumps are already operating and the water rises to a dangerous level. That is, depending on the input conditions there is a smooth changeover between control based mainly on pumps and control based mainly on gates; it can be seen that the control is cooperative. In addition, when the pump rpm is suppressed the gate opens slightly to increase the amount of inflow; it can be seen from Fig. 6.20 that the pump is operating at full power.

In this section we have shown by simulation that when fuzzy adaptive control is used for pump and gate control in a pumping station that cooperative control of pumps and gates is possible. We have also shown that cooperative control corresponding to rain water input conditions is possible with only a few easily designed control rules. This suggests that fuzzy adaptive control can be effective in solving the problem that in previous control methods such as PID, it is difficult to achieve cooperative control of 2 or more quantities in response to conditions.

6.6 In conclusion

Examples of fuzzy control carried out by Fuji Electric Co., Ltd. and the FRUITAX, the first general purpose fuzzy controller made in Japan, have been introduced against the background of the history of fuzzy control. Today, as we are in what is sometimes called the 2nd fuzzy boom, it is perhaps necessary to reaffirm that the use of fuzzy control is expanding, particularly in the process control field. It is also important to not treat this fuzzy boom as merely a momentary fad, but to search for those fields for which fuzzy control is best suited. As one example of an approach to this problem, we have explained fuzzy adaptive control in the water treatment field. I will be pleased if this discussion contributes even slightly to the design and introduction of fuzzy control and fuzzy controllers in the future.

References

1. Yagishita O, Itoh O, Sugeno M (1984) Application of fuzzy theory to control of chemical injection in a water treatment plant (in Japanese). Systems and Control 28 (10) : 597–604

2. Itoh O (1987) Application of fuzzy theory to control, Measurement and Control Technology Conference Abstracts (in Japanese). 2-6-1

3. Sugeno M, Itoh O, Yagishita O, Onizuka M (1985) A general purpose fuzzy control system (in Japanese). Fuji Electric J 58 (4) : 59–66

4. Miura K, Maruyama T, Simada Y, Yamamoto K (1988) Example of expert system development for automated propulsion (in Japanese). Collected Papers of Electrical Soc Jpn 108–C (4) : 238–245

5. Itoh O, Miyamoto A, Nakatani T (1990) Application of fuzzy theory to control (in Japanese). Fuji Electric J 63 (10) : 663–666

6. Sugeno M (1988) Fuzzy Control (in Japanese). Nikkan Kogyo Shimbun Sha.

7. Hasumoto R, Morimoto M, Itoh O, Yagishita O (1990) Fuzzy control of coagulant injection in a water treatment plant (in Japanese). In : Proceedings of the 6th Fuzzy Systems Symposium, pp 23–26

Chapter. 7

A COMBUSTION CONTROL SYSTEM FOR A REFUSE INCINERATION PLANT

7.1 Introduction — Fuzziness incorporated into a refuse incineration plant

As one field of application of fuzzy theory, fuzzy control has recently achieved great successes[1]. Since fuzzy control was first applied to cement kiln operation in 1980, many examples of its application have been reported, including process control[2–4], a water treatment plant[5], a steel mill[6] and chemical plants[7, 8]. The power of fuzzy control stems from the fact that it allows the vague but superior decision making ability of human beings to be implemented with a rule base, without need for a rigorous numerical model. The first application of fuzzy control was to a cement kiln. The reason that it then became widely applied in process control is that it has this advantage which makes it suitable for automating plant operations.

Combustion control in the incinerator of an urban refuse incineration plant (Fig. 7. 1) fulfills the roles of making sure that the target amount of refuse is incinerated, stably recovering energy from the exhaust gas and minimizing the thermal load on the main incinerator and the exhaust gas treatment device when refuse is incinerated. When the incinerator combustion temperature is increased, the amount of refuse incinerated increases and the unburned residue in the ash decreases, but, for example, the amounts of NO_x in the exhaust gas increase; the plant state variables and operating variables interact with each other in complicated ways. For this reason, the control system is a multiple-purpose system with multiple inputs and outputs, capable of treating complicated mutual interventions.

An incinerator is normally operated automatically by an Automatic Combustion Controller (ACC)[9], but fine adjustments during normal operation and restoration of the system in emergencies are performed by a specialist operator. In this operation, accurate judgment of the state of the refuse in the incinerator and appropriate adjustment of the controls based on that judgment are important; skill and experience are required of the operator.

Many of these judgments and operations can be described by linguistic empirical rules that incorporate vague expressions; for example, "the accumulation of refuse in the incinerator has increased somewhat and there are signs that the completeness of combustion is falling, so increase the air flow to the combustion and reduce the supply of refuse a bit".

Refuse combustion is accompanied by complicated phenomena, and whether combustion is good or bad is difficult to evaluate quantitatively. Attempts have been made to express the dynamic combustion characteristics as a physical model with a lumped parameter system [10,11], but in actual combustion, the dry region, combustion region and ash layer are distributed in 3 dimensions in a complicated manner; their boundaries are vague and difficult to determine clearly. In their present state of development the most that can be achieved with these models is to use them for qualitative understanding of the dynamic characteristics; they are considered difficult to use for on-line control.

In addition, the properties of refuse, which serves as the fuel for the incinerator, reflect the life style of the people in the surrounding area, and continually undergo both long- and short-term changes.

For example, after the oil shock in the mid-1970s the paper content of refuse decreased, leading to decrease of the amount of heat produced and making response to the deterioration of combustion an important facet of incinerator operation. Conversely, in the mid-1980s, the upward revaluation of the yen led to drastic reduction in the amount of used paper collected for recycling, causing the paper content of refuse to increase and along with it the amount of heat produced. The problem then became how to keep the temperature inside the incinerator at or below a certain level to prevent heat loss. Recently, the rapid increase in office automation has produced a large increase in waste paper, and the amount of plastics that produce a large amount of heat has increased, so that now high heat-producing combustion in incinerators has become a problem[12]. Thus, the properties of refuse are greatly affected by social changes, making it difficult to predict refuse properties from past statistical changes.

In addition, there are seasonal changes: dry winter refuse burns more easily than the rainy season refuse. Regionally, refuse properties change with differences in diet in the countries and cities where plants are located and with differences in the way trash is collected; in some areas different types of trash are collected separately, while in others trash is all lumped together. Even over the short term, each load of refuse loaded into the incinerator has different combustibility.

Refuse combustibility is determined by the caloric content and water content in a unit of weight, the apparent density, and the physical and chemical properties of the refuse, but it is difficult to quantitatively determine the combustibility of the refuse from these quantities. Changes in refuse prop-

Fig. 7.1 An urban refuse incineration plant.

erties act as disturbances to the incinerator combustion control, but it is difficult to express the relation between those properties and combustibility quantitatively, and nearly impossible to measure combustibility on-line.

Thus, a plant includes many vague factors such as the human factor, process characteristics and the disturbances of changes in refuse properties, making fuzzy control well suited for incorporation of operators' empirical rules and specialists' knowledge as linguistic control rules in automation of incinerator operation[13–15].

The remainder of this chapter discusses a fuzzy control system designed for the purpose of automating operation of an urban refuse incinerator.

First, I will explain the composition and functions of the sections of a refuse incineration plant; then I will explain the dynamic characteristics of refuse combustion.

Next, I will point out the problems of fuzzy inference in a multiple input-multiple output system when the operator's operation rules are expressed as skill-based or rule-based control rules, and the inference method which considers the dynamic priority relationships among control rules.

Then I will give the design concept and composition of the fuzzy rule system having the purpose of refuse combustion control. Here I will introduce the concept of the fuzzy sensor which is used in place of an operator's observations of conditions in the furnace, and the actual control rule composition.

The results of a demonstration test over about 1 year in an actual plant will be given, and the fuzzy combustion control system which was developed will be evaluated.

7.2 Characteristics of refuse incineration

The flow in a refuse incineration plant is shown in Fig. 7. 2. The refuse is transported from the refuse pit to a hopper by a crane, then loaded into the incinerator from the bottom of the hopper by the reciprocating movement of a feeder. Inside the incinerator the refuse is combusted by a fuel-air mixture blown in from below a stoker. After the exhaust gas that is produced is recovered, the harmful components are removed by a gas treatment device, then it is released into the air. Meanwhile, the combustion ash is removed from the ash pit at the bottom of the incinerator.

The crane, in addition to transporting the refuse, has the role of stirring the refuse inside the pit to keep the properties of the refuse uniform. The combustion ash is utilized for earth filling, or hardened and then reused as a road aggregate material.

The purposes of refuse combustion control include the following.

1. Combustion control: The exhaust gas temperature inside the incinerator or amount of steam produced in the waste heat boiler is kept within target limits to keep the amount of heat produced constant.
2. Combustion gas control: The oxygen concentration and nitrogen oxide (NO_x) concentration in the combustion exhaust gas are held at or below fixed values.
3. Control of amount of refuse processed: The target for the amount of refuse treated per day is maintained.
4. Start-stop control: The incinerator is started and stopped to heat and cool it in accordance with a predetermined schedule.

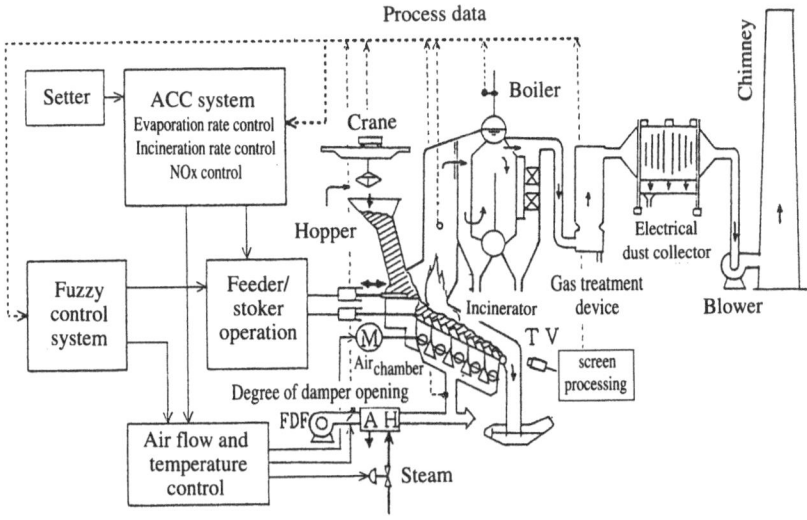

Fig. 7.2 Flow of action and control meters in an urban refuse incineration plant.

Combustion control, in which the amount of heat produced in the incinerator is held constant, maintains stable refuse combustion and permits the waste heat boiler to perform good quality heat recovery with minimum fluctuation. In addition, it prevents thermal degradation of the plant overall due to overheating and dying out of the flames due to reduction of heat produced, and reduces the dynamic load on the gas treatment equipment downstream from the incinerator.

Harmful gases such as NO_x , HCl and SO_x contained in the exhaust gas are normally removed by specialized gas treatment equipment using ammonia, slaked lime, bag filter, etc. Combustion gas control plays the role of pretreatment of the gas upstream from this treatment equipment, so that the gas treatment equipment can always operate under good conditions.

In control of the amount of refuse processed, the target value of amount of heat produced is varied so that the planned amount of refuse is burned each day.

In addition to these controls, it is also important to reduce the amount of unburned combustible refuse contained in the ash. This is done by monitoring the location of the boundary between the combustion region and the ash layer, that is, the "burn-off point"[9], within the refuse in the incinerator. The burn-off point is detected from the brightness on a TV screen used to monitor the inside of the incinerator by image processing.

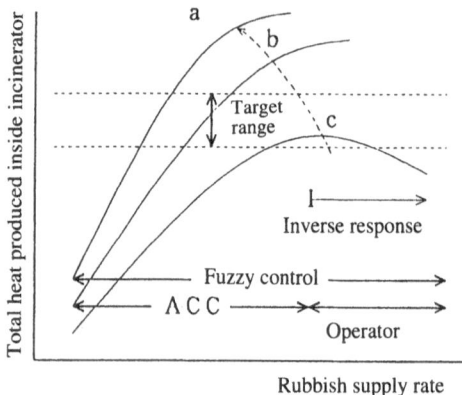

Fig. 7.3 Rubbish heat-producing characteristics, and range within which each system can respond.

The ACC, which keeps the amount of heat produced in the incinerator constant, detects either the amount of heat produced in the waste heat boiler or the temperature inside the incinerator, both of which have a fixed correlation with the amount of heat produced; then the supply of refuse, which is the fuel for the combustion, is increased or decreased by feeder on-off control in response to the deviation from the target value.

The following are among the most common causes of abnormal refuse combustion in an incinerator.

1. Reduction of heat production due to reduction of the amount of refuse being burned in the incinerator.
2. Reduction of the flames due to input of a large quantity of wet refuse.
3. Sudden combustion due to falling of a large quantity of dry refuse.
4. Sudden rapid combustion due to drying of wet refuse.

The amount of heat produced can be reduced by either too much or too little fuel; ACC treats both cases as insufficiency of fuel and continues operating the feeder to supply refuse. Consequently, if the reduction is due to too much fuel, there is a danger of positive feedback in which the continued supply of refuse results in increased rather than decreased reduction of heat production. This resembles the phenomenon, familiar in everyday life, of a bonfire cooling down when wet leaves are scooped onto it and then dying out completely when the quantity of wet leaves becomes large.

For this reason, a sequence is incorporated into ACC such that if the feeder operation signal continues to be generated continuously for more than a certain time, feeder operation is halted temporarily and the system waits for combustion to return to normal.

If the cause of abnormal combustion is 3) above, the amount of heat produced increases suddenly, then the combustion returns to normal relatively
quickly. Cause 4) occurs after cause 2); a great deal of steam is produced
after the relatively low heat production.

These qualitative relations between the amount of refuse supplied and
the amount of heat produced are shown in Fig. 7. 3. The parameter Qr
in the figure is the qualitative ease of combustion of the refuse, that is, a
vague quantity that indicates the refuse quality. For good quality refuse a
the relation between quantity supplied and heat produced rises to the right;
for poor quality refuse c increase of the amount supplied produces a negative
response in which the heat production decreases. Stopping the supply of
refuse permits refuse c to dry out, and the response curve progresses through
b to a.

7.3 Fuzzy control methods and problems

In this chapter, we will touch briefly on some inference formats that are
commonly used in fuzzy control, and discuss problems that are encountered
when they are applied to inference in a multi-input system and examples of
application to process control.

In addition, "inference by ordinal structure model" will be explained with
the aim of having the priority ordinal relations among control rules reflected
in dynamical inference.

7.3.1 Fuzzy inference methods

A number of fuzzy inference methods have been proposed for use in fuzzy
control; they are distinguished by, for example, inference format and method
of putting output into numerical form. Here, we will envision application to
process control and discuss two methods that are actually in use, and give
their inference procedures.

1) The condition part is taken to be a fuzzy set, and the conclusion part a
function of the input variables[17, 18].
The control rule $R^i(i = 1, \cdots, n)$ is given in the following format, where n is
the number of control rules.

$$R^i; \quad \text{IF } A_{i1}(x_1), A_{i2}(x2), \cdots, A_{im}(X_m)$$
$$\text{THEN } Y = f_i(x_1, x_2, \cdots, x_m) \quad (1)$$

Here $x_1, x_2, \cdots, x_m$ are input variables,
Y is the output variable,
$A_{ij}(x_j)(j = 1, \cdots, m)$ is a fuzzy set on X_j.

The If$\cdots$part of a control rule is called the condition part; the then$\cdots$part is called the conclusion part. $A_{ij}(x_j)$ means "x_j is A_{ij}", with A_{ij} as a fuzzy variable. A_{ij} has a linguistic meaning such as Big or Very Small. The comma in the condition part is a logical operator that means "and".

The function f in the operation part is often a linear or quadratic function of the input variables.

$$f = a_0 + a_1 x_1 + \cdots + a_m x_m \tag{2}$$

The outstanding feature of this method is that complicated relationships between input and output variables can be expressed in a relatively simple format. However, if the object of control has strongly non-linear characteristics, for example if there is a large hysteresis loop, and moreover it is changing with time, this method becomes difficult to apply. The coefficients $a_0, a_1, \cdots$ in equation (2) are determined by analyzing a skilled operator's operations and applying the least squares method to bring the inference results into approximate agreement with them. In a process having strong nonlinearity, it is not always possible to find coefficients that give good results, and for this reason measurements and data analysis often require a great deal of time. Also, it is difficult to use an operator's operation rules in a rule base expressed linguistically directly as the control rules.

2) Method in which both the condition part and the conclusion part are expressed as fuzzy sets[18, 19].

$$R^i; \qquad \text{IF } A_{i1}(x_1), A_{i2}(x_2), \cdots, A_{im}(x_m),$$
$$\text{THEN } Y \text{ is } B_i \tag{3}$$

Here, B_i is a fuzzy set on Y.

In this method, the conclusion part format is simple when there are many input variables and complicated input/ output relationships must be expressed; this in turn means that the division of input space becomes complicated, so that it tends to take a great deal of time to put the control rules together. However, as long as the operation rules and knowledge in an operator's or specialist's skill base or rule base that can be expressed linguistically are described as the control rules, this method is clear and easy to understand intuitively. For this reason, method 2) is used in the fuzzy controller described in this chapter.

Many other methods have also been proposed, including a special case of method 1) in which the inference rules are simplified, with the conclusion part being given in the form of constants[20]. Some other examples are presented in references [21–23].

When fixed inputs $(x_1, \cdots, x_m)$ are given, the degree to which the condition part satisfies each of the control rules $R^i (i = 1, \cdots, n)$ is called the degree of applicability of the condition part. It is written as β_i. The formula

$$\beta_i = \min\{A_{i1}(x_1), \cdots, A_{im}(x_m)\} \tag{4}$$

is often used to compute the β_i. Here $A_{i1}(x_1), \cdots, A_{im}(x_m)$ are the membership functions. A number of other methods have also been proposed for computing the degree of applicability, such as taking the algebraic product $A_{i1}(x_1), \cdots, A_{im}(x_m)$. The definition in equation (4) will be used in the remainder of this chapter.

When fixed inputs are given, the set of inference results from the control rules is called the solution set and written B^*. A number of methods for obtaining this solution set have been proposed. The methods given below are in general use. Note that the subscript i is the number of a control rule.

1) Compression operation is performed by multiplying the membership function by the degree of applicability.

$$B_i^* = \int_{y \in Y} \beta_i \cdot B_i(y)'y \qquad (5)$$

2) The conclusion part membership function is reduced by the degree of applicability:

$$B_i^* = \int_{y \in Y} \min(\beta_i, B_i(y))/y \qquad (6)$$

3) The operation part membership function is parallel- shifted downward by the value of the part of the degree of applicability which is less than 1:

$$
\begin{aligned}
B_i^* &= \int_{y \in Y} (\beta_i(y) - (1 - \beta_i))/y; \\
&\quad \text{for } B_i(y) \geq 1 - \beta_i, \\
&= \int_{y \in Y} 0/y; \\
&\quad \text{for } B_i(y) < 1 - \beta_i.
\end{aligned}
\qquad (7)
$$

These methods are shown in Fig. 7. 4.

The operation part solution set B* is formed by adding the individual solution sets, as follows:

$$B^* = B_1^* \cup B_2^* \cup \cdots \cup B_n^*. \qquad (8)$$

where $\cup$ means union of sets (linguistically "or"). Expressing equation (8) in terms of the membership $B^*(y)$, we obtain:

$$B^*(y) = \max_{y \in Y}\{B_1^*(y), \cdots, B_n^*(y)\} \qquad (9)$$

This method is shown in Fig. 7. 5.

The most commonly used method of determining the final output y^* from the composite solution set B^* (this is called putting the solution into numerical form, or defuzzification), is to use the y coordinate of the center of gravity of the composite solution set, the so-called "center of gravity method" [17,23]. When this "compositional rule of inference" is used, y^* is computed

X is A ⟶ Y is B

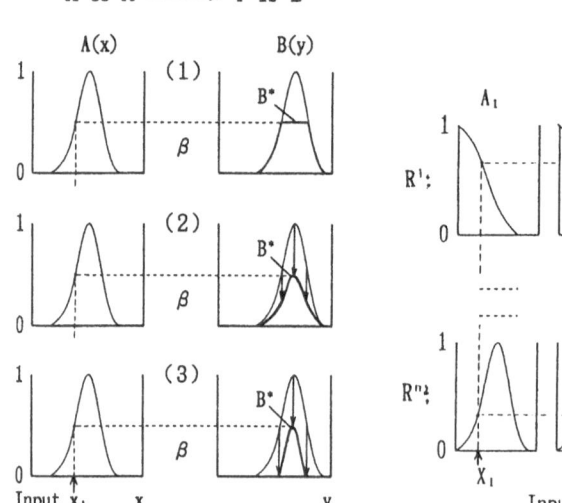

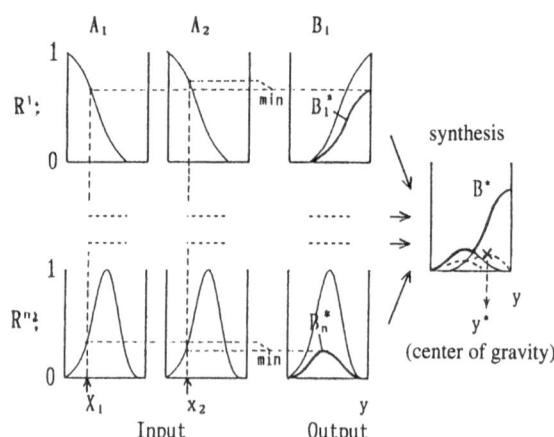

Fig. 7. 4 Inference results B*
obtained by various
methods.

Fig. 7. 5 Synthesis of inference results, and output by cen-
ter of gravity method.

by the following formula.

$$y^* = \frac{\int_{y \in Y} B^*(y) \cdot y dy}{\int_{y \in Y} B^*(y) dy}. \tag{10}$$

In fuzzy control, as can be seen from the above calculation weights are assigned to the conclusion parts according to the degree of applicability of each condition part of the control rule; the average value is then output and used.

7.3.2 Characteristics and problems of fuzzy inference

In this section we discuss some characteristics and problems of fuzzy inference, then consider the process of decision making on operator's operation quantities in an actual plant.

As discussed in the above sections, the inference process in fuzzy control can be roughly divided into the following 3 processes[17].

1. Calculation of the degree of applicability of the condition part of each control rule.
2. Calculation of the inference result (solution set) of each control rule.
3. Synthesis of the solution sets of each control rule, and calculation of output.

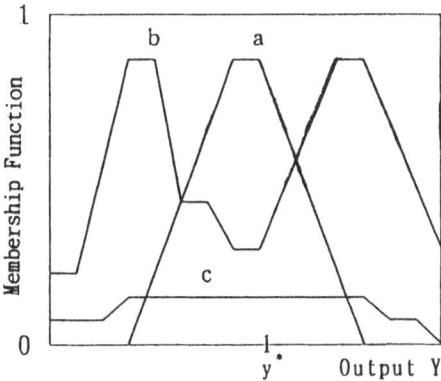

Fig. 7.6   Inference synthesis result pattern.

A number of methods have been proposed for inference in fuzzy control, but this process is shared in common by nearly all of the methods.

One characteristic of fuzzy control is that even in input in an intermediate area of an input space partitioned linguistically by the inference synthesis rule, approximate reasoning is performed by several control rules around it. This feature permits a fuzzy controller to have complicated input/output characteristics with a small number of control rules. In many specific applications which have been carried out up to the present, there have been anywhere from a few to, at most, several tens of control rules, 1 to 2 orders of magnitude less than are required by a conventional expert system.

Next, let us consider the characteristics in the case of application to a multi-input system with inference in format (3), "IF X is A, THEN Y is B".

Since a feature of this method is that both input and output are expressed as fuzzy sets, as was stated above operator's empirical rules and specialist's knowledge expressed in natural language can be easily incorporated into the controller as control rules. However, as can easily be guessed from the format of the condition part of the inference, as the input becomes multi-dimensional and the number of fuzzy variables increases, the number of control rules increases exponentially. For example, if the input is 4-dimensional and the input and output each have 5 fuzzy variables, the maximum number of control rules becomes $5^4 = 625$.

However, the number of control rules used in an actual plant is generally on the order of 20 for one output. The reason for this is that rules in the rule base possessed by operators and specialists that can be expressed in the format "IF $\cdots$ THEN $\cdots$" is generally not more than about this number.

In general the number of areas into which the input space is divided is equal to the number of control rules. If the maximum number of input spaces that can be devided is large and at the same time the actual number of control rules which are composed by an input space having fewer dimensions than the total number of input variables is small, then the synthesized inference result is apt to simultaneously have a solution that has different conclusions. There are three main types of synthesized inference result pattern[24], as shown in Fig. 7. 6. Inference result a in the figure shows the case in which there is only 1 adequately large peak; it can be expected that inference having adequately high reliability will be obtained through all of the control rules. b shows the case in which 2 separated peaks do not appear; in this case, rules having different operation parts can hold simultaneously. c shows the case in which pronounced peaks do not appear throughout the whole output region; in this case a highly reliable solution cannot be obtained from any of the control rules.

The final inference result of fuzzy control is expressed in a 2-dimensional pattern, as shown in Fig. 7. 6, but it is necessary for an actual controller to select just one reliable number from this result. In this case, even though the 3 synthesized results all have different meanings, they all give nearly the same output in a center of gravity calculation, for example using equation (9).

There are many methods of converting an inference result to a number including, in addition to the center of gravity method, for example one which gives a median or maximum peak value in the pattern of the solution set as the output value. Only the numerical processing methods are different; in substance the same problems occur.

Phenomena which occur in relation to synthesized inference results, such as the contradiction that occurs in case b or the occurrence of results of low reliability as in case c, occur frequently in fuzzy inference in multi-input systems having a large number of dimensions of input variables. This is thought to happen because there is a limit to the number of control rules received from operators and specialists, and it is difficult to find a combination of method of dividing input space and control rules such that contradictions absolutely will not occur.

If a contradiction or solution of low reliability arises from synthesized inference results, and the result of converting them to a numerical value does not match what human intuition would consider a reasonable result, in the conventional fuzzy inference method it was necessary to further subdivide the input space and correct the control rules. That is to say, in multi-input system fuzzy control, if one attempts to simplify the control rules contradictory interpretations and difficulty in matching the results to human intuition occur; conversely, when an attempt is made to force strictly logical compatibility with control rules, inevitably the number of control rules tends to increase.

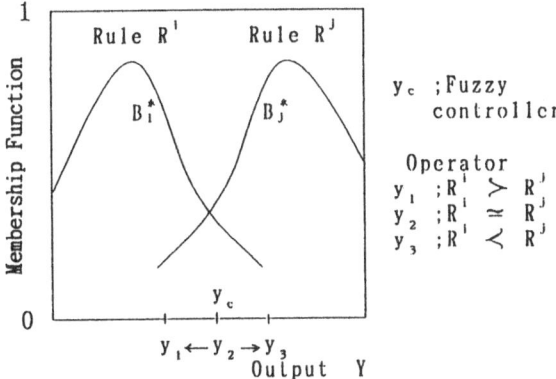

Fig. 7.7 Fuzzy inference and operator selection.

Let us now consider the response of an operator to a case such as that shown in Fig. 7. 7, in which 2 different peaks occur in the conclusion from the final inference results, that is, the synthesized solution set. Let R^i and R^j be the control rules that form these 2 peaks, respectively. In ordinary fuzzy inference, according to the inference procedure in the preceding section, according to equation (10) the approximate midpoint yc between the 2 peaks becomes the controller output. However, if the operator's assessment of the relative importance of these rules changes, it is expected that the operator will choose an output closer to the conclusion of the rule that he considers more important. The difference between the operator's selected output y and the output yc found by the center of gravity method in a case in which a contradiction arises in the synthesized solution set becomes larger as the degree to which one control rule is considered more important than the other becomes larger and as the degree of applicability of that rule becomes relatively larger; and output close to the conclusion of the priority rule is selected[18].

In a fuzzy controller having the purpose of automating the operator's operations in a plant, in many cases the action of the operator's skill base or rule base is formulated into control rules. The smallest unit of such a control rule possessed by an operator often has a simple structure with 1 input and 1 output (or at most 2 inputs). A human being possesses many such simple control rules, and intuitively and instantaneously determines operation quantities based on the degree of applicability with which these rules satisfy the condition part conditions and the difference in evaluation of the importance of these rules. Fig. 7. 8 shows a structural model used to determine an operator's operation quantities; p_i in this figure is a constant that indicates the degree of importance attached to the rule R^i. The differences between this and fuzzy inference in an ordinary multi-input system lie in the interpretation function in case of contradiction between control rules and the smaller

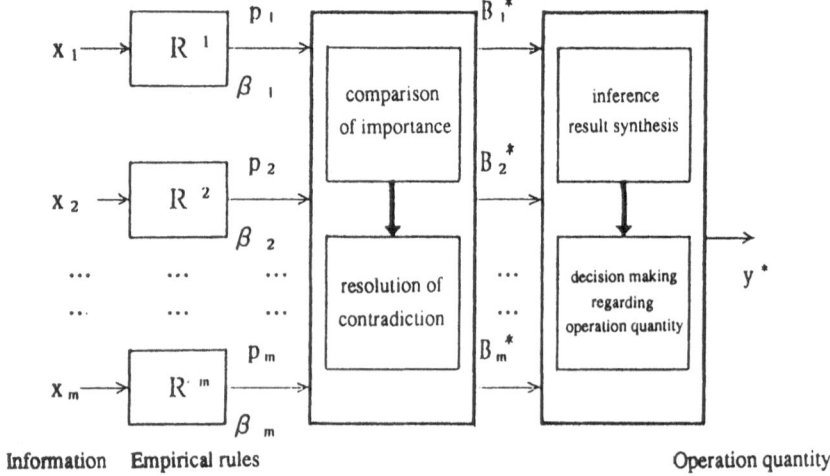

Fig. 7.8 Operator's operation quantity determination model.

number of input variables for each control rule.

In existing forms of fuzzy control, when it becomes necessary to interpret a contradiction that arises in a synthesized solution set, this has been taken care of by adjustments in the control rules themselves, for example by partition of the input space and changes in the set of operation parts. In order to obtain the subtle operations of an operator in a multi-input system, it is sufficient to provide a function that resolves the contradiction in accordance with the relative importance of the various rules. Use of such an inference method[24] makes it easy to describe the control rules; also, the number of control rules can be reduced; and control action that approximates an operator's subtle control action is obtained.

7.3.3 An ordinal structure model of control rules

In this section we discuss an "ordinal structure model of control rules" which has the purpose of resolving contradictions that arise in a synthesized solution set. The process by which an operator intuitively resolves such a contradiction has characteristics such as those given below. These apply when 2 rules having different operation parts are applied simultaneously.

Characteristic 1: The larger the degree of importance of a priority rule, the closer the output is to the conclusion of that rule.

Characteristic 2: The larger (relatively) the degree of applicability of a priority rule, the weaker the effect of the operation part of the other rule in determining the output.

Letting R^i and R^j be 2 control rules having mutually contradictory operation parts, the priority relations between them are expressed by such symbols as $<$, $\lesssim$, $>$ and $\gtrsim$. For example, if greater weight is to be attached to the conclusion of R_i when R_i and R_j are applied simultaneously, this priority relation is expressed as follows.

$$R^i \gtrsim R^j, \text{ or } R^i \gtrsim^p R^j. \tag{11}$$

Here p is an index that indicates degree of importance; if $p \geq 0$, then the larger p is, the greater the priority given to the rule on the left, in this case R^i. The linguistic meanings of p include, for example, the following.

$p = 0$: R^i and R^j have equal priority.
$p = 1/2$: R_i has slight priority over R^j.
$p = 2$: R_i has strong priority over R^j.

Let the set of control rules be R, and R_r the set of groups of elements (R^i, R^j) of R which have certain priority relationships, and the set of ordered elements (R^i, R^j) of R be R_r. If we also specify that the set of elements with the sequential relationship not considered does not belong to R_r, then 2 arbitrary elements R^i, R^j of R_r have the following properties.

1) One and only one of the following conditions holds.
   $R^i \gtrsim R^j, R^i < R^j, R_r \not\ni (R^i, R^j)$
2) If the relationships $R^i \gtrsim R^j$ and $R^j \gtrsim R^k$ hold, then $R^i \gtrsim R^k$.

Therefore, the set of control rules forms a semi-ordered set subject to these relationships.

Suppose that the rules R_i and R_j satisfy the relationship $R^i \, R^j$, and that β_i and β_j are the degrees of applicability of the respective condition parts of the rules R^i and R^j. With or being the binomial operation that projects the relative size relationship of β_j and β_i onto [0,1], the following properties hold.

1-a: $0 \leq \tau(\beta_i, \beta_j) \leq 1$
   $\tau(0, \beta_j) = 1, \quad \tau(\beta_i, 0) = 0$

1-b: τ is a continuous, monotonically increasing function with respect to β_j/β_i.
   That is, if $\beta_j/\beta_i \geq \beta_l/\beta_k$ then $\tau(\beta_i, \beta_j) \geq \tau(\beta_j, \beta_l)$

Next, we introduce the function $\Phi(p, \tau)$ which expresses the strength of the order relationship between 2 control rules. $\Phi(p, \tau)$ is:

$$\Phi(p, \tau); p \times \tau \to [0, 1]$$

and has the following properties:

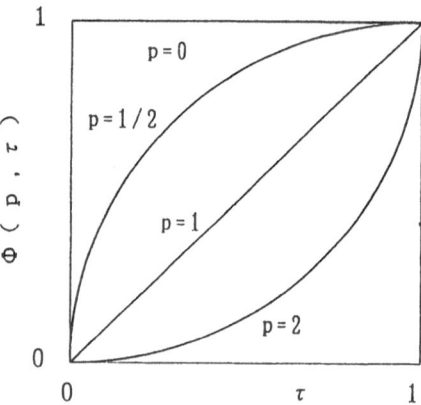

Fig. 7.9 The dynamic weighting coefficient $\Phi(p, \tau)$.

$$2 - a \quad p_1 \geq p_2 \rightarrow \Phi(p_1, \tau) \leq \Phi(p_2, \tau)$$
$$2 - b \quad \tau_1 \geq \tau_2 \rightarrow \Phi(p, \tau_1) \geq \Phi(p, \tau_2)$$

$\Phi(p, \tau)$ is a dynamic weighting coefficient which corrects the lower priority side degree of applicability β_j according to the relative magnitudes of the degrees of applicability of the control rules. Letting $\beta_i{}^*$ and $\beta_j{}^*$ be the degrees of applicability which have been newly corrected by the interpretation of the contradiction, β_i and β_j are replaced by:

$$\beta_i* = \beta_i$$
$$\beta_j* = \Phi(p, \tau)\beta_j \tag{12}$$

In this chapter, τ and $\Phi(p, \tau)$ which have the above properties are determined as follows:

$$\tau = \beta_j/(\beta_i + \beta_j) \tag{13}$$
$$\Phi(p, \tau) = \tau^p \tag{14}$$

(13) and (14) satisfy 1–a, b and 2–a, b, respectively. At this time equation (11) becomes as follows:

$$\beta_j{}^* = [\beta_j/(\beta_i + \beta_j)]^p \beta_j \tag{15}$$

The relation between τ and $\Phi(p, \tau)$ according to equations (13) and (14) is shown in Fig. 7. 9.

The expression in brackets in equation (15) is monotonically increasing with respect to β_j/β_i. As the degree of applicability β_i of the condition part of the priority side becomes large relative to the degree of applicability β_j of the lower priority side, the expression in brackets approaches 0; in the opposite case it approaches 1. Also, as p becomes large $\beta_j{}^*$ becomes small and the contribution of the lower priority side rule R^j to the computation of the output y^* becomes small. Consequently, this kind of computation satisfies "Property I" and "Property II".

7.4   A fuzzy control system

In this section, we will discuss the original basic concept in developing the fuzzy control system and the conditions for being able to put it into practical application. Next, we will discuss the functions and procedures used in creating the fuzzy sensors and controllers that comprise the system.

7.4.1   Composition of the fuzzy control system

The principal role of the fuzzy control system is to substitute for the repetitive operations of a skilled operator when combustion becomes abnormal, and prevent overheating and deterioration of the combustion. Consequently, the basic concept calls for the controller operation to resemble the operations of a skilled operator as closely as possible in each set of conditions.

The controller is meant to be used not only in a newly built plant but also in an existing incinerator. For this reason, it is necessary that it be possible to install the fuzzy control system without need for major modification of or addition to the existing plant instrumentation or development of special sensors, and without changes in the existing operation control system. If for some reason it should become necessary to disconnect the fuzzy controller from the system, this policy will make it possible to easily switch to operation by the ACC and a human operator. For this reason, the basic policy is that it must be easy to install and remove the controller.

According to analysis of the operator's operations, the control operations are not usually determined by trial and error, but rather on the basis of knowledge of conditions in the incinerator. For example, when the feeder speed is reduced, it is usually because of a sense that the combustion is worsening and there is a sign of reverse response by the ACC. This diagnosis is deduced from images from the monitor TV inside the incinerator and process data. It is necessary to have information which by nature is imprecise, such as the accumulation condition of refuse inside the incinerator and the quality of the refuse.

Consequently, in order to reproduce the operations of a skilled operator by fuzzy control, it is not sufficient to merely incorporate the operator's operation rules into the controller as fuzzy control rules; the controller must have the capability to make subtle judgments about conditions in the incinerator. For this purpose, it is necessary to be able to sense conditions inside the incinerator in a way that provides the same kind of imprecise information used by a human operator.

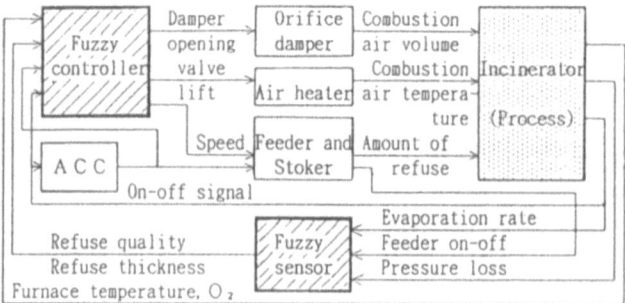

Fig. 7.10 Composition of a refuse incineration plant control system
(diagonal lines show fuzzy control system).

For these reasons, the fuzzy combustion control system consists of fuzzy
sensors and a fuzzy controller. Some of the operator's observations of quantities indicative of conditions inside the incinerator are done by the fuzzy
sensor. This information is input together with process data into the fuzzy
control section, where fuzzy inference is performed to produce control signals
which are output. The composition of this fuzzy control system is shown in
Fig. 7. 10.

7.4.2 Fuzzy sensors

When abnormal combustion occurs, the operator observes the refuse combustion condition and the condition of piling of refuse on the stoker, judges
the cause and decides what operation to perform. However, these observation results are probably in imprecise form, and cannot be measured with an
ordinary sensor. Meanwhile, in fuzzy control a certain amount of imprecision
is allowed in the input signals. From this point of view the fuzzy sensors
infer and test-prepare imprecise data which the operator uses to determine
operation quantities, including the refuse quality, refuse layer thickness and
combustion condition. The methods of sensing refuse quality and refuse layer
thickness, which are used directly in actual fuzzy control, are described below.

(1) Refuse quality: Refuse quality means the ease of burning in a qualitative
    sense. The operator calls refuse that burns easily "good quality refuse".
    The refuse quality sensor, like a skilled operator, infers the refuse quality
    Qr from the time T_1 after refuse input is started until the incinerator
    temperature drops and the rate $C\theta$ of the subsequent temperature increase.
    Qr is a projection from $T_1 \times C\theta$ to $[0,1]$; the closer to 1, the better the refuse
    quality. There are 9 inference rules, in a format such as the following.

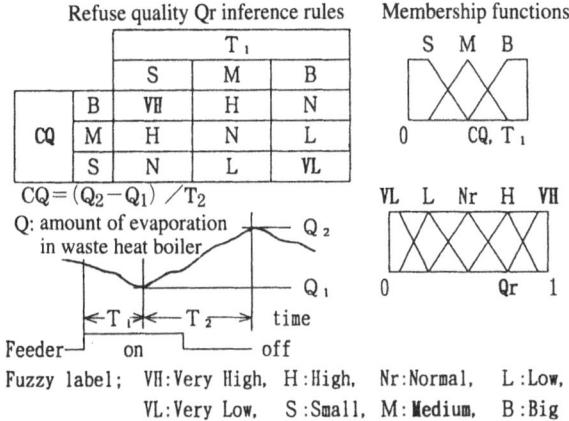

Fig. 7.11 Refuse quality inference rules.

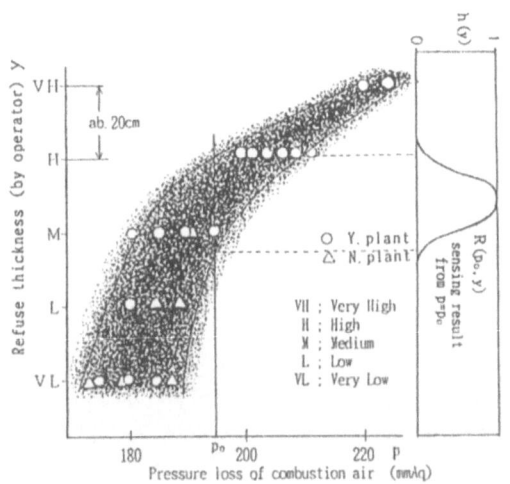

Fig. 7.12 The fuzzy relation between pressure loss and refuse layer thickness.

IF T_1 is big and $C\theta$ is small, then Qr is Bad.

This inference rule is shown in Fig. 7. 11.

(2) Refuse layer thickness: The greater the amount of refuse piled up on the stoker, the greater the loss of pressure in the air that blows through it. Consequently, the refuse layer thickness can be estimated very roughly from the fuzzy relationship $R(p, y)$ between pressure loss p and refuse layer

thickness y. An example of actual measurement is shown in Fig. 7. 12. The refuse layer thickness in the figure was estimated visually by a worker by comparison with the brick wall of the incinerator. Letting $h_R(p, y)$ be the membership function of the relation $R(p, y)$, when the non-fuzzy input $p = p_0$ is given, the result of refuse layer thickness inference is given by $h_R(p_0, y)$, as shown in Fig. 7. 12. In actual control, if A is the refuse layer thickness subset and $h_A(y)$ is its membership function, the degree of appliciability β of the proposition "y is A" is given by the following formula.

$$\beta = \max_{y \in Y} [\min\{h_A(y), h_R(p_0, y)\}] \tag{16}$$

In addition, the combustion range and the position of the boundary between the burning refuse and ash are important in judging the combustion condition. These are automatically sensed by performing image processing on the television image used to monitor the inside of the incinerator (Fig. 7. 13) and performing fuzzy inference.

Fig. 7.13 View on TV screen that monitors inside of incinerator (the horizontal line is the computed position at which the refuse is completely combusted).

It has been confirmed in actual incinerator tests that these sensor inference results agree qualitatively with a human operator's judgment and incinerator conditions [13, 15]. In conventional PID control, it has generally been considered that this type of sensor, which makes inferences indirectly from process data, is not practical because of errors and reliability problems.

The fuzzy sensor concept introduced here is effective in such cases as when creating control rules when it is difficult to infer operation quantities directly from process output, and when the output is to be used for monitoring incinerator conditions.

7.4.3 Fuzzy control rules

Both the condition parts and conclusion parts of the control rules, indicated by equation (3) in section 7.3.1, were expressed as fuzzy sets. Since this method describes an operator's empirical knowledge that can be expressed linguistically, it is a simple and clear method that is easy to understand intuitively and has been used successfully many times in this type of fuzzy control.

In the actual controller, the degrees of applicability of the control rules are found from equation (4) and the solutions of the various control rules are given by equation (6) as algebraic products of the set of operation parts and the degree of applicability. The center of gravity method of equation (10) is used to put the output in numerical form.

The actual control rules were created from skilled operators' empirical rules and the knowledge of combustion technology specialists. The following are examples of this kind of knowledge.

(1) If the oxygen concentration in the hot exhaust gas is stable, normally the NO_X concentration will also be stable, so stabilizing the oxygen concentration can be used as an indirect means of stabilizing the NO_X concentration.
(2) When the refuse quality on a given day is poor, if the combustion air temperature is increased a bit the temperature inside the incinerator will stabilize.
(3) If the deviation of the amount of evaporation in the boiler and its rate of change are negative, the combustion is deteriorating, so the amount of air should be increased.
(4) If the feeder continuous operation time is long and the refuse layer thickness is increasing, it means that ACC reverse response is occurring, so the refuse supply rate must be drastically decreased.

These rules of thumb were refined by giving an operator various process condition data and video tapes of combustion conditions inside an incinerator and analyzing his judgments and operations in each set of conditions, and by seeking the opinions of process specialists and control technicians.

The fuzzy controller uses 8 inputs to infer 3 outputs: the degree of opening of the air damper, the feeder operation speed and the combustion air temperature. There are a total of 46 control rules. The degree of opening of the air damper, which determines the combustion air flow rate, is inferred from 4 variables: deviation from the target value of evaporation rate in the

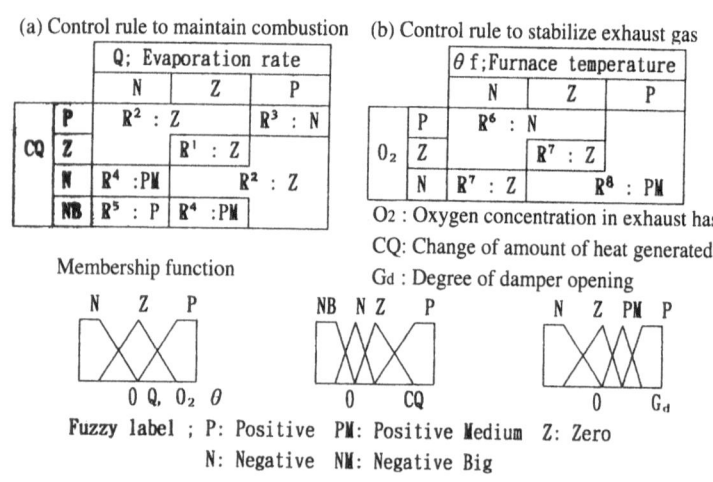

Fuzzy label ; P: Positive PM: Positive Medium Z: Zero
N: Negative NM: Negative Big

Fig. 7.14 Control rules for degree of damper opening.

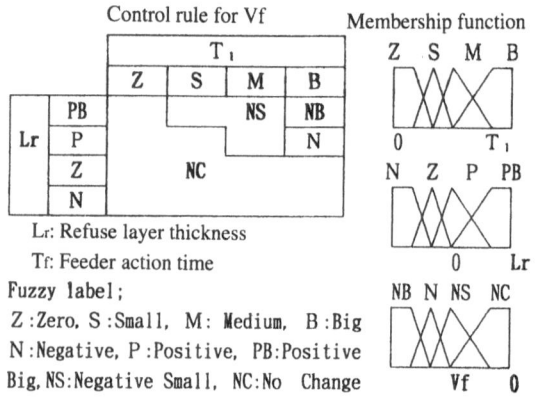

Lr: Refuse layer thickness
Tf: Feeder action time
Fuzzy label ;
Z :Zero, S :Small, M: Medium, B :Big
N :Negative, P :Positive, PB:Positive
Big, NS:Negative Small, NC:No Change

Fig. 7.15 Control rule for feeder speed

waste heat boiler and its rate of change, temperature inside the incinerator, and oxygen concentration in the exhaust gas. The relevant control rules are shown in Fig. 7. 14.

The feeder operation speed, which determines the refuse supply rate, is inferred from the time the feeder operation signal, which is an ACC output, remains continuously ON, and the refuse layer thickness, as shown in Fig. 7. 15. Feeder speed control is applied in response to worsening of the combustion; it returns to its initial value every time the ACC signal goes

θc control rule

		Qr		
		VL, L	Nr	II, VII
	N	PB	P	Z
θf	Z	P	Z	N
	P	Z	N	NB

Membership function

VL L Nr II VII

0   Qr 1

N   Z   P

0   θf

NB N Z P PB

0   θc

θc: combustion air temperature
Qr: refuse quality
θf : temperature inside incinerator
Fuzzy label:VL;Very Low L;Low
Nr;Normal, II;High, VII;Very High
NB;Negative Big, N;Negative,
Zero, P;Positive, PB;Positive Big

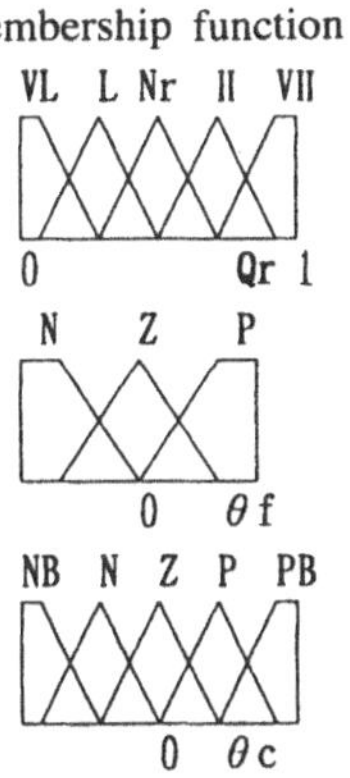

Fig. 7.16 Control rule for combustion air temperature.

OFF. As can be seen in Fig. 7. 15, there is no change of speed for a short time after the feeder operation is started. This is the time during which a human operator would observe the course of combustion in the incinerator without operating the feeder; the rule is based on the operator's empirical rule.

Combustion air temperature is inferred from refuse quality, which is a fuzzy sensor output, and temperature inside the incinerator. These control rules are shown in Fig. 7. 16. These rules are for the purpose of keeping the combustion gas temperature constant and are created mainly from the control technician's knowledge.

The inference rules for the degree of opening of the air damper, shown in Fig. 7. 14, consist of control rules that have the evaporation rate deviation and its rate of change as inputs, and control rules that have the oxygen concentration in the exhaust gas and the temperature inside the incinerator as inputs. The former rules are to maintain the combustion, the latter to stabilize the exhaust gas properties. In the process of synthesizing these rules, if rules for which the operation parts are different hold simultaneously, dynamic weighting is done according to equations (12) and (15) for rules that are in a sequential relationship with one another, with regard to the degree of importance attached to each. This sequential relationship is shown in Fig. 7. 17. The sequential relationship for the air damper control rules is created based on the operator's empirical rules, as follows.

(1) If combustion is stable, the exhaust gas properties will stabilize, so the degree of damper opening is held as is regardless of the temperature inside the incinerator and the oxygen concentration. This rule R' has a very strong priority ($p = 4$):

$$R^1 \gtrsim^4 R^j \quad (j = 6, 8)$$

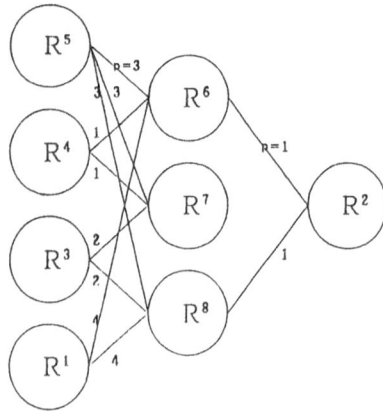

Fig. 7.17 Ordinal stracture model of control rule for degree of damper
opening (control rule numbers are as in Fig. 7. 14).

(2) When the deviation Q of the evaporation rate from the target value and
its rate of change CQ are negative, namely in the case of poor combustion,
when Q and CQ are positive but the combustion is excessive, the operation
that will return the combustion to normal has priority corresponding to the
degree of imbalance:

$$R^5 \gtrsim^3 R^j \qquad (j = 6, 7, 8)$$
$$R^4 \gtrsim^1 R^j \qquad (j = 6, 7)$$
$$R^3 \gtrsim^2 R^j \qquad (j = 7, 8)$$

This inference method according to an ordinal structure model has the
advantage that, by inserting the interpretation of the contradiction into the
inference process dynamically, even when the number of input variables is
large the number of dimensions of the input variables for each control rule
is reduced and the number of control rules is reduced. The purpose of this
method is to obtain a skilled operator's intuitive operation with relatively
simple control rules.

7.5   An actual incinerator test

Fuzzy control systems were installed in 2 existing commercial incinerators
handling 200 tons/day, and an operational test was carried out over almost
a full year. The fuzzy controller assumes that the process time constant is
1 to 2 minutes. Input data are taken in every second, and every 10 seconds
the input data are averaged and operation values inferred. The control rules
that are applied and the synthesized inference results are displayed on-line
as shown in Fig. 7. 18. This kind of capability can also be used to monitor

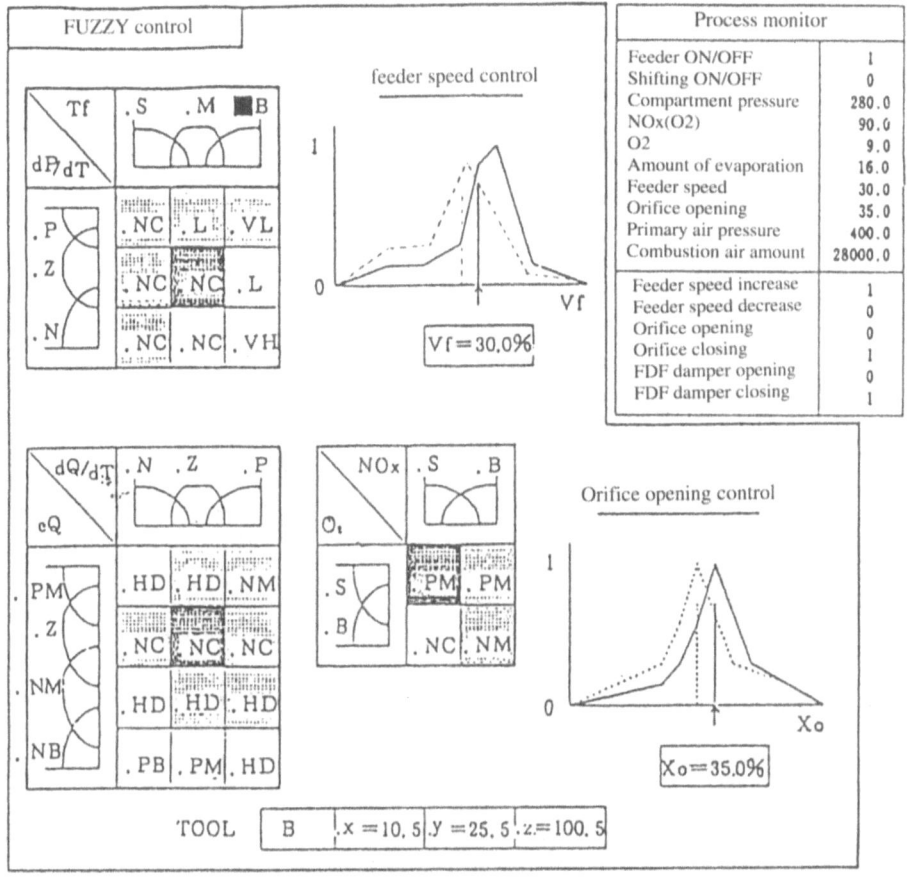

Fig. 7.18 Example of controller monitor screen.

incinerator operation.

On-site adjustment of a fuzzy controller was principally done by changing the support sets of input and output variables. First, the support set of input variables was changed in an open loop while comparing the fuzzy controller inference results to the judgments of operators and control technicians. Next, fine adjustment was performed while observing conditions in a closed loop. In the initial stage, the combustion air temperature control function was not incorporated; it was added in a subsequent improvement. This was mainly to respond to a change of refuse quality to refuse that burns hot that occurred between the time the planning for the test was started and the time the actual test started, as mentioned in section 1. In the initial stage there were 25 control rules, and this control system was evaluated as follows.

1) In normal operation, both the temperature inside the incinerator and the evaporation rate were controlled about as well as with a human operator.
2) When poor quality refuse containing a great deal of water is input, fuzzy control gives more rapid response than a human operator and normal combustion is restored more quickly.
3) With regard to the specific operations used to restore normal combustion, whereas a human operator mainly adjusts feeder speed, fuzzy control mainly adjusts air flow.

In subsequent improvements, a combustion air temperature control function was added; and the control rules were changed somewhat, considering 3) above. At the end continuous automatic operation was continued for 2 months, and the following operation side evaluations were obtained for the fuzzy control system.

1) In normal operation, stable combustion was obtained, with both the amount of heat produced and the temperature inside the incinerator deviating from the target range far less frequently than with a conventional system. In addition, there was almost no change in the quality of control obtained from one incinerator to another or with seasonal change in refuse quality. Variations in NO_X and O_2 concentrations in the exhaust gas were about the same as with a human operator.
2) During abnormal combustion, in which refuse quality worsens and combustion deteriorates, normal combustion is restored more quickly than with a human operator.

Measured examples of frequency distributions of deviations of waste heat boiler evaporation rate and temperature inside the incinerator for both operation by fuzzy control and operation by a human operator are shown in Figs. 7.19 and 7.20, respectively. These show 2 weeks of data sampled once every minute during continuous operation. With fuzzy control, the frequency with which evaporation rate and temperature inside the incinerator fell outside their target ranges was reduced to $1/2 - 1/3$ that with conventional control. This is very important in reducing thermal wear on the incinerator, waste heat boiler, gas treatment unit, etc., and in obtaining stable heat recovery from the combustion exhaust gas; and is the item requiring the most serious attention from the operator in incinerator operation.

An example of the characteristics during worsened combustion due to input of waterlogged refuse is shown in Fig. 7. 21. Arrows in the figure show times when waterlogged refuse was input; the fuzzy controller detected reduction in the heat produced and increase of the refuse layer thickness, and reduced the refuse supply rate and increased the air flow; and in about 10 minutes combustion was restored to normal.

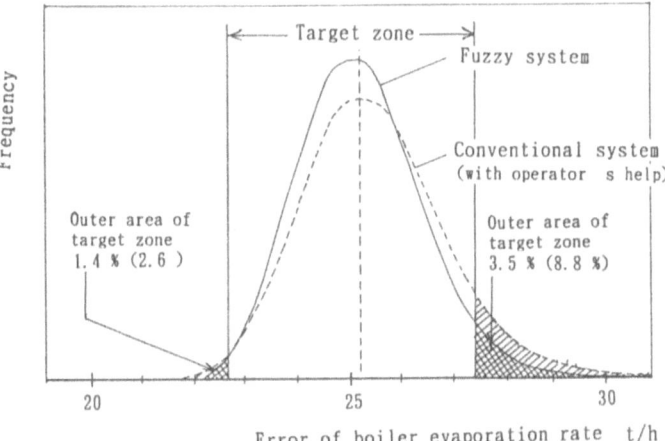

Fig. 7.19 Waste heat boiler evaporation amount frequency distribution (percentages are frequencies of falling outside the target range).

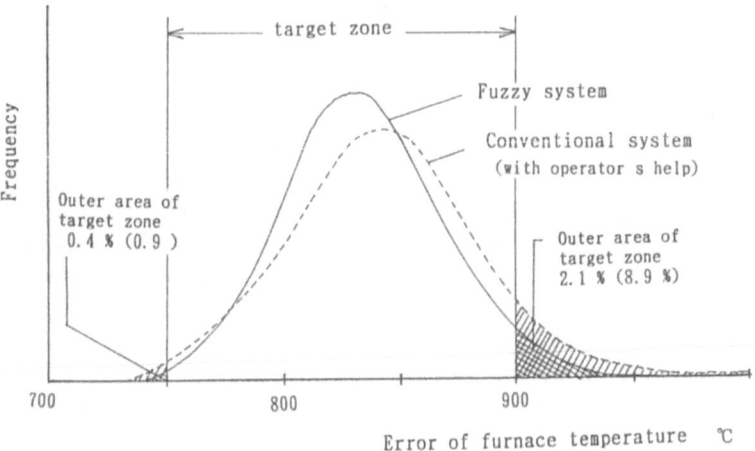

Fig. 7.20 Frequency distribution of temperature deviation inside incinerator (percentages are frequencies of falling outside the target range).

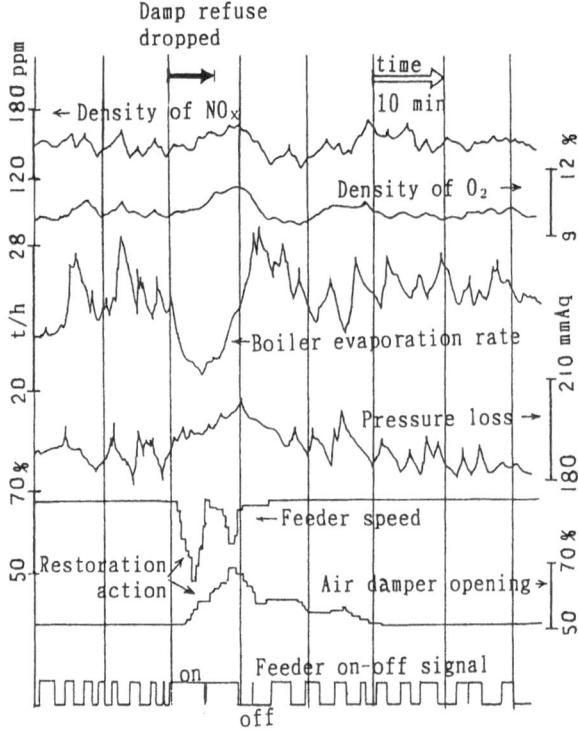

Fig. 7.21 Example of plant response when fuzzy control is used (arrows indicate that refuse of high water content has been input).

The reason that fuzzy control was able to respond quickly to abnormal combustion is believed to be that the fuzzy controller was able to respond faster than a human operator when abnormal conditions first appeared. Normally 2 or 3 incinerators are operating in parallel in a plant. Normally an operator is not monitoring data from only 1 incinerator, and does not respond until after a controlled quantity deviates outside the target range and an alarm sounds; the timing varies with the work squad, the individual operator and the conditions at that moment. It is believed that the principal reason that the fuzzy control system was able to give better results than a human operator in obtaining stable combustion in an incinerator was that the fuzzy controller is superior to the human operator in both response speed and reproducibility..

7.6   In conclusion

In fuzzy control of a refuse incineration plant, as described in this chapter, a fuzzy sensor, which infers imprecise observed quantities such as refuse quality and pile thickness which an operator uses to determine operation quantities, is inserted between the process variables and the fuzzy controller, and provides part of the input for inferring operation quantities. This concept is believed to be useful inasmuch as, in general, fuzzy control of a rule base is widely applied for process control.

In operations in a plant, in general it is not common for humans to instantaneously determine an operation method using a multi-dimensional input space; it is more common to have many simple control rules using input having few dimensions and then to intuitively determine operation quantities by weighting these control rules as to their importance.

In the inference methods used in previous fuzzy control, inference considering the ordinal relationship corresponding to the degree of importance of each control rule has not been directly performed. If a contradiction arises between synthesized inference results, and in addition inference by the center of gravity method does not agree with human intuition, it becomes necessary to respond to the problem by subdividing the input space and increasing the number of control rules. The inference method introduced here, in which the control rules are treated as a semi-ordered set model corresponding to the degree of importance of each rule, has the purpose of simplifying division of the input space and duplicating a human operator's intuitive operation quantity determination process.

In order to make good use of the characteristics of fuzzy theory and apply it as an effective control system, it is not sufficient to merely use straightforward simple methodology such as putting skilled operators' and specialists' experience and knowledge into the form of fuzzy control rules and performing necessary adjustments with membership functions. It is important to do what has been introduced in this chapter, to consider knowledge special to the specific plant and customize the system based on it.

Results of a test in an actual plant confirmed that the fuzzy combustion control system gives better stability of combustion and better responsiveness to sudden changes than a conventional control system with a human operator. As of this writing, at the end of 1991, actual operations using this system are starting at 4 plants with 11 incinerators, in Toyohashi, Sapporo, Kanazawa and Sagamihara, and it is planned to use this system widely in plants.

Applications of fuzzy systems are not limited to process control such as combustion control in an incineration plant. It is hoped that the future will see further developments in controller learning and optimization [20, 26], diagnosis of abnormal operation, supervision, etc. [27, 28]. Perhaps more sophisticated systems will provide us with even more advanced concepts and methods.

References

1. Hirota K (1989) Trends in application of fuzzy control, measurement and control (in Japanese). J Soc Instrument and Control Engineers 28 (1) : 28–33

2. Larsen PM (1980) Industrial application of fuzzy logic control. Int J Man-Machine Stud 3–10

3. Umber IG (1980) An analysis of human decision-making in cement kiln control. Int J Man-Machine Stud 12 : 11–23

4. Holmblad LP et al (1982) Control of cement kiln by fuzzy logic.
    In: Gupta mm, Sarchez E (eds) Fuzzy information and decision-making process. North-Holland, pp339–399

5. Yagishita O, Tanaka Y (1983) Estimation of amounts of chemicals to be added by fuzzy inference (in Japanese). 34th Nationwide Waterworks Research Society Papers, Instrumentation Division : 365–367.

6. Ueyama T, Esaki K, Hirayama S, Niidome T (1986) Application of imprecise control to cold rolling setup (in Japanese). 2nd Intelligent Engineering Res Soc Papers 15–18.

7. Nakamura T (1988) Application of fuzzy control to glutamic acid fermentation process (in Japanese). Instrumentation 31 : 7

8. Nakamura K (1989) Sugar extraction process control (in Japanese). Measurement Technol 221 (17, 9) : 47–51

9. Okada M, Yanagisawa Y, Kubota S, Takahashi M, Kobayashi K , Ono H (1985) Development of an automated system for a refuse incineration plant (in Japanese). Mitsubishi Heavy Industries Technical Reports 22, 6, 926–930

10. Goromaru T, Hanabusa H, Yonezawa H, Ito R, Takano O, Iwakawa N (1989) A model of the refuse combustion process in an oscillating flame lattice type refuse incinerator in Japanese. Collected Papers of J Soc Instrument and Control Engineers 25 (1) : 62 – 82

11. Goromaru T, Hanafusa H, Yonezawa H, Iwakawa N (1990) A model of the refuse combustion process in an oscillating flame lattice type refuse incinerator (in Japanese). Collected Papers of J Soc Instrument and Control Engineers 26 (8) : 894–901

12. Sato H (1989) The refuse problem and spaceship earth (in Japanese). Waste Matter 15 (6) : 54–83.

13. Oonishi T, Ono H, Terada Y (1986) A proposal for a combustion system in an urban refuse incinerator using fuzzy control (in Japanese). Abstracts of the 29th Combined Lectures on Automatic Control, 545–546

14. Ono H, Oonishi T, Okada M, Takahashi Y, Miura S (1987) Development of a fully automated system for an urban refuse combustion plant (in Japanese). Mitsubishi Heavy Industries Technical Reports 24 (6) : 563–568

15. Ohnishi T, Ono H, Terada Y (1989) Combustion control of refuse incineration plant by fuzzy logic. Fuzzy Sets and Systems 32 (2) : 193–206

16. Oonishi T (1989) Combustion control in an urban refuse incinerator (in Japanese). Measurement Technol 17 (9) : 40–46

17. Sugeno M (1985) An industrial application of fuzzy reasoning methods. Information Sciences 36 : 59–83

18. Sugeno M (1988) Fuzzy control (in Japanese). Nikkan Kogyo Shimbun Sha

19. Mandani EH, Assilian S (1975) An experiment in linguistic synthesis with a fuzzy logic controller. Int J Man-Machine Stud 7 : 1–13

20. Maeda M, Murakami S (1988) A self-adjusting fuzzy controller (in Japanese). Collected Papers of J Soc Instrument and Control Engineers 24 (2) : 181–197

21. Mizumoto M (1988) Recent fuzzy theory and approximate inference (in Japanese). Numerical Science (special ed) : 69–75 Science Sha

22. Mizumoyo M (1987) Comparison of various fuzzy reasoning methods. Preprint of Second IFSA Congress pp2–7

23. Mizumoto M (1989) A fuzzy inference method suitable for fuzzy control (in Japanese). Measurement and Control 28 (11) : 28–33

24. Oonishi T (1990) A method of inference by an ordinal structure model of fuzzy control rules (in Japanese). J Jpn Soc Fuzzy Theory and Systems 2 (4) : 125 – 132

25. Oonishi T (1991) Fuzzy control of an urban refuse combustion plant (in Japanese). Collected Papers of J Soc Instrument and Control Engineers 27 (3) : 326–332

26. Procyk TJ (1978) A linguistic self-organizing fuzzy controller. Automatica 15 : 15–30

27. Furuta H (1987) An expert system for damage assessment of reinforced concrete bridge deck. Preprint of Second IFSA Congress pp160–163

28. Mizumoto M (ed.) (1990) Special edition on fuzzy expert systems (in Japanese). J Jpn Soc Fuzzy Theory and Systems, Fuzzy Society 2 (2) : 114–173

Chapter. 8

FUZZY CONTROL FOR JAPANESE *SAKE* — Fuzzy decision controller and fuzzy simulator for Japanese *sake* fermentation

8.1 Introduction

8.1.1 Background

Japanese *sake* brewing technology has a history of more than 1,000 years. Many predecessors of the present-day master-hands went through numerous trials to develop the technology, tested their methods and made improvements. These technologies have been based on experience and intuition; until now the master brewer who perfected a technique has passed it on to his successors. However, with the revolution in industrial organization after World War II, there has been a drastic drop in the number of young people who want to work in *sake* breweries, and aging of the master brewers and technicians has become a problem. The *sake* industry must resolve this problem if it is to ensure the continuation of *sake* brewing, one of Japan's distinctive food industries.

In order to help solve this problem, the present authors have investigated the feasibility of replacing empirical process control technology in the *sake* fermentation process with fuzzy control by computer, with the eventual aim of producing a *sake* brewing technology expert system based on the experience and knowledge possessed by master-hands in *sake* brewing [1–5]. An outline of the results is presented here.

8.1.2 On the *sake* brewing process

Here, before discussing the research results, we simply explain the *sake* brewing process. An outline of the *sake* brewing process is shown in Fig. 8.1. Part of the steamed rice is used for solid culture of *Aspergillus oryzae*. The resulting molded rice becomes a source of liquefying and sacchairifying enzymes and proteases, and is called *koji*. Seed yeast is prepared in the mixture mash of *koji*, steamed rice and water containing lactic acid, and the whole mash named *moto* is used as inoculum for the following unrefined

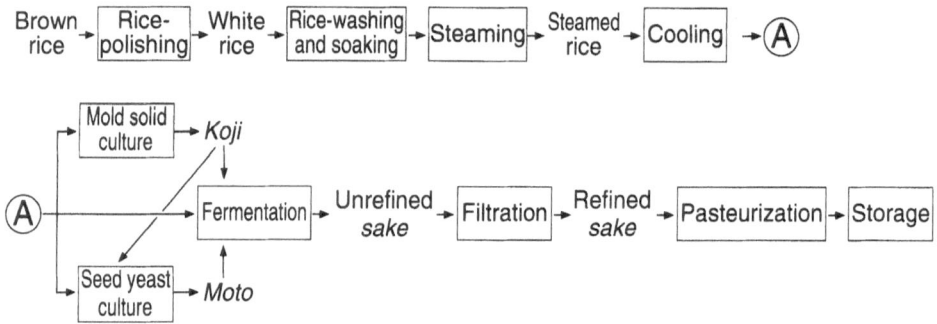

Fig. 8.1 Outline of Japanese *sake* production process.

sake fermentation: inoculum size is around 7% in commercial scale *sake* production. Because of the great amount of solid materials, steamed rice and *koji*, they are supplied to a fermentor in 3 or 4 stages; the 2nd supply is done 2 days after the 1st one with inoculation, while the 3rd one is done the next day. Fresh *koji* and steamed rice are prepared each time or the day of supply. The day of the 3rd supply is defined customarily as the beginning day (= the 1st day) of the unrefined *sake* fermentation (see Fig. 8. 2 and the following figures), although the fermentation has started 3 days before. When producing a thick full-bodied *sake*, Japanese *sake* brewers perform a 4th supply of steamed rice partly liquefied and saccharified by *koji* and/or commercially available enzymes before filtration.

The object of this study is the unrefined *sake* production process. This process usually takes 2 to 3 weeks, and is greatly affected by the type of rice and yeast used, and the condition of *moto* and *koji* prepared in previous steps. Therefore, it requires well-balanced control of enzymatic liquefaction and saccharification and microbiological alcohol fermentation, which are carried out simultaneously in the fermentor. The aim is to make sure that the unrefined *sake* reaches the target range in alcohol content and Baumé degree, both of which greatly affect the taste, after a given period of fermentation. In order to perform this, each and every master brewer measures temperature, Baumé degree and alcohol content during fermentation, and determines the optimum temperature for the following days through overall evaluation based on his own experience. We have to emphasize that the *sake* fermentation process is not isothermal, differing from ordinary fermentation in microbial industries. The principle of the process is to change brewing temperature to harmonize enzymatic reactions and yeast fermentation. The present research aims to construct a fuzzy inference controller that can replace human decision-making and to verify the inference with a fuzzy simulator.

8.2   Developing a fuzzy dicision system to perform Japanese *sake* fermentation control

8.2.1   Analysis samples

In order to analyze a skilled *sake* brewer's unrefined *sake* process control technology in the frame of fuzzy control theory, we first selected an average-sized *sake* brewery in Hiroshima Prefecture as a model brewery, and obtained process data of 70 runs in 1985–1986 under nearly identical conditions. All of these fermentations were conducted by the same master brewer under the following condition: The rice used was Nakate-Shinsenbon strain, polished to 70% for *moto* and *koji* production and to 73% for *sake* fermentation; the total amount of polished rice including that used for *moto* and *koji* preparation was 2,600kg. The yeast strain used was *Saccharomyces sake* (*s. cerevisiae*) KYOKAI 601 strain (a mutant depressed in foaming from the strain of KYOKAI 6; KYOKAI = Japanese Brewers' Association), and the seed yeast mash, *moto*, was prepared using 156kg of rice. The total amount of *koji* in the 1st to 3rd supplies was 520kg as polished rice. At each material supply, the solid substrates, *koji* and steamed rice, were mashed with water in the ratio of 135 parts water (w/w) to 100 parts polished rice by weight in the fermentor.

8.2.2   Result of brewing unrefined *sake* in the model brewery

Changes in fermentation temperature, Baumé degree and alcohol content during the brewing period for the 70 fermentations under the conditions mentioned above are shown in Fig. 8. 2. The mash temperature after the 3rd material supply varied from 8 to 11°C. It is believed that the fermentation temperature was controlled so that it increased 0.5 to 3°C per day from the initial temperature until it reached the maximum temperature of 17 to 19°C, then remained constant or fell less than 1°C per day. It must be mentioned here that the maximum temperature cannot be stringently planned before the fermentation starts, although its approximate value and the approximate day on which that value will be reached are well-known. When the temperature is increasing, whether it should be caused to decrease is decided through an overall estimation by the master brewer based on alcohol content and Baumé degree. When he decides to decrease the temperature, it has reached its maximum in that fermentation. The fermentation period was 17 to 18 days for all of the samples. The middle and bottom graphs of Fig. 8. 2 show that even though the brewing conditions were nearly identical, the Baumé and alcohol values during fermentation period did not vary along identical paths, but drifted within fixed limits. However, all of the unrefined *sake* samples were within the target ranges (Baumé degree 0.1±0.2, alcohol 19±0.5%) by the last day.

Table 8.1 Basic control technologies for unrefined *sake*
with 4th supplies of *moto* in the model *sake* brewery

1.	The initial temperature is about 10 (±1) °C .
2.	The fermentation period is 18 (±1) days until the 4th supply.
3.	Culture dilution to activate yeast is performed until the 7th day, if necessary.
4.	The temperature increases 0.5°C per day or more until the maximum temperature is reached.
5.	The temperature increase is not to exceed 3°C per day until the maximum temperature is reached.
6.	The maximum temperature is reached around the 6th (±1) day.
7.	The maximum temperature is 18 (±1) °C , and is not to exceed 19°C.
8.	Temperature higher than 17 °C is to be maintained for 3(±1) days.
9.	After the maximum temperature has been defined, the temperature is not to be increased.
10.	After the maximum temperature has been defined, the temperature is not to be decreased by more than 1 °C per day.
11.	The temperature (or temperature change) from the 7th day to the 13th day is to be determined by referring to the Baumé degree and alcohol content.
12.	When the Baumé degree and alcohol content are thought to be reasonable, the temperature is to be decreased 0.5°Cper day until the 11th day, and 0.8°C per day after that.
13.	If the Baumé degree is lower or the alcohol content higher than desirable between the 7th day and the 13th day, the temperature is to be decreased more rapidly.
14.	If the Baumé degree is higher or the alcohol content lower than desirable between the 7th day and the 13th day, the temperature is to be decreased less rapidly.
15.	Every time the Baumé degree is thought to be slightly higher or lower than a reasonable value between the 7th day and the 13th day, the necessary operation is to be performed gradually.
16.	Temperature control on and after the 15th day is to be based mainly on the Baumé degree.
17.	If the alcohol content is normal and after the 15th day, temperature control is to be based solely on the Baumé degree.
18.	If the alcohol content is not normal on or after the 15th day, temperature control is to be performed referring to both Baumé degree and the alcohol content.
19.	If the alcohol content becomes 19% and the Baumé degree is 0.2 or less on the 17th day, the 4th supply is immediately to be performed.
20.	The 4th supply is to be performed on the 19th day.

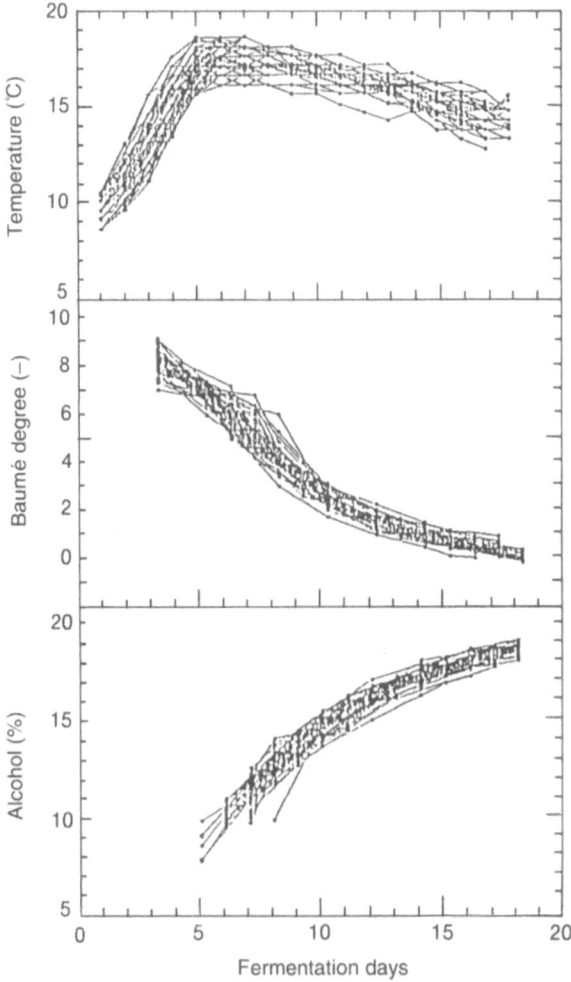

Fig. 8.2 Result of unrefined *sake* fermentation in the model *sake* brewery.

The basic points of unrefined *sake* process control technologies of the master brewer in the model brewery were determined referring to the results presented in Fig. 8. 2 and interviews with the master. The results are summarized in Table 8. 1. The control technologies during the time required to reach the maximum temperature (5 to 7 days) are given as 1 to 7; the technologies from then until the 13th day are given as 8 to 15; the technologies on and after the 15th day are given as 16 to 20.

Table 8.2   Conversion of the principal unrefined *sake* control technologies in the model *sake* brewery to the form of fuzzy production rules

1. IF FD<7	THEN FO=0
2. IF FD$\geq$7 AND BE is HIGH	THEN BFO$_{(H)}$=$k_1 \cdot$B$_{(H)}$
3. IF FD$\geq$7 AND BE is NORMAL	THEN BFO$_{(N)}$=$k_2 \cdot$B$_{(N)}$
4. IF FD$\geq$7 AND BE is LOW	THEN BFO$_{(L)}$=$k_3 \cdot$B$_{(L)}$
5. IF FD$\geq$7 AND AL is HIGH	THEN AFO$_{(H)}$=$k_4 \cdot$A$_{(H)}$
6. IF FD$\geq$7 AND AL is NORMAL	THEN AFO$_{(N)}$=$k_5 \cdot$A$_{(N)}$
7. IF FD$\geq$7 AND AL is LOW	THEN AFO$_{(L)}$=$k_6 \cdot$A$_{(L)}$
8. IF FD$\geq$7 AND FD<13	THEN FO=(BFO+AFO)/2
9. IF FD$\geq$13 AND AL is not LOW	THEN FO=BFO
10. IF FD$\geq$13 AND AL is LOW	THEN FO=(BFO+AFO)/2
11. IF FD=16 AND BE is LOW	THEN END
12. IF FD=17 AND BE is not HIGH	THEN END
13. IF FD=18	THEN END

FD: Fermentation days
BE: Baumé degree
AL: Alcohol content
B$_{(H)}$, B$_{(N)}$, B$_{(L)}$: Membership values relative to the Baumé degree
A$_{(H)}$, A$_{(N)}$, A$_{(L)}$: Membership values relative to the alcohol content
BFO$_{(H)}$, BFO$_{(N)}$, BFO$_{(L)}$, BFO (= BFO$_{(H)}$ + BFO$_{(N)}$ + BFO$_{(L)}$):
  Fuzzy output variables derived from the Baumé degree
AFO$_{(H)}$, AFO$_{(N)}$, AFO$_{(L)}$, AFO (= AFO$_{(H)}$ + AFO$_{(N)}$ + AFO$_{(L)}$):
  Fuzzy output variables derived from the alcohol content
FO: Output variables Defuzzified
k_1 – k_6: Proportionality constants

8.2.3   Conversion to fuzzy control rules

After fermentation has started, the principal process temperature control technologies presented in Table 8. 1 is that the optimum rate of temperature increase (0.5 to 3°C) is maintained until the maximum temperature is reached. After that, the optimum temperature profile is determined from experience, based on the master brewer's evaluation of the Baumé degree and alcohol values. The unrefined *sake* process control technologies after the maximum temperature is reached were expressed as fuzzy control rules in IF-THEN production form (Table 8. 2). Here, the membership functions of the fuzzy input variables which express the subjective linguistic evaluations of "high", "normal", and "low" Baumé degree and alcohol included in the condition part of the fuzzy control rule were prepared for each fermentation day based on the distributions of Baumé degree and alcohol values in the samples. Some of those shapes are shown in Figs. 8. 3 and 8. 4.

In Fig. 8. 3, the range within which the membership function for "normal" becomes 1 narrows gradually as the fermentation pregresses, and is set to reach the target range on the last day. However, since the object of control

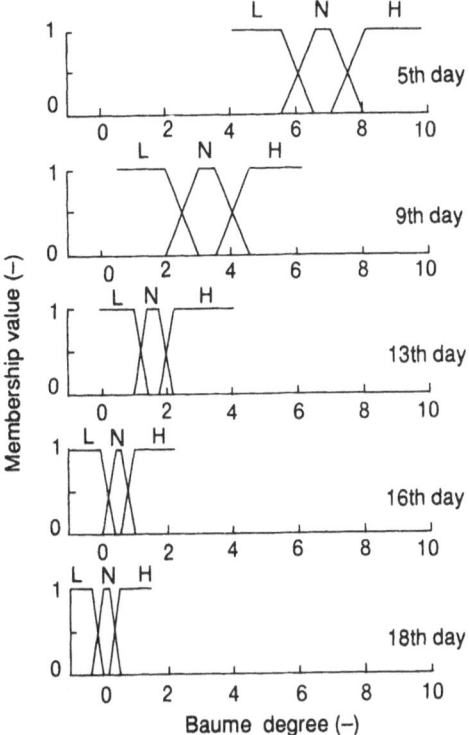

Fig. 8.3 Membership functions which indicate subjective evaluations with respect to Baumé degree on each fermentation day; L, N and H indicate low, normal and high evaluations, respectively.

after the 15th day is shifted to the adjustment of Baumé degree on the last day (Table 8. 1), the range of "normal" in alcohol becomes relatively wide after the 15th day (Fig. 8. 4).

The consequent of the fuzzy control rule presented in Table 8.2 is set from Table 8. 1 as in the following equations; defuzzification of the output variables in the consequent is done at the same time.

$$
\begin{aligned}
BFO &= BFO_{(H)} + BFO_{(N)} + BFO_{(L)} \\
&= k_1 \cdot B_{(H)} + k_2 \cdot B_{(N)} + k_3 \cdot B_{(L)} \\
AFO &= AFO_{(H)} + AFO_{(N)} + AFO_{(L)} \\
&= k_4 \cdot A_{(H)} + k_5 \cdot A_{(N)} + k_6 \cdot A_{(L)}
\end{aligned}
$$

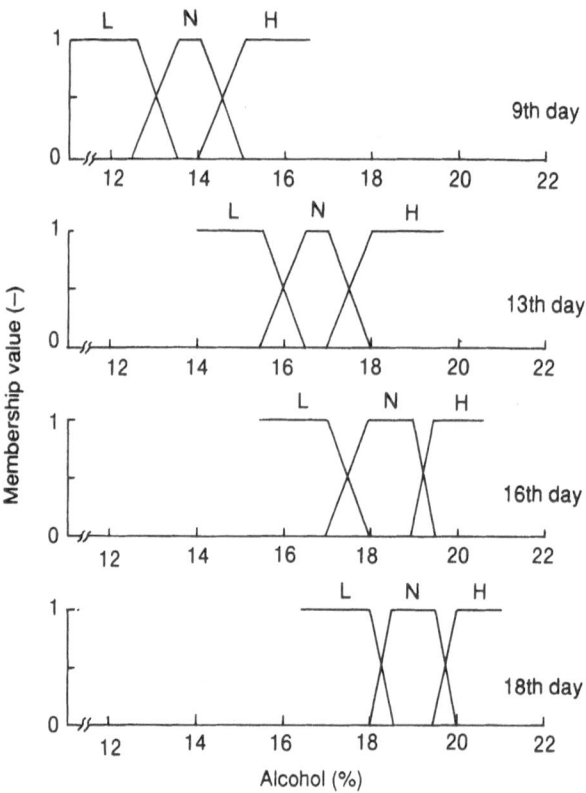

Fig. 8.4 Membership functions which indicate subjective evaluations with respect to alcohol content at each fermentation days; L, N and H indicate low, normal and high evaluations, respectively.

Where:

BFO: Output variable derived from Baumé degree
AFO: Output variable derived from alcohol content
$BFO_{(H)}$, $BFO_{(N)}$, and $BFO_{(L)}$: Fuzzy output variables decided by $B_{(H)}$, $B_{(N)}$, and $B_{(L)}$, respectively
$AFO_{(H)}$, $AFO_{(N)}$, and $AFO_{(L)}$: Fuzzy output variables decided by $A_{(H)}$, $A_{(N)}$, $A_{(L)}$, respectively
$B_{(H)}$, $B_{(N)}$, and $B_{(L)}$: Values of membership functions for "high", "normal", and "low", respectively, as to Baumé degree
$B_{(H)}$, $B_{(N)}$, and $B_{(L)}$: Values of membership functions for "high", "normal", and "low", respectively, as to alcohol content
k_1, k_2, k_3, k_4, k_5, k_6: Proportionality constants

The constants k_1 – k_6 are set as follows, according to Table 8. 1 and Fig. 8. 2.

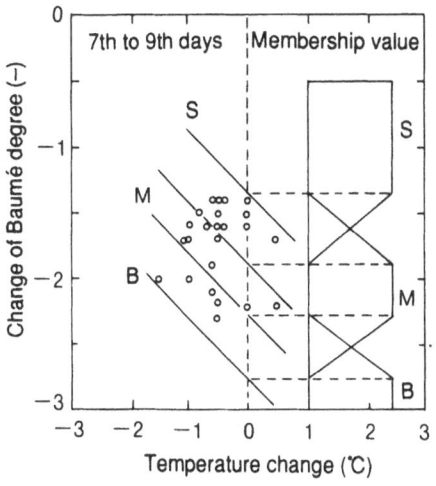 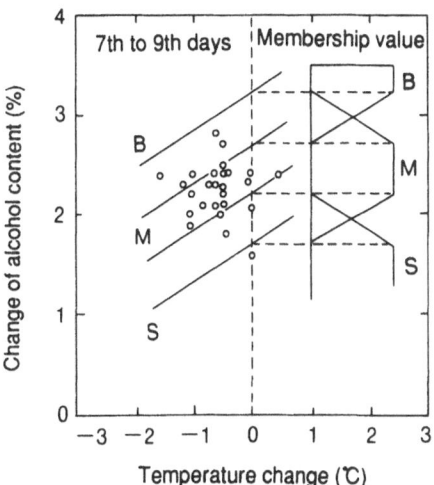

Fig. 8.5 A subjective evaluation by the master brewer of the relation between changes of fermentation temperature and Baumé degree from the 7th to the 9th day. The membership function in the figure is for the case of 0 temperature change.

Fig. 8.6 A subjective evaluation by the master brewer of the relation between the changes of fermentation temperature and alcohol content from the 7th to the 9th day. The membership function in the figure is for the case of 0 temperature change.

7th, 9th days: $k_1 = -1.5, k_2 = -0.5, k_3 = 0.5, k_4 = -1.5, k_5 = -0.5, k_6 = 0.5$
11th, 13th days: $k_1 = -1.6, k_2 = -0.8, k_3 = 0, k_4 = -1.6, k_5 = -0.8, k_6 = 0$
15th, 16th, 17th, 18th days: $k_1 = -0.8, k_2 = -0.4, k_3 = 0, k_4 = -0.8,$
$\quad k_5 = -0.4, k_6 = 0$

A number of methods have been reported for defuzzifying fuzzy output variables[6]. Here all fuzzy output variables in the consequents of the rules are devised to be scalars and defuzzified with the use of Sugeno's method to 0th order [6]. The final output is basically determined as the average of the two; if $A_{(L)} \neq 1$ on and after the 15th day, the Baumé degree fuzzy output variable is used as is as the final output variable.

8.2.4 Fuzzy simulator construction

The fuzzy rules thus translated from Table 8. 1 are believed to be equivalent to the technologies and to be capable of reproducing the *sake* fermentation conditions produced by the master brewer in the model *sake* brewery. However, further investigations were conducted into a fuzzy simulator to obtain even more appropriate fuzzy control.

Table 8.3 One example of fuzzy simulator calculations compared to actual *sake* fermentation results at the model *sake* brewery

	Fermentation days							
	7	9	11	13	15	16	17	18
Temperature (°C)								
calculated value		18.0	17.3	16.4	15.3	15.6	14.9	14.2
measured value	18.5	18.0	17.3	16.4	15.9	15.5	15.0	14.2
Baumé degree								
calculated value		3.1	2.00	1.15	0.64	0.37	0.12	0.03
measured value	4.9	3.2	1.99	1.10	0.60	0.30	0.10	0.00
Alcohol content (%)								
calculated value		14.6	16.4	17.8	18.5	18.7	18.8	19.2
measured value	12.3	14.6	16.5	17.5	18.3	18.5	18.9	19.0

When temperature control quantities are computed by fuzzy inference based on Baumé degree and alcohol values on predetermined days (the 7th, 9th, 11th, 13th, 15th, 16th, 17th and 18th days), it is necessary to evaluate the changes in Baumé degree and alcohol content that will take place after those control operations are performed. For this purpose, the relations between the temperature change between each pair of measurement days and the corresponding changes in Baumé degree and alcohol shown in Fig. 8. 2 were analyzed using fuzzy set theory. Since those relationships depend in a complicated manner on many factors including the glucose content, alcohol content yeast concentration and temperature of the unrefined *sake*, rigorous estimates are difficult to obtain. Therefore the relation between the control quantities and the resulting changes in Baumé degree and alcohol was framed based on subjective evaluation by the master brewer (high, normal, low): Some of those results are illustrated in Figs. 8. 5 and 8. 6, respectively. Figs. 8. 5 and 8. 6 summarize the membership functions of the subjective predictions (big, medium, small) of the changes in Baumé degree and alcohol content from the 7th day to the 9th day for different temperature changes: (only in the case of temperature change 0) the shapes of membership functions are shown. Note that the menbership function graphs should actually be perpendicular to the plane of the paper, the third axis being value of the membership function. The parallel lines at left are loci of points on the alcohol content change vs. temperature change graph at which membership functions for different subjective evaluations take the values 0 or 1. Note that these loci are straight lines.

It is plausible that the Baumé degree and alcohol changes for a given temperature change at each fermentation day are normally within the range in which the membership function expressing a subjective prospect of "medium" in Figs. 8.5 and 8.6 has the value 1. First, using measured values of temperature, Baumé degree and alcohol on the 7th day as input values, the temperature control quantity believed to be appropriate until the 9th day was calculated using the fuzzy control rules that had been created. Next, using the results in Figs. 8.5 and 8.6, this output value was used to predict the

Table 8.4 Comparison of fuzzy simulator calculations compared to the actual *sake* fermentation results (70 samples) at the model *sake* brewery

	Fermentation day						
	9	11	13	15	16	17	18
Average Temperature (°C)							
calculated value		17.5	16.9	16.1	15.3	15.1	14.9	14.3
measured value		17.4	16.9	16.3	15.5	15.1	14.9	14.5
Average of absolute values of differences							
between calculated and measured values	0.39	0.31	0.28	0.38	0.21	0.24	0.32
Average Baumé degree							
calculated value		3.4	2.1	1.32	0.83	0.59	0.34	0.18
measured value		3.4	2.1	1.35	0.78	0.56	0.35	0.15
Average of absolute values of differences							
between calculated and measured values	0.22	0.14	0.07	0.09	0.08	0.04	0.05
Average alcohol content (%)							
calculated value		14.0	16.0	17.2	18.2	18.5	18.7	19.0
measured value		14.0	15.9	17.1	18.0	18.4	18.7	19.0
Average of absolute values of differences							
between calculated and measured values	0.19	0.18	0.19	0.16	0.10	0.11	0.10

Baumé degree and alcohol values on the 9th day. Using the same procedure, the measured values of temperature, Baumé degree and alcohol on the 9th day were then used to predict the values on the 11th day. This process was repeated until the last day. An example of calculated results is presented in Table 8.3.

The same kind of calculation was performed for all of the process data shown in Fig. 8.2. The averages of those calculated results, measured values and the differences between calculated and measured values with respect to temperature, Baumé degree and alcohol were examined for the 70 samples for each fermentation day (Table 8.4). Large differences between the averages of calculated and measured values of temperature, Baumé degree and alcohol for each fermentation day were not found; in general they agreed. In addition, the averages of differences between calculated and measured values of temperature, Baumé degree and alcohol for each fermentation day for all of the fermentation results in the model brewery were in the ranges of 0.2 to 0.4°C, 0.04 to 0.22 and 0.1 to 0.2, respectively, so the calculated and measured values can be considered to be essentially in agreement for each of the brewing results.

As explained above, it was believed that the fuzzy control rules for the unrefined *sake* fermentation process in the model brewery could reproduce the basic process control technologies of the master brewers to reasonable accuracy. To further examine the capability of these fuzzy control rules, test brewing using 45kg of rice was then performed.

8.3   Test brewing using a pilot plant

8.3.1   Brewing conditions

Rice of Nakate-Shinsenbon strain harvested in Hiroshima Prefecture in 1987 was used again in the test brewing. The rice was refined to the extent of 70%, and washed with tap water at 12°C. It was soaked for 18 hours, and the water drained for 5 hours. Then the rice was steamed for 50 minutes. The molded rice, *koji*, was prepared at one time in advance; after drying it was stored at 4°C and then used as necessary. The yeast used was *S. sake* KYOKAI 601 strain. The recipe of materials used is shown in Table 8.5: the total amount of rice was 45kg. The target range of the final Baumé degree was ± 0.1.

Table 8.5   Recipe of materials for test brewing

	1st supply	2nd supply	3rd supply	4th supply	Total
Total rice (kg)	7.5	14.2	23.3		45
Steamed rice (kg)	5.2	11.7	19.1		36
Molded rice (kg)	2.3	2.5	4.2		9
Water (liters)	10.5	17.3	30.0	2.2	60
Yeast (liters)	0.3				0.3
Lactic acid (liters)	0.035				0.035

A schematic diagram of the equipment used for the test brewing is shown in Fig. 8.7. Two 150 liter sealed stainless steel tanks (made by Yabuta Industries), each with agitator (60rpm), cooling jacket on a side surface and heating jacket on the bottom , were used. A thermistor (sensitivity = 0.1°C) was inserted into the central portion of each tank. The temperature inside each tank (the fermentation temperature) was read out on a digital indicating controller (DIC: Yamatake Honeywell SDC 350). Warm water (about 20°C) and cool water (about 10°C) were used to control the temperature. The fermentation temperature and the temperature setting were input every 10 minutes into a control computer (Fujitsu FMR-60HD) through a communication controller (CMC: Yamatake Honeywell CMC 300B); the average temperature was printed out once an hour. New temperature settings were set by transmission in the reverse direction to the digital indicating controller.

8.3.2   Test brewing by manual operation

To examine the appropriateness of the fuzzy control rules developed from the model brewery, test brewing was done both manually, based on the process control technologies used in the brewery, and by computer control using the fuzzy control rules; and the progress of the unrefined *sake* fermentation, values obtained from analysis of the brewed *sake* and the various yields were compared.

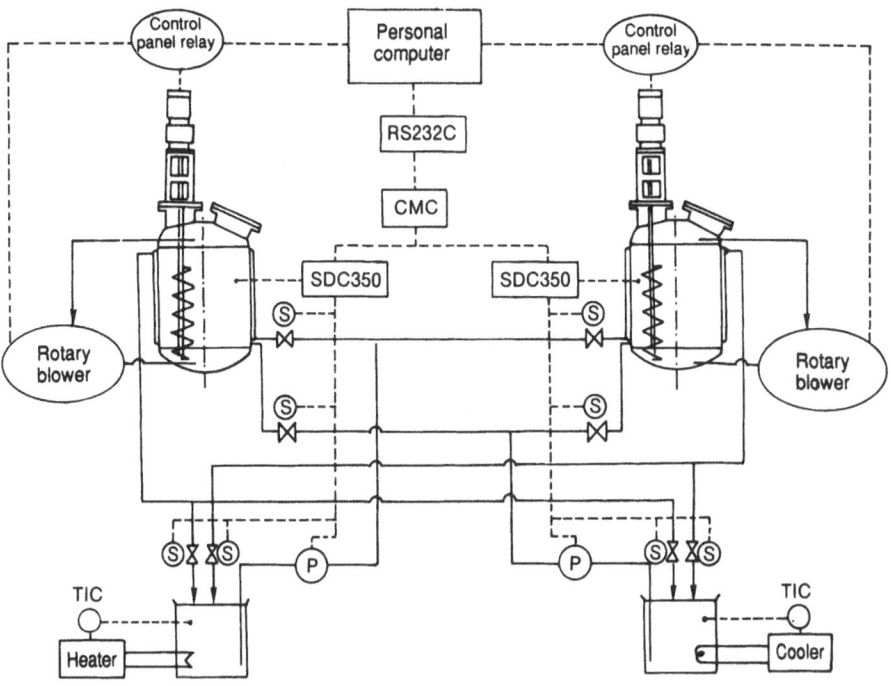

Fig. 8.7 Schematic diagram of equipment used in test brewing. Dotted lines are input/output lines for the control computer; solid lines show the flow of warm and cold water used to control the fermentation temperature.

Manual brewing was done following the basic process control technologies in the model *sake* brewery as closely as possible, with 2 batches . The temperature, Baumé degree, alcohol, acidity and amino acidity changes for each fermentation day are shown in Fig. 8.8. In the 2 batches the maximum temperatures were 17.5°C (manual batch #1) and 18.5°C (manual batch #2). The fermentation days required were 17 days and 16 days, respectively. Even though the Baumé degree and alcohol values on the 7th day were 5.0 and 11.0% in both cases, the effect of the subsequent course of temperature change caused the two batches to behave differently, resulting in a 1-day difference in the time required to reach the target Baumé degree. Comparing these values with the results obtained in the model brewery, the time required to produce the desired unrefined *sake* was 1 to 2 days shorter and the final alcohol content was about 1% less. These differences largely resulted from the fact that in this test brewing the mask of seed yeast, *moto*, was not used; instead precultured yeast broth was used. However, in both cases the fermentation proceeded smoothly until the final day. It was confirmed that if the temperature is controlled in accordance with the measured temperature,

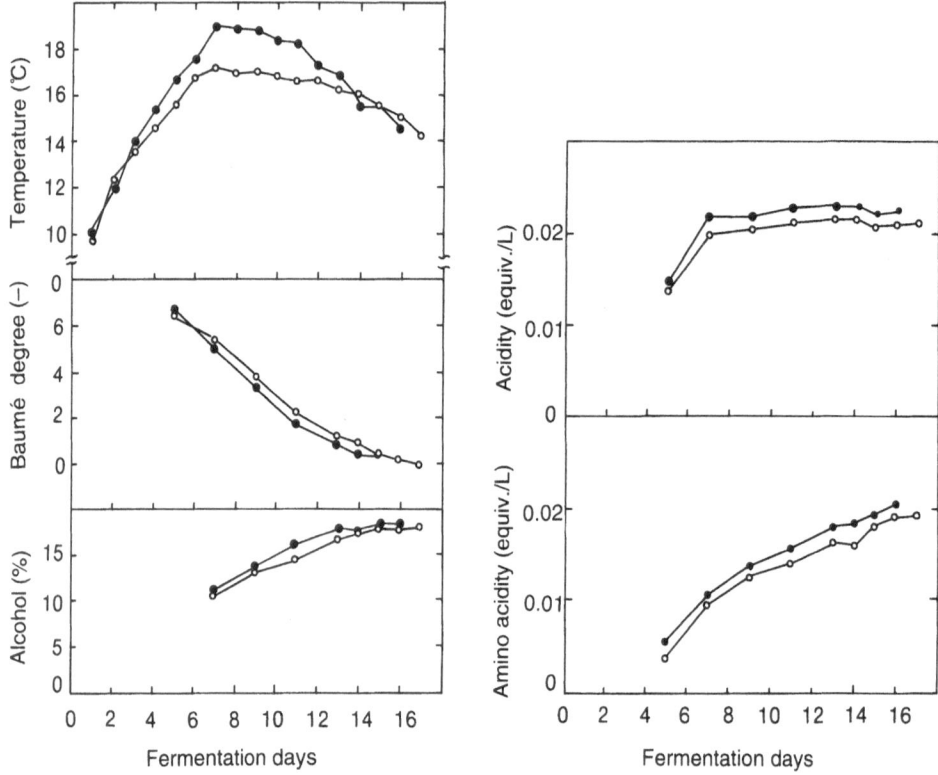

Fig. 8.8   Results of test brewing by manual operation.

Baumé degree and alcohol for each fermentation day, with the process control technologies, the time required to produce the desired unrefined *sake*, final Baumé degree and alcohol content can be adequately controlled.

8.3.3   Test brewing using fuzzy control

Referring to the results of the brewing by manual operation, the fuzzy control rules mentioned earlier were corrected to reduce the number of days needed to brew unrefined *sake* by 1, and the results of decision and inference from the fuzzy control rules were applied in computer-controlled brewing. The fuzzy control rules after the correction are given in Table 8. 6. Baumé degree and alcohol values determined off-line on the 7th, 9th, 11th, 13th, 14th, 15th, 16th and 17th day were input by keyboard interrupt. Based on these values, temperature control values believed to be appropriate were inferred by fuzzy control rules to determine the temperature setting pattern until the time of the next key input. In addition, every time new test brewing results were obtained, the average values until then of the Baumé degree and alcohol content for each fermentation day were employed to tune the membership functions as follows: the center of normal range for the Baumé degree and

Table 8.6 Fuzzy control rules used in test brewing

1. IF FD<7	THEN FO=0
2. IF FD≥7 AND BE is HIGH	THEN $BFO_{(H)}=k_1 \cdot B_{(H)}$
3. IF FD≥7 AND BE is NORMAL	THEN $BFO_{(N)}=k_2 \cdot B_{(N)}$
4. IF FD≥7 AND BE is LOW	THEN $BFO_{(L)}=k_3 \cdot B_{(L)}$
5. IF FD≥7 AND AL is HIGH	THEN $AFO_{(H)}=k_4 \cdot A_{(H)}$
6. IF FD≥7 AND AL is NORMAL	THEN $AFO_{(N)}=k_5 \cdot A_{(N)}$
7. IF FD≥7 AND AL is LOW	THEN $AFO_{(L)}=k_6 \cdot A_{(L)}$
8. IF FD≥7 AND FD<13	THEN FO=(BFO+AFO)/2
9. IF FD≥13 AND AL is not LOW	THEN FO=BFO
10. IF FD≥13 AND AL is LOW	THEN FO=(BFO+AFO)/2
11. IF FD=16 AND BE is LOW	THEN END
12. IF FD=17 AND BE is not HIGH	THEN END
13. IF FD=18	THEN END

FD: Fermentation days
BE: Baumé degree
AL: Alcohol content
$B_{(H)}$, $B_{(N)}$, $B_{(L)}$: Membership values relative to the Baumé degree
$A_{(H)}$, $A_{(N)}$, $A_{(L)}$: Membership values relative to the alcohol content
$BFO_{(H)}$, $BFO_{(N)}$, $BFO_{(L)}$, BFO $(=BFO_{(H)} + BFO_{(N)} + BFO_{(L)})$:
　Fuzzy output variables derived from the Baumé degree
$AFO_{(H)}$, $AFO_{(N)}$, $AFO_{(L)}$, AFO $(=AFO_{(H)} + AFO_{(N)} + AFO_{(L)})$:
　Fuzzy output variables derived from the alcohol content
FO: Output variables defuzzified
k_1 – k_6: Proportional constants

alcohol content were reset to the averages without changing the shape of the previously framed function. The new membership functions were used in the next test brewing, as seemed appropriate.

The changes of temperature, Baumé degree, alcohol content, acidity and amino acids in fuzzy-controlled brewing are shown in Fig. 8.9. The temperatures until the 7th day followed different courses in each batch. However, in every batch the maximum temperature was defined within the range 17 to 18.5°C, and the total fermentation times were all 16 to 17 days, the same result obtained in manual brewing. The Baumé degree and alcohol content also behaved differently in each batch, but by the last day always reached about the same values that were obtained in manual brewing.

In fuzzy-controlled brewing, the fermentation of the unrefined *sake* always progressed smoothly and there were not great differences in time required for fermentation or composition on the last day from manually brewed unrefined *sake*. Comparisons of the final compositions and the resultling various yields after filtration are given in Table 8.7. In Table 8.7, each of the *sake*s brewed had a Baumé degree of −0.08 to 0.15 and alcohol content of 17.0 to 18.0%, and there also are not great differences in total sugar, acidity, amino acidity or color.

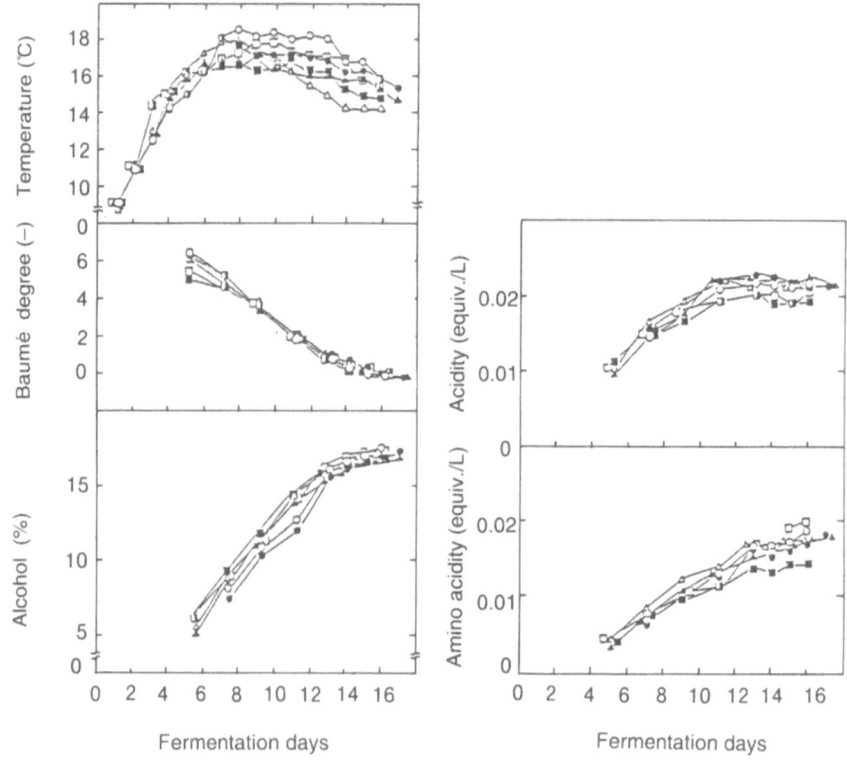

Fig. 8.9 Results of test brewing using fuzzy control.

Table 8.7 Comparison of compositions and yields of refined *sake*

Item	Manual brewing		Fuzzy controlled brewing					
Brewing runs	1	2	1	2	3	4	5	6
Baumé degree (−)	−0.08	0.02	0.10	0.14	0.10	0.00	0.15	0.10
Alcohol (%)	17.9	17.8	17.8	17.5	17.1	17.0	17.7	17.0
Acidity (equiv./L)	0.021	0.022	0.021	0.022	0.022	0.022	0.023	0.020
Amino acidity (equiv./L)	0.019	0.020	0.019	0.018	0.018	0.018	0.021	0.018
Total sugar (%)	3.59	3.59	3.74	3.89	4.29	4.10	4.45	4.43
Reducing sugar (%)	0.88	0.85	1.00	0.84	0.91	0.87	1.07	0.83
Color intensity (OD_{420})	0.020	0.025	0.035	0.028	0.035	0.030	0.040	0.046
Iron (ppm)	0.08	0.16	0.12	0.07	0.09	0.07	0.06	0.07
Sake production yield (L/L-unrefined *sake*)	0.906	0.905	0.891	0.889	0.908	0.906	0.898	0.897
Cake yield (kg/kg-rice)	0.230	0.227	0.237	0.230	0.238	0.236	0.264	0.282
Amount of pure Alcohol yield (L/kg-rice)	0.346	0.338	0.338	0.326	0.339	0.329	0.345	0.325

Table 8.8 Results of organic evaluation of refined *sake*

Brewing runs		Average value	Examiners' comments
Manually controlled brewing	1	2.2	Good in taste, sharp but fair in fragrance
	2	3.2	Bad in taste, poor in fragrance, a little smell came from pyruvate, acid-stimulating
Fuzzy controlled brewing	1	3.7	Bad in taste, poor in fragrance
	2	2.3	A little bad in taste, fair in fragrance, well-matured
	3	3.0	A little bad in taste, well–matured, not sharp in fragrance
	4	1.5	Good in fragrance and taste, well-matured
	5	1.3	Good in frangance and taste, well-matured
	6	1.3	Clisp in taste, good in fragrance, too sharp but good

The sensory evaluation was done by 6 examiners on a scale of 5 (1: good, 3: fair, 5: bad).

The yields in each batch were of the same order except for cake yields. The cake yield was 22.7% for manually-brewed batch #2 and 28.2% for fuzzy-controlled batch #6, for a difference of 5.5%. This is believed to be mainly because, in temperature control up until the 7th day, the temperature in fuzzy-controlled batch #6 rose higher than that in manually-brewed batch #2.

The results of sensory tests of each *sake* alter filtration are given in Table 8.8. There was not a great difference among manually brewed batches #1 and #2 and fuzzy-controlled batches #1 to #3, but fuzzy controlled batches #4 to #6 were evaluated as having better taste and fragrance than manually brewed batches #1 and #2. This is suspected as possibly being a side effect of tuning of the membership function by learning.

As explained above, good results were obtained in test brewing using a pilot plant. Next, as the final demonstration step, commercial-scale brewing was carried out at Chugoku Jozo K.K. (Hatsukaichi City, Hiroshima Prefecture) (1988 and 1989). In this test, different yeast and molded rice were used in each batch, and the control was applied to brewing of unrefined *sake* for the *sake* of extra full body requiring not only 4th but also 5th supply.

8.4 Commercial scale application

8.4.1 Brewing conditions

The rice used for the test commercial scale brewing was also Nakate–Shinsenbon strain, but Hattan–nishiki strain was sometimes used for *koji* preparation. The polishing extent of rice was 70 and 73% for *koji* and other preparations, respectively. The yeast used was *S. sake* KYOKAI 701 (a mutant depressed in foaming from KYOKAI 7). The total amount of rice was 2,000kg; 22.5% of rice was converted to the molded, *koji*. A ratio of 127.5 parts water to 100parts rice was used to make a mask of steamed rice and *koji*. The target values of the fermentation period to produce the desired unrefined *sake* and Baumé degree on the last day were 20 ± 1 days and −15 ± 3, respectively.

In a test run, ten 9-kiloliter sealed tanks (made by Fuji Hohro, Inc.) were used. The fermentation temperature was measured using a thermistor (sensitivity = 0.01°C) inserted from the top of the tank and a digital indicating controller (Yamatake Honeywell SDC 350). The temperature was maintained at the set value by circulating cooling water (about 10°C) prepared in advance through a cooling mat (Daiichi Kogyo CR-670) wound around the outside of the tank side surface and controlling the flow through it using a proportional valve (Yamatake Honeywell VCT-M944B). Other specifications were the same as for the pilot plant.

8.4.2 Test results

The fuzzy control rules used in this test were based on those created previously and modified based on new interviews with the master brewer and his assistants involved in controlling the brewing process of a brewery different from the model one mentioned earlier. In the initial stage of fermentation (until the 8th day) the Baumé degree was subjectively judged as "high", "normal" or "low", and the temperature control quantity was correspondingly set to "big", "medium" or "small" in order to maintain the extent of liquefaction of the steamed rice approximately constant. Then, in and after the mid-period of fermentation (8th day and after), in order to maintain good fermentation progress (decrease of Baumé degree and alcohol formation) a linguistic expression for the temperature control quantity believed to be appropriate (big, somewhat big, medium, somewhat small, small) was determined by fuzzy inference from subjective evaluation (high, normal, low) for the Baumé degree and the alcohol content. The center of gravity method was used for defuzzification.

Using fuzzy control rules created in this manner, 28 batches were test-brewed. All batches completed fermentation in 20 ± 1 days and the final Baumé degrees were in the range from 1.3 to 2.0, with an average value of 1.6, so that the progress of fermentation was excellent for each batch. The *sake* production, cake, and alcohol yields for the batches using the fuzzy control method (number of samples: 14) were on the same order as for batches brewed by the conventional method (number of samples: 16). In addition, in the sensory evaluation there was not a great difference between the resulting *sake* from the two fashions; the final *sake* was of quality equal to or better than that obtained by the conventional method.

8.5 Summary

An investigation was conducted into application of fuzzy control theory to *sake* brewing in a model brewery, based on the experience and intuition of a master brewer, and the fuzzy method was applied to computer control of unrefined *sake* brewing. The results showed that it is possible to construct a fuzzy control method that faithfully reproduces the basic process control technology.

Next, the appropriateness of the fuzzy control rules for controlling the brewing of unrefined *sake* was investigated by performing test brewing on a pilot plant scale. The results of comparing the fuzzy control with manual control showed that for each number of fermentation days the changes in temperature, Baumé degree and alcohol content were about the same, and, in addition, no great difference was observed in the final refined *sake* composition or various yields, fragrance composition or sensory tests, so that refined *sake* of about the same quality was obtained.

Based on the above results, as the final demonstration test the fuzzy control rules were introduced into the unrefined *sake* production process for producing ordinary *sake* on a commercial scale (2,000kg). The results showed that, even though different types of yeast and molded rice were used in each batch, fermentation progressed satisfactorily in all of the test batches; the fermentation period required to produced unrefined *sake*, final Baumé degree, final alcohol content and the various yields were about the same as obtained with the conventional method. In addition, in the sensory evaluation, there was no clear difference in the resulting filtered *sake* produced by the fuzzy control method and the conventional method; the filtered *sake* was evaluated as being of about the same quality.

From these results, we believe that it is possible to put a fuzzy inference and decision system based on the experience and intuition of master brewers into general use. The process control rules for producing unrefined *sake* differs depending on the fermentation conditions, brewing methods of different master brewers and proprietors, and types of rice, yeast and molded rice used. However, in computerizing those methods it is easy to construct fuzzy control rules based on the brewing results and policies at each brewery, making this one of the most suitable methods available.

Finally, we thank Mr. Jun Hara, Mr. Kazuhiko Takatsuna and Mr. Toshiyuki Fukuhara of Chugoku Jozo K.K., which provided the necessary facilities and assistance, and Senior Researcher Yoshiharu Teshima of the Hiroshima Prefectural Food Industry Technology Center for his assistance.

References

1. Tsuchiya Y, Koizumi J, Suenari K, Teshima Y, Nagai S (1990) Creation of fuzzy rules and a fuzzy simulator for unrefined *sake* process control for Hiroshima brewers (in Japanese). J Fermentation Technol 68 : 123–129

2. Suenari K, Tsuchiya Y, Teshima Y, Koizumi J, Nagai S (1990) A *sake* production test using a fuzzy control method (in Japanese). J Fermentation Technol 68 : 131–136

3. Tsuchiya Y (1990) Fuzzy control of the *sake* brewing process (in Japanese). J Jpn Brewing Assoc 85 : 300–308

4. Tsuchiya Y, Suenari K, Teshima Y (1990) Fuzzy control of the *sake* brewing process (in Japanese). Food Technol 33 (22) : 32–42

5. Hara J, Takatsuna K, Tsuchiya Y, Suenaga K, Teshima Y, Koizumi J, Nagai S (1990) Fuzzy control for the unrefined *sake* brewing process (in Japanese). Jpn Fermentation Engineering Society Conference Abstracts : 49

6. Terano H (1985) Introduction to systems engineering (a challenge to imprecise problems) (in Japanese). Kyoritsu, p 304

Chapter. 9

ELEVATOR CONTROL USING A FUZZY RULE BASE

9.1 Introduction

Recently fuzzy theory has been attracting attention as a powerful means of giving computers the capability to make human-like judgments. In the previous methodology, a condition could only be judged by the computer as "correct" or "incorrect"; but this theory permits the computer to make the same kind of subjective judgments that humans make in judging something to be "somewhat correct" or "a little bit in error". In the last few years, a great deal of research has been conducted on applications of this theory to real systems [1–3].

In this chapter, we introduce the example of application of fuzzy theory to control of a group of elevators. This control has the purpose of increasing the efficiency of vertical transportation in a building by systematically operating several elevators [4]. In order to improve the efficiency of control it is indispensable to make use of human knowledge, and fuzzy theory is a very effective way of doing this [5, 6].

First, in section 9.2, we discuss elevator group control systems and outline the conventional method. In section 9.3 we explain control of a group of elevators using a fuzzy rule base [5, 6]. Then in section 9.4 the effectiveness of this system is investigated by simulation examples.

9.2 Outline of elevator group control

9.2.1 What is a group control system?

In order to discuss the role of an elevator group control system, first let us look at an elevator system from the passenger's point of view.

Fig. 9. 1 depicts an elevator which is controlled by a group control system (in the business this is called a cage) and an elevator hall inside a building.

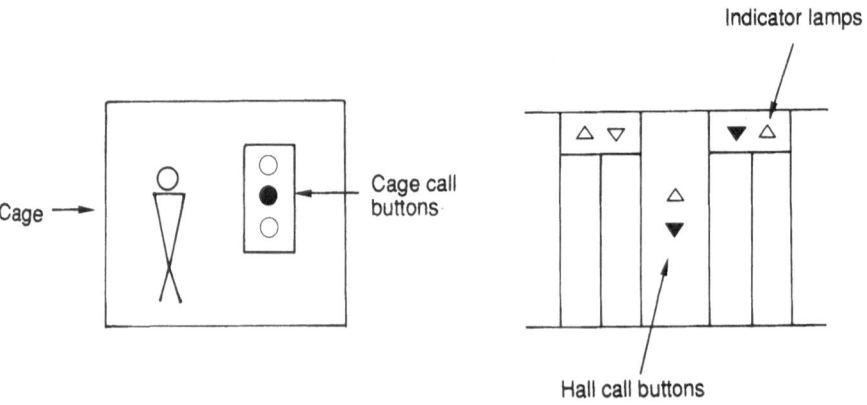

Fig. 9.1 Outline of elevator group control

Normally in a building 3 to 8 cages are controlled by an elevator group control system as 1 group of elevators.   When a passenger on any floor presses the elevator call button, the cage in the group which is judged to be most appropriate (called the assigned cage) is assigned to answer the call. When the cage to be assigned is determined, a guide lamp and chime on that floor inform the passenger of which cage is arriving. The passenger waits in front of the designated cage; after getting in he presses the button corresponding to his destination (called the cage call button) and is transported to the floor to which he wishes to go.

Thus, seen from the passenger's point of view an elevator system is a simple transportation system. However, in order to transport the passenger quickly and comfortably to the destination floor, many detailed considerations are necessary. An outline of the functions possessed by the group control system is given below.

Operation supervision function

Each cage transports passengers through repetitions of a few operations: starting, running (up or down), stopping, and opening and closing the door. This series of cage operations requires a number of judgments. For example as shown in the example in Fig. 9. 2 the cage must stop on the cage call floor (the floor to which the passenger travels), but as necessary it might have to pass through a floor to which it has been called by a waiting passenger. It is necessary for a judgment to be made as to whether or not the cage should stop on a given floor and whether or not it should reverse direction at a floor at which it stops.

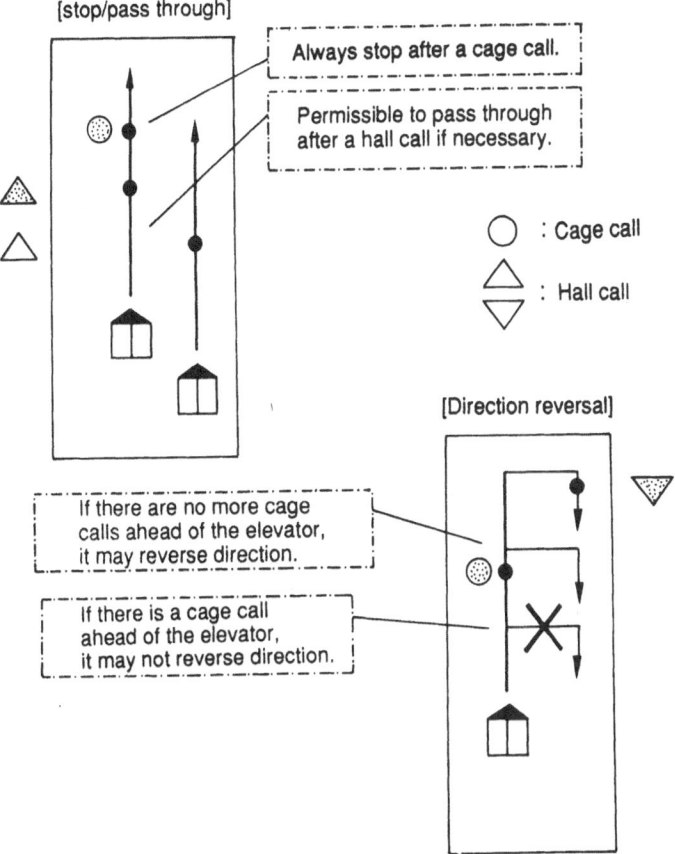

[stop/pass through]

Always stop after a cage call.

Permissible to pass through after a hall call if necessary.

○ : Cage call

△▽ : Hall call

[Direction reversal]

If there are no more cage calls ahead of the elevator, it may reverse direction.

If there is a cage call ahead of the elevator, it may not reverse direction.

Fig. 9.2 Operation control functions

Response to hall calls

In response to a hall call, it is necessary for a judgment to be made as to which cage is to be assigned. This selection must be made based not only on the passenger waiting time, but on a number of factors as shown in Fig. 9. 3 to make the selection that is optimum for the building as a whole. The details of this cage assignment process will be discussed below.

Responsiveness to fluctuating traffic conditions

As shown in Fig. 9. 4, the traffic demand within a building fluctuates greatly with time of day. For this reason it is necessary to have a number of response functions for use at different times. A method by which the system learns the traffic demand, in order to understand and predict this kind of traffic pattern, has been developed [7].

Evaluation item	Content of evaluation
Waiting time	Average waiting time and proportion of long waits
Elevator completely full	Frequency of times people are unable to board or elevator passes through a hall call because it is full.
(Service elevator indicator)	Indicator timing, prediction accuracy (prediction error rate)
Departure control	Maintaining boarding/departing time; fast departure
Priority service	Relative waiting time
Uniform service	Relative prediction accuracy
Maintenance, management	Fluctuation of start frequency
Energy saving	Electric power consumption
Service completion time	Average and maximum boarding and departing time; average and maximum service completion time

Fig. 9.3 Principal evaluation items in group control

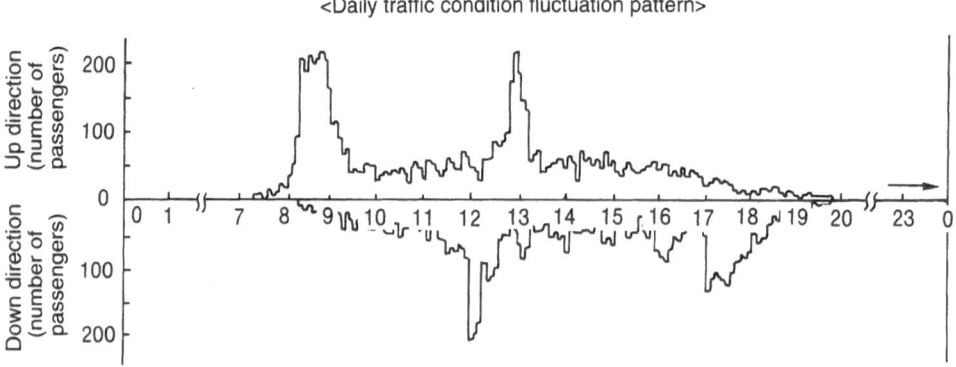

Fig. 9.4 Traffic condition fluctuations

<Traffic flow characteristics and response functions>

Time range	Traffic flow characteristics	Response functions
Commuting to work time	Passengers preparing to ride to all floors congregate on the first floor in a short time.	Up-down operation divided into 2 parts (shorten round-trip time and increase transport capacity)
Lunch time <first half>	Many passengers ride to dining room floor and first floor.	Cages are distributed evenly among floors.
Lunch time <second half>	Passengers preparing to ride to all floors congregate on the first floor and dining room floor in a short time.	Concentrate cages mainly on the first floor and the dining room floor (to reduce the number of passengers left behind).
Work leaving time	Many passengers ride from all floors to the first floor.	Distribute cages evenly among all floors (to reduce the number of times fully loaded elevators have to skip floors).
Building closed time (night, holidays)	Traffic is very light.	Only cages that are needed are used; others are idle.
Normal time (most of the working day)	Up and down passengers are in approximate balance. There is much traffic but not as much as at rush hours.	Equal service is provided to all floors.
Time of ending of exhibition or gathering	Passengers from a certain floor temporarily increase.	Degree of crowding is automatically judged and cages concentrated where they are needed (to reduce the number of passengers left behind).

Fig. 9.4 *(continued)*

Indicator function

This function, shown in Fig. 9. 5, provides the waiting passenger with information as to when a cage will arrive, which cage will arrive and how long it will be until it arrives.

9.2.2 Procedure for determining which cage to assign

The most basic and important of the functions of the group control system discussed above is the determination of which cage will respond to the calls which are being made all the time. In a high rise building, the selection as to which cage will respond greatly effects the efficiency of transportation in the building. Let us now explain how this selection is made in detail.

The control procedure which is followed in determining which cage to assign will be explained with reference to Fig. 9. 6.

Transmission of information on traffic conditions

When a hall call is issued, the call conditions at that time and the operational status of each cage are transmitted from each hall control unit and each elevator control unit through communication channels to the group control system.

Prediction and evaluation operations

Based on the information on conditions throughout the building, first it is assumed that each cage in turn is assigned to respond, and then predicted values for each of the following items are computed. 1) Time of arrival of each cage at each hall 2) Number of passengers on each cage at each hall

Let us explain this referring to Fig. 9. 7. As can be seen in the figure, cage #1 is already scheduled to stop at the 3rd, 5th and 7th floors. Let us suppose that the predicted arrival times at those floors are T3, T5 and T7, respectively. At this time a call comes in from the 4th floor. If cage #1 were to respond to this call, the predicted arrival time at the 3rd floor would remain the same, but the arrival times at the 5th and 7th floors will naturally be delayed from T5 and T7. Similarly, the number of passengers at each floor will be affected. The prediction operation makes this kind of prediction for each cage in response to each call.

Next, based on these prediction results, evaluations are performed of the waiting time for each passenger and the probability of the prediction being in error, as shown in Fig. 9. 3.

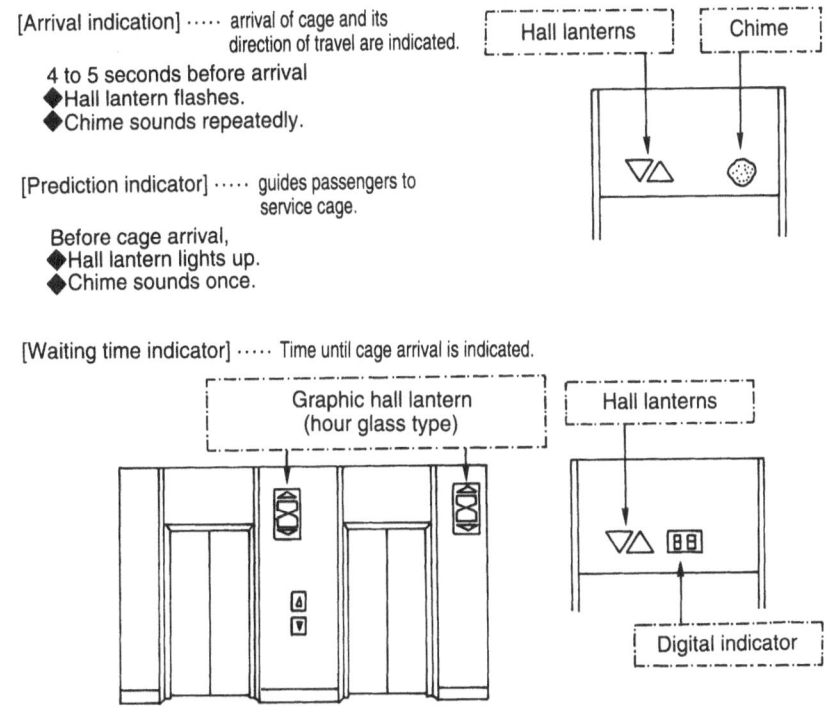

[Arrival indication] ····· arrival of cage and its
                          direction of travel are indicated.

4 to 5 seconds before arrival
◆Hall lantern flashes.
◆Chime sounds repeatedly.

[Prediction indicator] ····· guides passengers to
                            service cage.

Before cage arrival,
◆Hall lantern lights up.
◆Chime sounds once.

[Waiting time indicator] ····· Time until cage arrival is indicated.

Fig. 9.5 Indication functions

Determination of which cage to assign

Based on the results of the above prediction and evaluation computations, an appropriate cage is assigned, and assignment and announcement instructions are given to the elevator and hall control units.

The problem is how to decide which cage to assign through this series of steps. A specific explanation of this will be given referring to Fig. 9. 8. For simplicity we consider the case of 2 cages. #1 has a cage call for the 5th floor and #2 has a hall call for the 6th floor, so both of them are moving upward. Which cage will be assigned if at this time a call for an upward traveling cage is issued on the 4th floor? #2 can get to the 4th floor faster. For this reason assigning #2 will shorten the waiting time of the passenger on the 4th floor. However, if #2 stops on the 4th floor its arrival at the 6th floor will be delayed, so the waiting time of the passenger on the 6th floor will become longer. Consequently, in a case like this it is often optimum to assign cage #1.

Determination of which cage to assign in a case like this must be based on consideration of service to the whole building, not only the latest hall call.

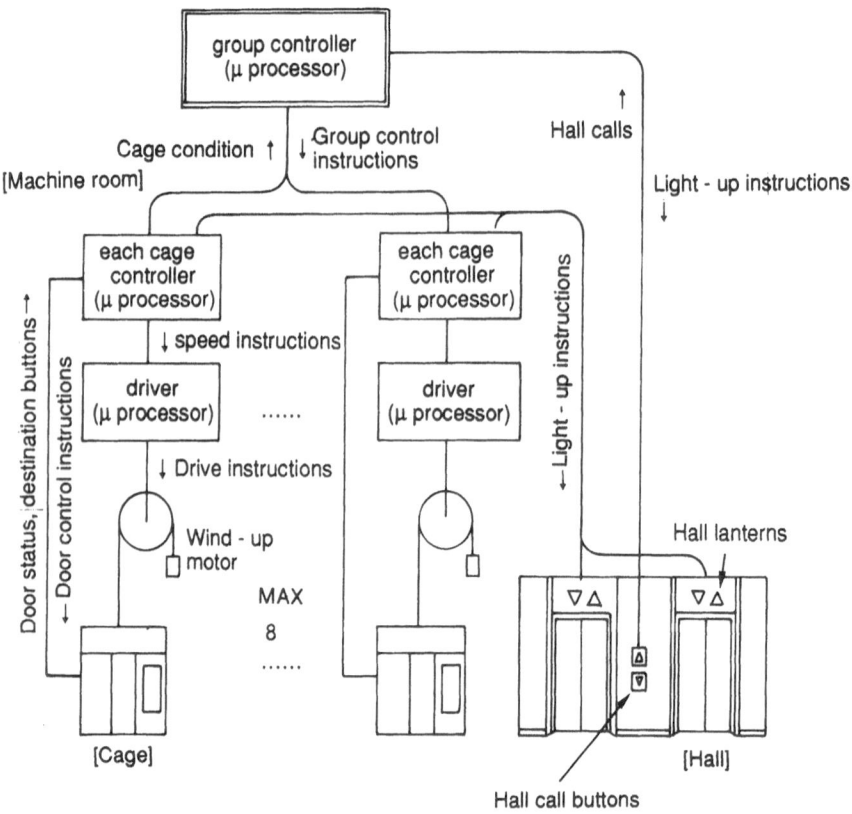

Fig. 9.6 Control instructions in group control

In a conventional group control system, in general an evaluation function is provided for determining which cage to assign. The service of the whole group is evaluated numerically on the assumption that each cage is assigned. This is expressed in general form by the following equations.

$$J(e) = \min\{J(1), J(2), \cdots, J(n)\}$$
$$J(i) = f(x, y, z \cdots) \tag{1}$$

N: Number of cages
e: Assigned cage
x, y, z: Evaluation items

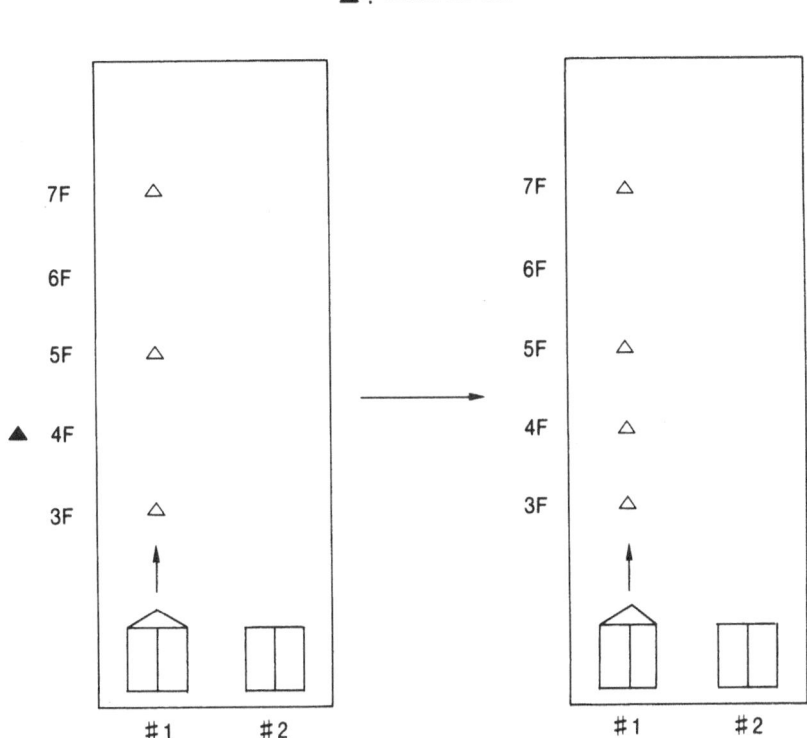

△ : Registered hall calls
▲ : Latest hall call

Fig. 9.7 Example of prediction operation

9.3 An elevator group control system using a fuzzy rule base

9.3.1 System construction concept

As was stated in the previous section, in a conventional group control system an evaluation function is used to determine which cage to assign.

A common defect of all of the evaluation functions that have been used is that they lack flexibility and cannot necessarily respond to the constantly fluctuating traffic conditions in the building. For example it is difficult to provide flexible control such that when a hall call arrives, in one case the cage that can arrive soonest will be assigned, while in other cases the relative positions of all cages or a new call which will be issued with high probability are considered to determine which cage to assign.

In order to provide this kind of flexible control, it is indispensable for the knowledge and judgment of a human expert to be reflected in the way the system operates. However, to express such human knowledge in quantitative form as an evaluation function is very hard work. For this reason, even

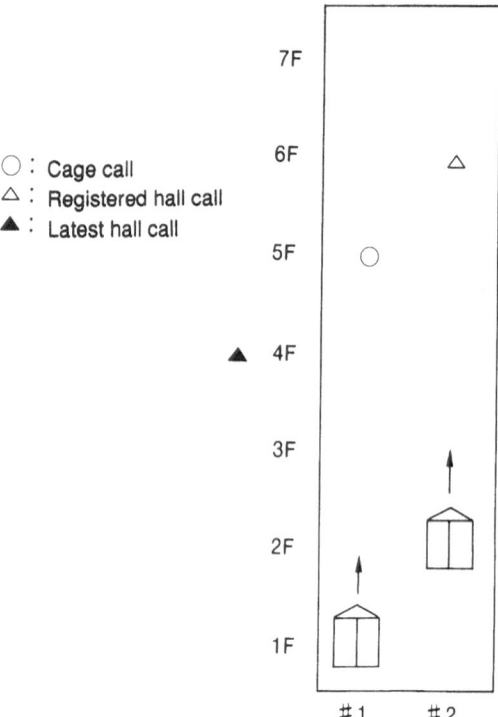

Fig. 9.8   Example of hall call assignments

though the developers (experts) of elevator group control systems possess a great deal of valuable experience, that experience could not be adequately reflected in the systems in the past.

A method that has been proposed to solve this problem and make it possible for expert knowledge to be applied actively is to use a knowledge base, as shown in Fig. 9. 9. A characteristic of this system is that the system possesses several control rules (logic for determining which cage to assign) and assigns a cage to respond to a hall call based on the control rule that is appropriate for the traffic condition at that time. This makes it possible to respond flexibly to the fluctuating traffic conditions in the building. However, in practice the following problems must be solved in order to construct a system based on this concept. 1) Many different traffic conditions must be conceived of and control rules which are appropriate for them devised. 2) It is necessary to have a mechanism for determining which of the many control rules is appropriate every time a hall call is issued. 3) In order to perform real time control, it is desirable for the system configuration to be kept as simple as possible. 4) The knowledge and judgment possessed by a human must be reflected in the system.

For this reason the fuzzy concept has been introduced into group control systems. A system has been developed in which the control rules needed

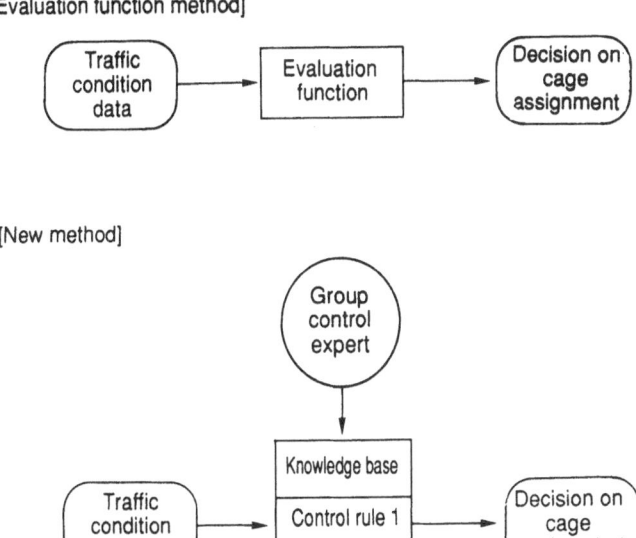

Fig. 9.9 Comparison of control concepts

for group control are described as "fuzzy rules" in "IF-THEN" format, and this group of fuzzy rules is used to perform group control. This system is described, and the advantages of introducing fuzzy theory discussed, below.

9.3.2 Fuzzy rule base construction and action

The system composition is shown in Fig. 9. 10.

When a new hall call is issued, the traffic condition inside the building is transmitted to the system. Then, based on the results of computations in the prediction computation section and evaluation item computation section, the cage assignment determination section selects and executes a suitable rule from the group control rule base, and determines which cage to assign. The control rules that are needed for group control are stored in the group control rule base. In general, each rule is described in the following format.

IF

    ([condition 1]
    and ([condition 2] or [condition 3])
    ......

)
THEN
    ([procedure 1]
    [procedure 2]
    ......

)

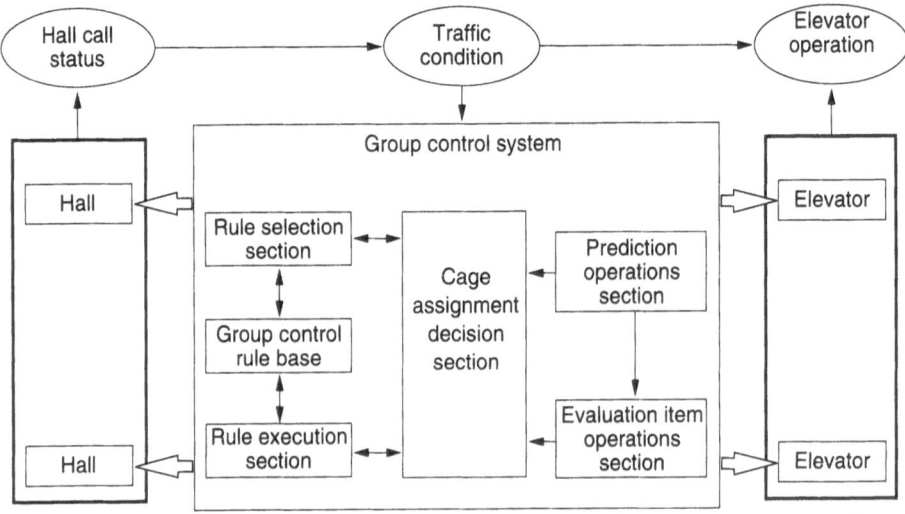

Fig. 9.10 System composition

```
(Rule Rm)
IF
    (   (A new call is issued on an upper floor.)
    and (There exists an elevator (X)
            (that is already headed upward)
        and (if it is assigned, a long wait wil not occur.)
        )
    )
THEN
    (Make elevator (X) a candidate for assignment. )
    (The elevator to be assigned is selected from
    among the candidates by evaluation function f1.)
```

Fig. 9.11 Example of rule

An example of a linguistic description of a rule is shown in Fig. 9. 11. The condition part of the rule prescribes the traffic condition in which that rule should be used; this is described using a number of conditions linked by "and" and "or". Each [condition] is prescribed as a fuzzy set by the membership function shown in the example in Fig. 9. 12.

For example, the condition "#if cage #1 is assigned then a long wait will become necessary" is described as:

$$\text{max\_wait}(1) \text{ is LG} \tag{2}$$

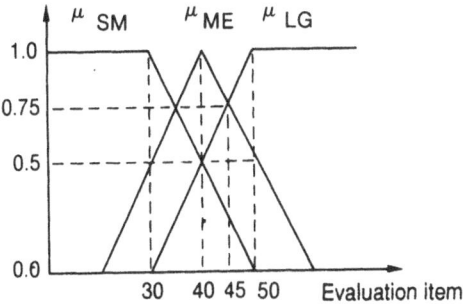

Fig. 9.12 Example of membership function

Here max_wait (1) indicates the maximum waiting time predicted for the passenger if cage #1 is assigned. LG is a fuzzy set that indicates "large".

The operation part of the rule describes the procedure to be executed when that rule is selected.

An outline of a group control procedure using this fuzzy rule base is shown in Fig. 9. 13. This is explained in detail below.

Computation of degree of applicability

First the degrees of applicability of the condition parts of all rules are computed. For example, if a certain condition as described in equation (2) is:

$$max\_wait(1) = 45$$

then, using the membership function shown in Fig. 9. 12, the degree of applicability (membership function value) for that [condition] becomes 0.75.

The degree of applicability of the rule condition part is computed from the degree of applicability for each [condition] and the min, max operations corresponding to each linking "and" and "or".

Suppose, for example, that the condition part of rule R_n is:

$$(\quad [\text{condition n1}]$$
$$\text{and } ([\text{condition n2}] \text{ or } [\text{condition n3}])) \tag{3}$$

If the degrees of applicability for these conditions are C_{n1}, C_{n2} and C_{n3}, then the degree of applicability C_n for rule R_n is computed from the following formula:

$$C_n = \min\{C_{n1}, \max\{C_{n2}, C_{n3}\}\} \tag{4}$$

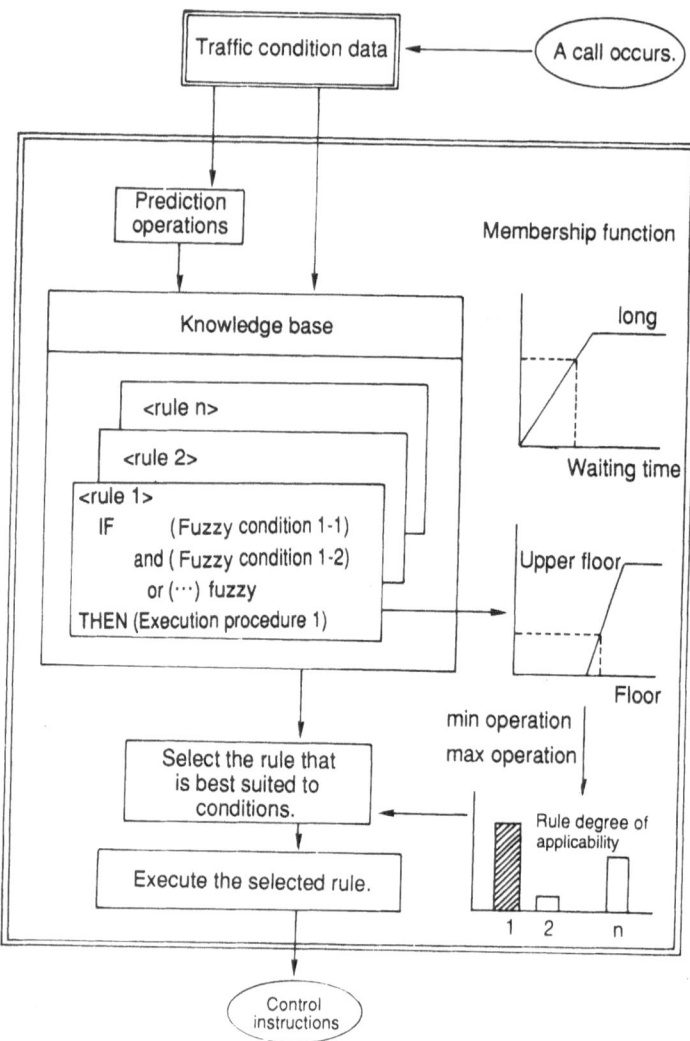

Fig. 9.13 Fuzzy rule base method control procedure

Rule selection

After the degrees of applicability are computed for the condition part of each rule, the rule having the maximum degree of applicability is selected. The rule that has this maximum degree of applicability is the rule that is most suitable for the traffic condition existing at the time the hall call is issued.

Rule execution

The procedure described in the operation part of the selected rule is

executed to determine which cage is to be assigned. This selection procedure
has 2 parts: selection of candidate, cages and determination of which cage
is to be assigned from among the candidates. These candidate cages are
determined by eliminating cages which would obviously be inappropriate from
all of the cages in the group.

For example, the rule in Fig. 9. 11 (Rule Rm), "make elevator (X) a
candidate", means that cages X (X = 1, ..., N) which satisfy the condition parts of the rule are candidates for assignment. Specifically, letting the
degrees of applicability of conditions 2 and 3 in the condition part for cage
X be $C_{m2}(X)$ and $C_{m3}(X)$, cages X that satisfy the following inequality are
candidate cages:

$$\min\{C_{m2}(X), C_{m3}(X)\} > C_X \qquad (5)$$

C_x: Threshold value

When candidate cages are selected in this manner, an evaluation function
such as the one in equation (1) is used to perform an overall evaluation such
as the waiting time and determine which cage is to be assigned.

9.3.3 Computation of rule degrees of applicability, and execution examples

In this section, we discuss the procedure for computing degrees of applicability of rule condition parts, and then a specific example of operation part
execution, using the rule shown in Fig. 9. 11. For simplicity we assume that
the system has 4 cages.

Computation of degree of applicability

First, note that the condition part of the rule in Fig. 9. 11, as shown in
Fig. 9. 14, consists of the 2 conditions p1 and p2 linked by "and". The degree
of applicability for p1, "a new call was issued on an upper floor", can easily
be computed. The membership function that indicates "upper floor" is, for
example, as shown in Fig. 9. 14; assume that the call is issued on the 9th
floor. At this time the membership function has the value 1.0, and this value
itself becomes the degree of applicability of condition p1.

However, condition p2 is a bit more difficult. This condition, "There
exists an elevator", means that (at least) one out of all of the cages in the
group satisfies a condition. As shown in Fig. 9. 15, this condition can be
decomposed using several "or"s as follows: "p21: #1 is headed upward, if
assigned to the latest hall call a long wait will not occur or p22: #2 is .....
or ..... p24: #4 is ..... ". Consequently it is sufficient to compute the degree
of applicability for each of the conditions p21, ..., p24.

These conditions for each cage can be further decomposed into 2 condi-

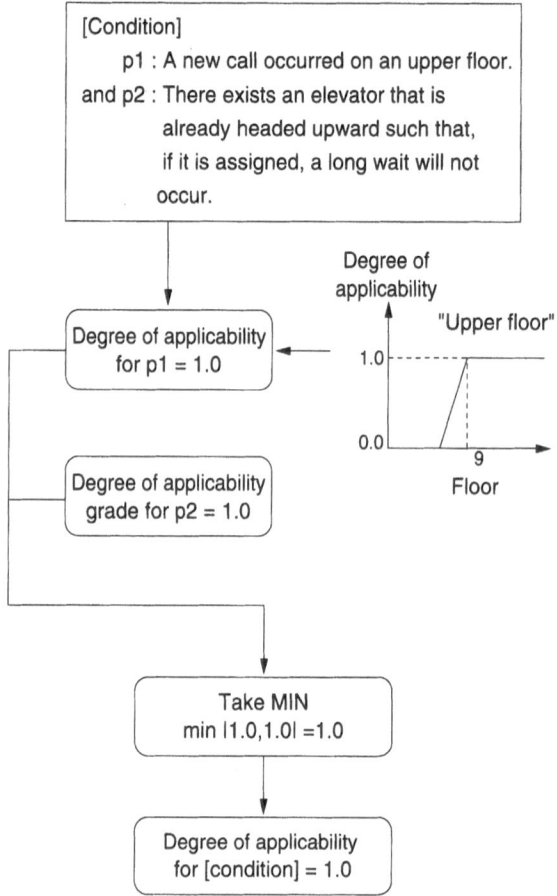

Fig. 9.14 Example of calculation of degree of applicability (1)

tions, as shown in Fig. 9. 16. Condition p21 can be decomposed into condition p211: "#1 is headed upward" and condition p212: "if #1 is assigned to the latest hall call, a long wait will not occur". The degrees of applicability for these 2 conditions can be computed from the membership functions. For example, suppose that #1 is scheduled to proceed upward to the 11th floor and that the maximum anticipated passenger waiting time if it is assigned to respond to the latest hall call is 50 seconds. Then, assuming that the membership functions for "upper floor" and "waiting time is short" are as shown in Fig. 9. 16, the degrees of applicability for p211 and p212 are 1.0 and 0.8, respectively.

The degree of applicability for condition p21 can be computed from the degrees of applicability for these 2 conditions p211 and p212. As shown in equations (3) and (4), when 2 degrees of applicability are linked by "and", the min should be taken. Accordingly, as shown in Fig. 9. 16 we have:

$$\min\{1.0, 0.8\} = 0.8$$

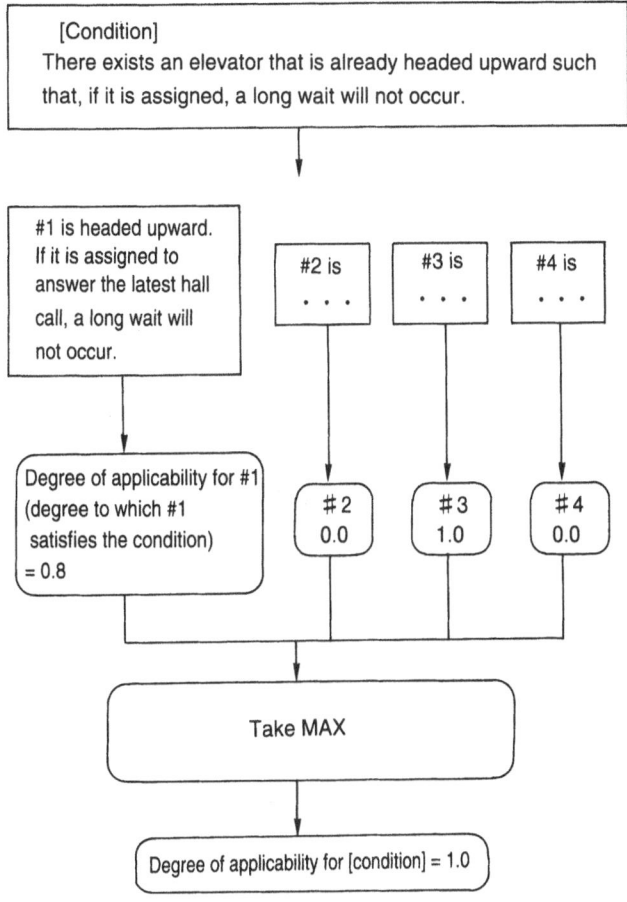

Fig. 9.15 Example of calculation of degree of applicability (2)

so that the degree of applicability for condition p212 becomes 0.8. The degrees of applicability for conditions p22, ..., p24 can be computed similarly.

Next, let us compute the degree of applicability for p2. Condition p2 consists of conditions p21, ..., p24 linked by "or"s. For this reason, the overall degree of applicability is found by taking the max. Supposing that the degrees of applicability for p22, ..., p24 are 0.0, 1.0, 0.0, respectively, the degree of applicability of p2, as shown in Fig. 9. 15, becomes:

$$\max\{0.8, 0.0, 1.0, 0.0\} = 1.0$$

We have found that the degrees of applicability for p1 and p2 are both 1.0. Finally, referring to Fig. 9. 14, we find the degree of applicability for the condition part of this rule by taking the min corresponding to the linking "and", which is naturally 1.0.

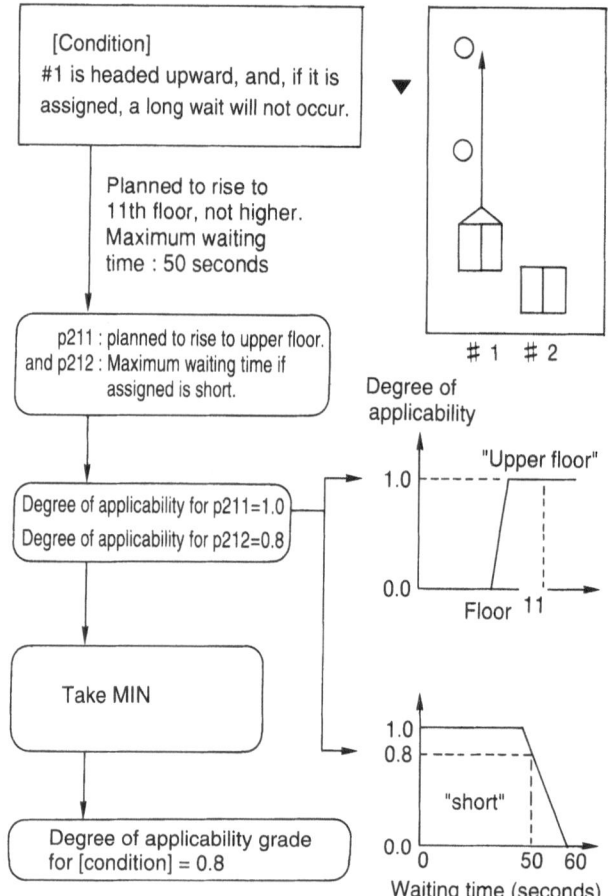

Fig. 9.16   Example of calculation of degree of applicability (3)

Execution of the operation part

As stated in section 9.3.2, after the degrees of applicability of all rule condition parts have been computed, the rule having the maximum degree of applicability is selected. Then the operation part of that rule is executed. Here we proceed on the assumption that the rule shown in Fig. 9. 11 has been selected.

The execution procedure for this rule starts with selection of the candidate cages. From the above computations, the degrees of applicability for the conditions for each cage are 0.8, 0.0, 1.0 and 0.0, respectively. Assuming that the threshold value Cx in equation (5) is 0.7, cages #1 and #3, which have degrees of applicability larger than the threshold value, as shown in Fig. 9. 17, are selected as candidate cages.

Then the evaluation function value for each cage is computed as shown in Fig. 9. 17. In this case the best evaluation value (here the smallest evaluation

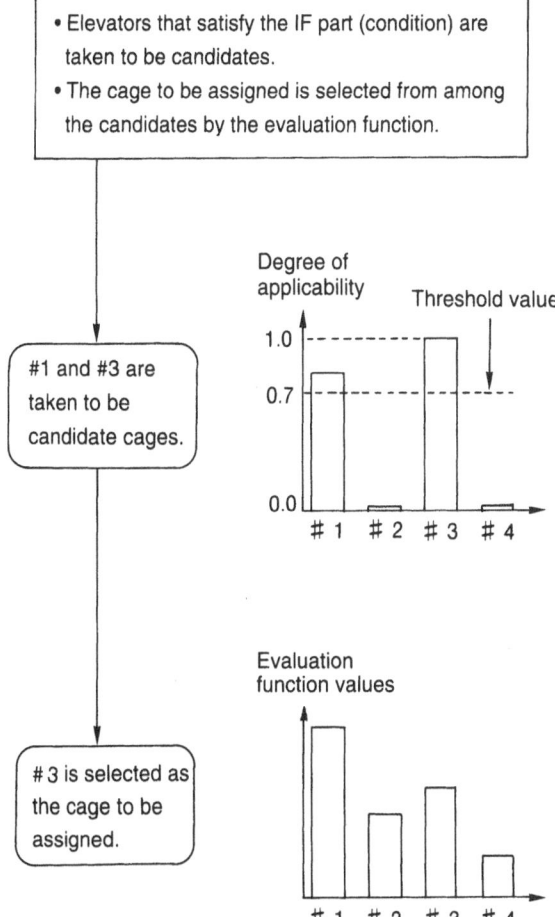

Fig. 9.17 Example of rule execution procedure

value is taken to be the best) is for cage #4, but cage #4 is not a candidate cage, so it is not selected as the assigned cage. Among the candidate cages #1 and #3, #3 has the better evaluation value, so it is determined to be the assigned cage.

9.3.4 Rule extraction

The method described above is simple, and the introduction of the concept of fuzzy sets into a group control system produces the following advantages. 1) By using fuzzy logic operations to compute the degree of applicability of each rule condition for the actual traffic condition, a suitable rule can be selected. 2) By introducing the concept of the degree of applicability, it is not necessary to describe all of the different possible traffic conditions rigorously as rules; it is sufficient to describe one example. Consequently, the number of rules can be greatly reduced, simplifying the system. 3) Since it is easy to convert a linguistic control rule into a fuzzy rule, the subjective knowledge and judgment methods possessed by humans can be easily reflected in the system.

It is clear that these correspond to system formation conditions (2), (3) and (4), which were discussed in section 9.3.1. What remains is to set rules (1); the following method is used to do this.

Optimum operation analysis [8]

A quantity of traffic data were prepared based on past data, and the optimum operation of an elevator group for those data was analyzed mathematically. For this analysis the S · A (Simulated Annealing) method, which is a powerful method of analyzing large-scale combinational problems, was used. The differences between the results of this analysis and cage assignment by a conventional system were extracted, and problems with the conventional system investigated.

The simulation environment [9]

As a tool for system development, a simulation environment was independently constructed on the EWS in order to consistently perform operations from rule input to demonstration of effectiveness.

Obtaining experts' knowledge

Experts were questioned as to both their general knowledge and proposals for improvement with regard to the problems extracted from the mathematical analysis discussed above. Then this information was organized and formed into rules, and the above-mentioned simulation environment was used to demonstrate their effectiveness.

As a result of combined use of these methods, a group of rules that are useful were extracted to form a fuzzy rule base (group control rule base).

9.4   A simulation example

As an example, consider the situation shown in Fig. 9. 18. There are 4 cages in a 12 story building. The principal floor, which is the most crowded, is the 1st floor. Suppose that cage #1 has cage calls for the 8th floor and the 11th floor and is moving upward. Similarly, #3 has a cage call for the 12th floor and is moving upward. Cages #2 and #4 are stopped at the 2nd floor and the 6th floor, respectively. At this time a new hall call (downward) is issued on the 10th floor. Which cage should be assigned to respond?

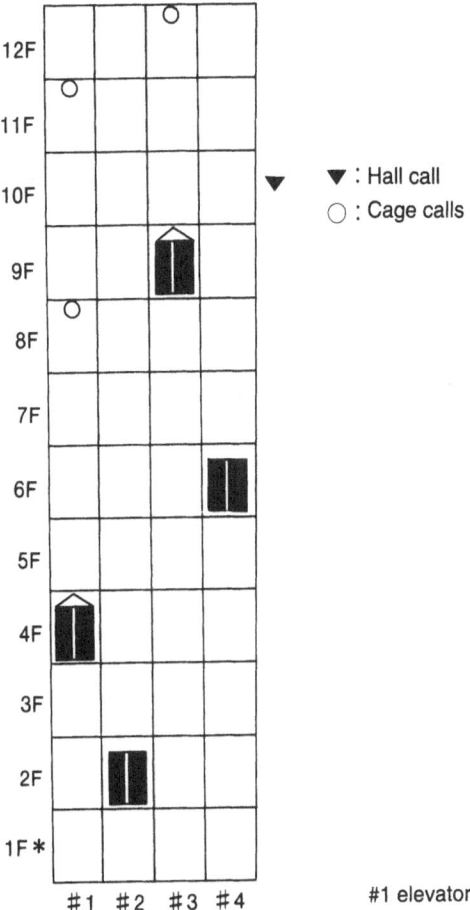

Fig. 9.18 Example of hall call distribution

The cage that can arrive at the 10th floor most quickly is #4. Since there is no hall call except from the 10th floor, there is no danger that by a fully loaded elevator passing through, a prior notification will have to be changed or will not match the actual elevator movement. Consequently, using the conventional evaluation method, normally cage #4 would be selected to be assigned. However, if #4 is assigned to respond to the hall call on the 10th floor, #1, #3 and #4 will all be moving upward. For this reason, if there are new hall calls from the 1st floor, which has the most traffic, or from intermediate floors, #2 will have to service all of them until the three cages that are moving upward turn around and return downward. It is easy to anticipate that in such a case service on the lower floors will be adversely affected. In fact, there is a high probability that just this kind of situation will occur.

#1 elevator

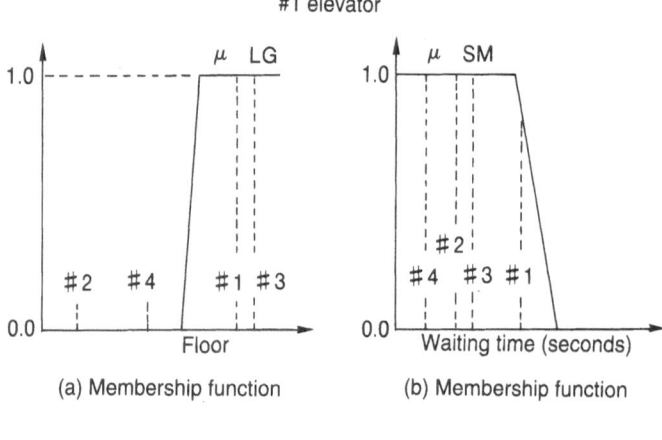

(a) Membership function   (b) Membership function

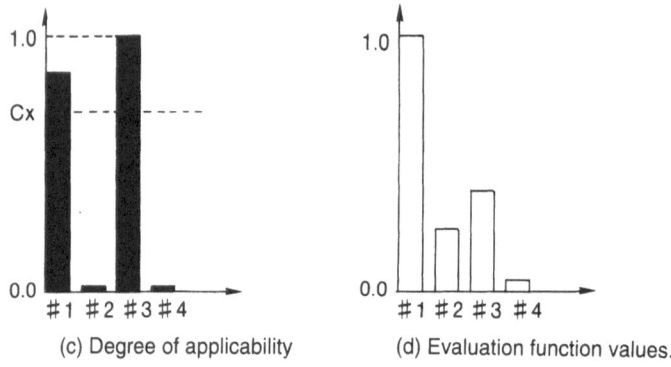

(c) Degree of applicability   (d) Evaluation function values.

Fig. 9.19 Examples of functions when rules are used

It is proposed that if a hall call such as shown in Fig. 9. 18 occurs, an appropriate solution is to reserve #2 and #4 for servicing the lower floors, and assign either #1 or #3 to respond to the call, considering such factors as the passenger's waiting time. In this case Rule Rm in Fig. 9. 11 is effective. Examples of membership functions and evaluation function values for this rule are shown in Fig. 9. 19. Figs. 9.19 (a) and (b) show the membership functions for the rule's 2nd and 3rd conditions. Here μLG and μSM mean "large (upward direction)" and "small (short time)", respectively. When computations are performed from these using the same procedure as in section 9.3.3, the degree of applicability of this rule becomes 1.0. Fig. 9. 19 (c) shows the degree of applicability with which each cage satisfies the condition in the rule condition part. From these results #1 and #3 are selected as candidate cages, in accordance with equation (5). Among these 2, from the result of applying the evaluation function (d) #3 is finally determined as the assigned cage.

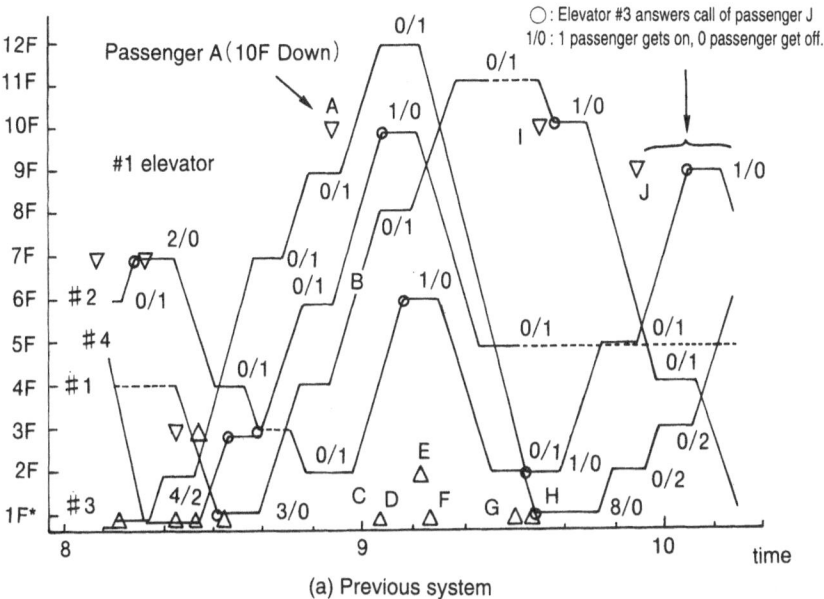

(a) Previous system

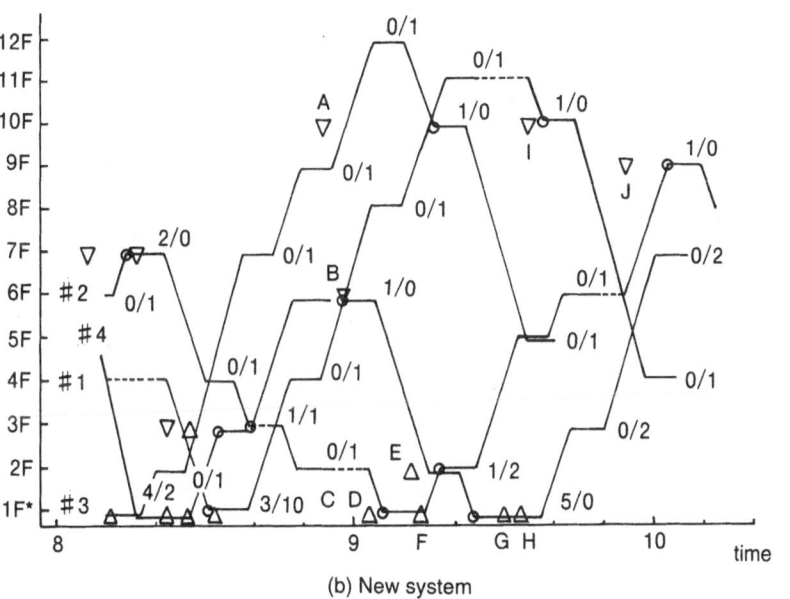

(b) New system

Fig. 9.20 Simulation results: operation records

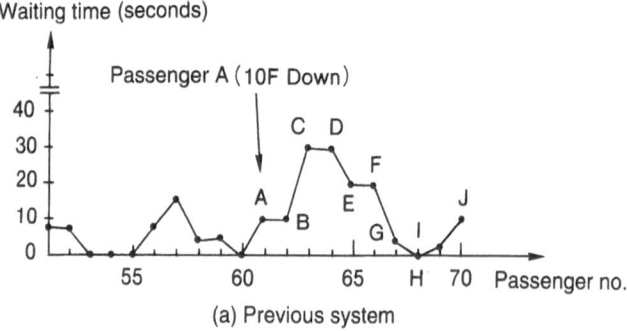

(a) Previous system

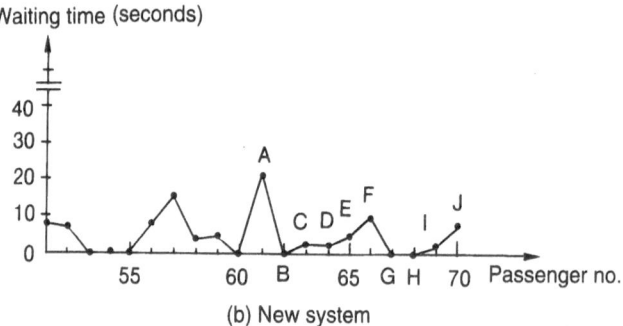

(b) New system

Fig. 9.21 Simulation results: passenger waiting time

Table 9.1 Simulation results

	Average waiting time	Proportion of passengers who wait 60 seconds or more
Previous system	18.0s	3.3%
New system	15.9s	1.7%
(Improvement rate)	(11.7%)	(48.5%)

4 elevators, standard office building with 12 floors Normal mid-morning time
(600 passengers/hour)

Figs. 9.20 and 9.21 show examples of the operation status of each cage
and the passenger's waiting time after #3 and #4, respectively, are assigned
to respond to a hall call.

In case (a), in which #4 is assigned, passenger A who issues the first call
from the 10th floor has the shortest waiting time, but servicing of subsequent
calls (C to H) is poor because #1, #3 and #4 were all moving upward at
the same time. In contrast, in case (b), in which #3 is assigned, the waiting
time for the 1st call on the 10th floor is a bit longer, but service after that is

better.

A comparison of simulations using this method and the conventional method is presented in Table 9.1.

The improvement rate in the table differs depending on the simulation conditions, but overall the average waiting time is reduced by 10% to 15% and the probability of having to wait a long time is reduced 30% to 50%.

9.5 In conclusion

In this chapter an elevator group control system using a fuzzy rule base has been introduced. This system has been installed in the <AI-2100> artificial intelligence group control system [10]. The AI-2100 has been operating in a number of major buildings, including the Mido Suji Grand Building in Osaka and the new Tokyo City Hall, since 1989.

This system can be considered an example of successful application of fuzzy theory. However, in the development stage of this system, it was not decided to definitely use fuzzy theory from the beginning; when problems occurred fuzzy theory was tried as a way to solve them and it worked.

Fuzzy technology has very broad potential applications; and it can be expected that it will find applications in many fields. However, there are many ways to use it; consequently, when it is applied to an actual system, it is important to investigate how (or in what part of the system) it should be applied. A great deal of work remains to be done on methods of applying fuzzy theory.

References

1. Hirota K (1987) Recent studies of and trends in fuzzy expert systems (in Japanese). J Information Processing Soc Jpn 22 (8) : 1065–1074

2. Yasunobu S, Miyamoto S, Ibara H (1984) A predictive fuzzy control for automatic train operations (in Japanese). Systems and Control 28 (10) : 605–613

3. Yagishita O, Itoh O, Sugeno M (1984) Application of fuzzy resoning to the water purification process (in Japanese). Systems and Control 28 (10) : 597–604

4. Watanabe H, Kamahara T (1978) Speed control and group supervisory system for high speed elevator (in Japanese). Systems and Control 22 (6) : 316–223

5. Hikita S, Tsuji S (1988) An elevator group-supervisory control system using fuzzy rule-base (in Japanese). J Information and Control Engineers 27 (10) : 925–926

6. Hikita S, Komaya K (1989) A new elevator group-supervisory control system using a fuzzy rule-base (in Japanese). Trans Soc Information and Control Engineers 25 (1) : 99–104

7. Araya S, Tsuji S, Umeda Y, Kamahara T, Kamaike H (1982) An elevator group control system having a traffic pattern learning function (in Japanese). Proc 8th Systems Symposium, pp225–239

8. Hikita S, Tsuji S, Komaya K (1986) The limits of elevator service-optimal operation analysis by the SA method (in Japanese). Proc Annual Conference Institute of Electrical Engineers of Jpn, pp1931–1932

9. Komaya K, Hikita S (1988) Development of an intelligent simulation environment for elevator group control (in Japanese). Proc Annual Conference Institute of Electrical Engineers Jpn, pp1956–1957

10. Ujihara H, Tuji S (1988) The revolutionary AI-2100 elevator-group control system and the new intelligent option series (in Japanese). Mitsubishi Electric ADVANCE 28 : 5–8

Chapter. 10

A HIGHWAY TUNNEL VENTILATION CONTROL SYSTEM USING FUZZY CONTROL

10.1 Introduction

In automobile highway tunnels, in order to maintain a suitable environment for drivers and traffic, visibility in the tunnel must be maintained, and the concentrations of poisonous substances including carbon monoxide must be kept at or below allowable levels. For this reason, in long tunnels and tunnels with heavy traffic, ventilation facilities are installed. When the ventilation facilities are run at full capacity, the environment in the tunnel is obviously adequately maintained, but this consumes a great deal of electric power. Consequently, a central problem in highway tunnel ventilation control systems is to keep the pollution concentration at or below the allowable level, and thus provide a safe environment for traffic, while consuming as little electricity as possible.

Here we will focus on longitudinal flow ventilation systems, which have recently become the mainstream, and discuss the application of fuzzy control to them.

10.2 Outline of a longitudinal flow ventilation system

In a longitudinal flow ventilation system, air is blown along the length of the tunnel to dilute the pollutants. Typical examples of longitudinal flow tunnel ventilation systems are shown in Figs. 10.1 to 10.3.

Fig. 10.1 A jet fan longitudinal flow tunnel ventilation system. Fig. 10.2 A jet fan longitudinal flow tunnel ventilation system with dust collectors. Fig. 10.3 A jet fan longitudinal flow tunnel ventilation system with vertical exhaust shafts and dust collectors.

In Fig. 10.1, only jet fans are used as ventilators; they blow as much air through the tunnel as necessary. This is a basic longitudinal flow ventilation system. Fig. 10.2 shows a system in which electrical dust collectors are installed at points along the tunnel. Air inside the tunnel is taken in to

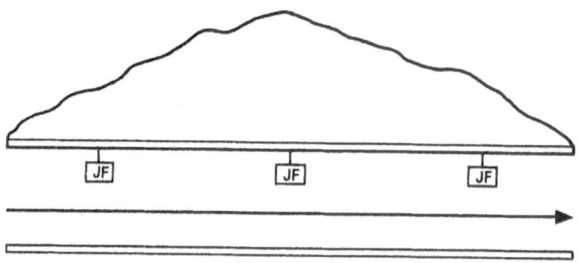

Fig. 10.1 A tunnel with jet fan longitudinal flow ventilation.

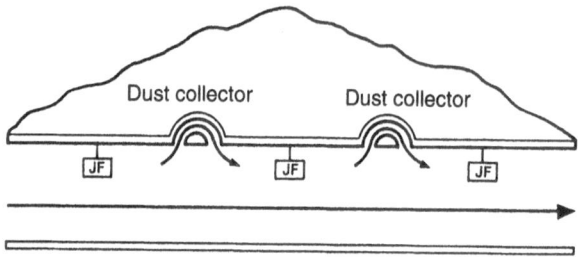

Fig. 10.2 A tunnel with jet fan longitudinal flow ventilation and dust collectors.

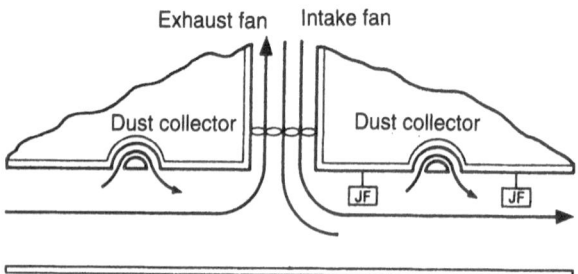

Fig. 10.3 A tunnel with jet fan longitudinal flow ventilation,
vertical intake and exhaust ducts, and dust collectors.

these, and returned to the roadway after soot and smoke are removed to
maintain visibility. In the system in Fig. 10.3, in addition to the jet fans and
electrical dust collectors, there are vertical shafts to remove air from inside
the tunnel to outside. Exhaust fans installed in these shafts remove polluted
air to the outside, and intake fans draw outside air in.

In a tunnel ventilated by longitudinal flow, the soot and smoke emitted
by passing vehicles can be regarded as being transported along the inside
of the tunnel approximately in accordance with the following 1-dimensional

advection and diffusion equation.

$$\frac{\partial C(x,t)}{\partial t} = -V_r \cdot \frac{\partial C(x,t)}{\partial x} + D \cdot \frac{\partial^2 C(x,t)}{\partial x^2} + \frac{q}{A_r} \tag{1}$$

Here $C(x,t)$ is the soot and smoke concentration, x is distance in meters along the tunnel in the long direction, t is time in seconds, V_r is the air flow speed over the roadway in m/s, D is the diffusion coefficient in $\mathrm{m^2/s}$, A_r is the cross-sectional area of the roadway in square meters and q is the amount of soot and smoke emitted per unit of distance and unit of time, in $[\mathrm{m^3/(m \cdot s)}]$.

As can be seen from equation (1), the pollution condition inside the tunnel is greatly affected by the air flow speed V_r. The air speed over the roadway is found from equation (2) below; it is affected by the ventilation due to the force of passing vehicles $\Delta P_t(\mathrm{Pa})$, ventilation due to natural wind $\Delta P_n(\mathrm{Pa})$, resistance in the tunnel $\Delta P_r(\mathrm{Pa})$, increased pressure caused by ventilation equipment $\Delta P_k(\mathrm{Pa})$ and so on.

$$\rho \cdot L \cdot \frac{dV_r}{dt} = \Delta P_t + \Delta P_n + \Delta P_r + \Delta P_k \tag{2}$$

Here ρ is the density of air in $\mathrm{kg/m^3}$ and L is the tunnel length in meters.

In the case of ordinary traffic, the force ΔP_t of the passing vehicles is always in the direction that aids ventilation, and there are even times when they cause a steady air flow velocity V_r over the roadway that is strong enough so that ventilation equipment does not have to be run. However, when there is traffic in both directions, the flows of air forced by traffic moving in both directions cancel each other, so it is easy for the air flow in the tunnel to become unsteady and difficult to maintain the pollution at the desired level. Consequently, in ventilation control in a tunnel with longitudinal flow ventilation and traffic moving in both directions it is important to stabilize the air flow over the roadway.

10.3 A ventilation control system using fuzzy control

As stated above, the ventilation process in a tunnel ventilated by longitudinal air flow is complicated; a great deal of time and effort are required to create even a linear model of the process. In particular, in the case of a highway tunnel it is first of all necessary to maintain the safety of the vehicles passing through, so naturally there are many restrictions on the tests that can be conducted after the tunnel opens for traffic, and it is not easy to create a linear model. Consequently, it is very difficult to apply conventional control theory based on a linear model. In addition, as stated above, in ventilation control in a tunnel with 2-way traffic, in order to obtain stable control it is effective to include air flow speed as well as pollutant concentration as a control quantity. In controlling such a multi-input system, it is conceivable that one would want to apply a multi-variable control theory such as

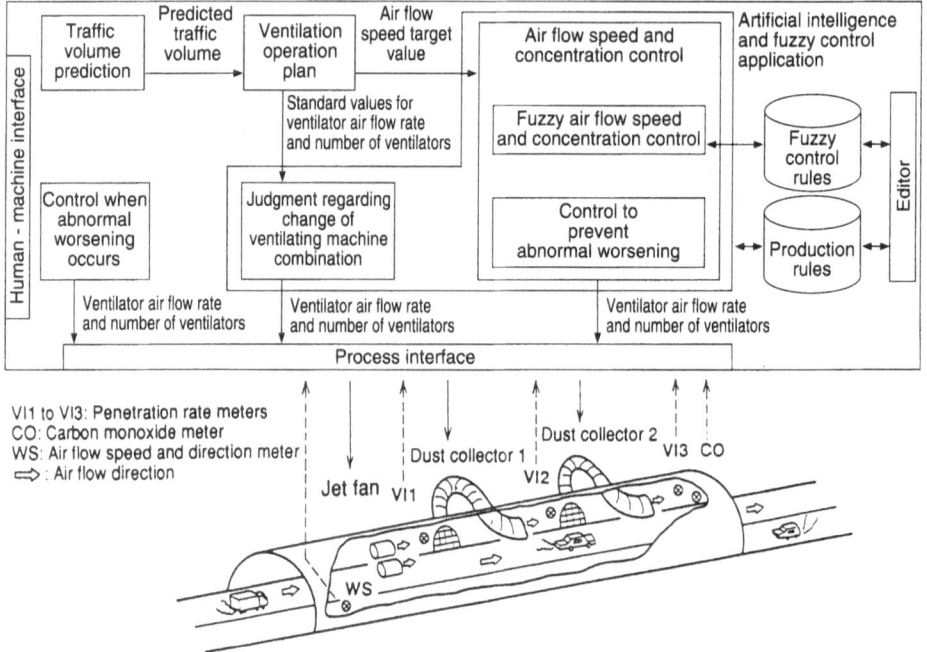

Fig. 10.4   Example of ventilation control system composition, for a tunnel with jet fan longitudinal flow ventilation and dust collectors.

optimum regulator theory or non-interference control theory. However, the tunnel ventilation process is described not as a concentrated system of constants but rather as a distributed system of constants, and it is difficult to obtain a highly accurate process model, so it is difficult to apply these control theories.

Against this background, we applied fuzzy control having the following characteristics to tunnel ventilation control. 1) A rigorous mathematical formula model is not necessary; qualitative knowledge which people have gained from experience is applied in the form of rules. 2) It is relatively easy to construct a multi-input multi-output control system. 3) It is relatively easy to change the control specifications. An example of the composition of a ventilation control system to apply to a jet fan longitudinal flow tunnel ventilation system with dust collectors is shown in   Fig.   10.4.   Fig.   10.5 shows a generalization of it for application to various types of longitudinal flow tunnel ventilation systems. A specific system matching the ventilation facilities to be used was constructed based on the system in Fig. 10.5.

These systems have the following functions. (1) Traffic volume prediction. (2) Ventilation equipment operation planning. (3) Judgments regarding changing the combination of ventilation equipment in use. (4) Fuzzy air flow speed and concentration control. (5) Level control. (6) Control when abnor-

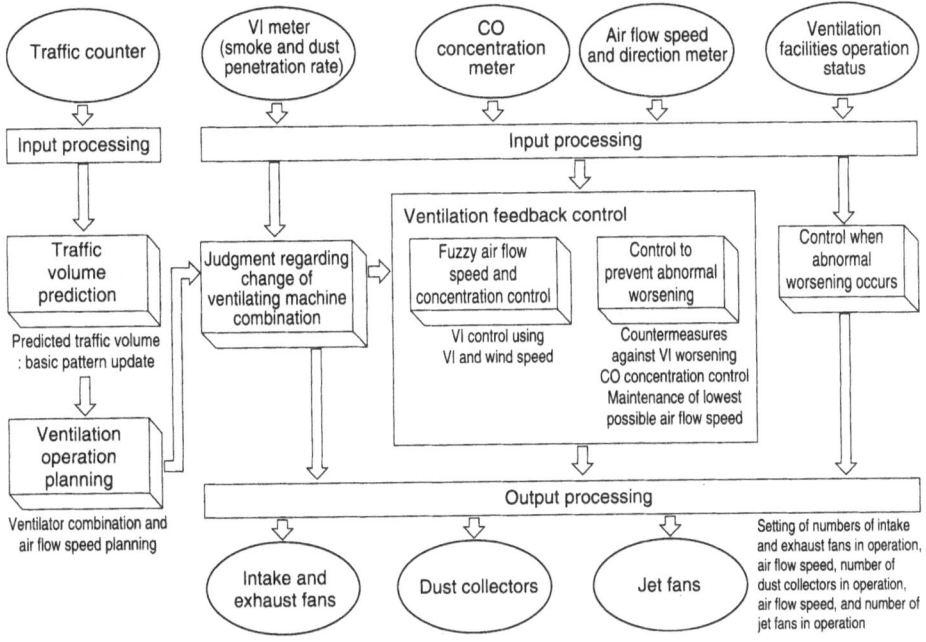

Fig. 10.5 Composition of a ventilation control system for all tunnels with longitudinal flow ventilation.

mal worsening of conditions occurs.

In ventilation control in a highway tunnel, it is important to appropriately set the combination of ventilation equipment being operated, number of machines in operation and air flow transport to correspond to the trend of traffic volume. For this reason, it is first of all necessary to predict the traffic volume, which is the basis for predicting the load on the ventilation equipment (10.3.1 Traffic volume prediction). Next, based on the predicted traffic volume, the combination of ventilating machines to be run and the air flow transport that will maintain the prescribed safety level while minimizing energy consumption is computed (10.3.2 Ventilation operation planning). Then, in proceeding to the stage of execution of this ventilation operation plan, the plan execution timing is determined and the plan values corrected considering various conditions such as the ventilation condition inside the tunnel at that time, the operating loads on the ventilation machines and protection of facilities with respect to the ventilation machines (10.3.3 judgment concerning change of the ventilation machine combination).

The above 3 functions, (1), (2) and (3), are ventilating machine operation functions based on the traffic volume. In a 1-way traffic tunnel, where the traffic volume in the tunnel can be predicted relatively accurately, these are executed every 5 to 10 minutes. However, in a 2-way traffic tunnel, the ventilation load is greatly determined not only by the amount of traffic in each direction but also by the ratio between the two, so even if a plan is prepared for ventilation control operations at short intervals such as every 5 to 10 minutes, good accuracy cannot be expected. Consequently, the plan is prepared for intervals on the order of 1 hour.

Once the basic operating load of each ventilating machine has been set in accordance with the predicted traffic volume, feedback control based on measurements by sensors installed in the tunnel then becomes necessary. The purposes of feedback control are to correct for errors in the ventilation operating plan and to respond to sudden external disturbances. The heart of the control system is wind speed and concentration control, which maintains the soot and smoke concentration at or below the allowable level while maintaining a steady flow of air through the tunnel (10.3.4 Fuzzy air flow and concentration control). This control is normally executed at intervals on the order of 5 minutes. In addition, feedback control is also applied in parallel with the air flow and concentration control, at intervals on the order of 1 minute, to deal with increase of the carbon monoxide (CO) concentration or a sudden drop in VI (VI is an index of visibility in the tunnel, also called the smog penetration rate, expressed on a scale from 0 to 100%) (10.3.5 Level control). There is also control applied at intervals of tens of seconds to deal with sudden increases in pollutant concentration that cannot be adequately handled by the air flow and concentration control or the level control (10.3.6 Emergency control).

In addition, there are also other controls added, particularly in 2-way traffic tunnels, for example to suddenly suppress the air flow in the tunnel, prevent smoke diffusion by keeping the air flow near zero, and secure driver escape routes, in case of a fire.

As stated above, this system can be divided into 2 blocks. The first prepares an operation plan for each ventilating machine based on the predicted traffic volume in the tunnel, and executes it (functions corresponding to 10.3.1, 10.3.2 and 10.3.3 above). Second, there is the part that monitors the actual ventilation condition in the tunnel and provides feedback corrections to the operation plan (corresponding to 10.3.4, 10.3.5 and 10.3.6, above). Artificial intelligence (AI) is used in the former block; the latter functions are based mainly on fuzzy control. Thus, this ventilation control system is a combination of functions using artificial intelligence and functions using fuzzy control.

The individual functions are explained below.

10.3.1   Traffic volume prediction

In general, the traffic volume varies through the day but often in a definite pattern (let us call it the standard daily traffic pattern). This function predicts the traffic volume (vehicles/hour) 2 hours ahead once per hour. In making the prediction, the count on a traffic counter and the above - mentioned standard daily variation pattern are used.

The standard daily traffic pattern is expected to differ from one day to another, so several are stored in the system, for example weekday, holiday, day before holiday, day after holiday. The standard traffic pattern is automatically updated every day based on past statistics. The variation from the standard pattern is predicted with an autoregression model. The coefficients in the autoregression model are set by on-line successive estimation using a Karman filter to account for the secular variation of traffic volume.

10.3.2   The ventilation operation plan

The principal purposes of the ventilation operation plan are to:

1) determine the combination of ventilating machines to be used,

2) determine standard values for the air flow transport and number of ventilating machine that is operated, so as to maintain the prescribed driving environment in the tunnel with minimum electric power consumption.

In a tunnel ventilated by jet fan longitudinal flow ventilation with dust collectors, as shown in Fig. 10.2, there are not many combinations of ventilating machines, as shown in Fig. 10.6, so for each combination, the optimum air flow transport and standard number of ventilating machines ("standard value" as used here means the set value before feedback control is applied) are computed, and from among those possibilities the combination that minimizes electric power consumption is extracted.

In the computation of the optimum air flow transport and standard number of ventilating machines for each combination, linear planning methods (multiplier method and conjugate slope method) are used so as to satisfy the constraint on pollution concentration and minimize the power consumption by the ventilating machines.

When the tunnel being ventilated is a large - scale tunnel, such as the tunnel ventilated by a jet fan longitudinal flow ventilation system with vertical exhaust shafts and dust collectors shown in Fig. 10.3, the number of possible combinations of ventilating machines becomes large, so often combinations of ventilating machines are selected first, then the optimum air flow transport and number of ventilating machines are computed within the limits of possibilities offered by those combinations. Specifically, the amounts of pollutants that will be emitted are estimated 1 hour ahead and 2 hours ahead based on the predicted traffic volume; then after an appropriate operation combination (ON/OFF) of ventilating machines is selected, the optimum air flow transport and standard number of ventilating machines for each ventilating machine are computed by the nonlinear planning method.

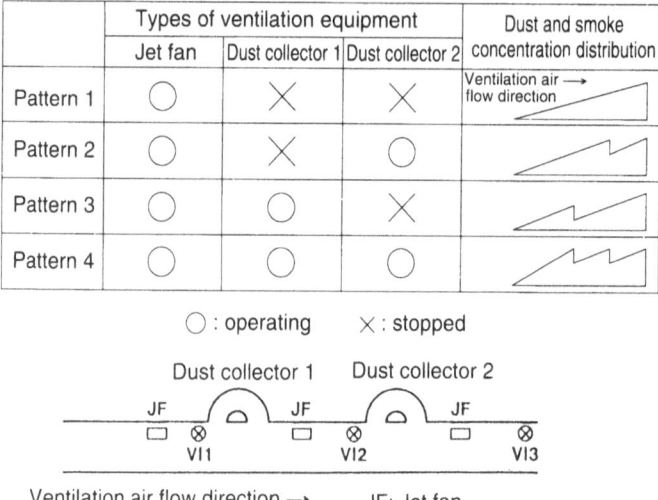

	Types of ventilation equipment			Dust and smoke concentration distribution
	Jet fan	Dust collector 1	Dust collector 2	
Pattern 1	○	✕	✕	Ventilation air → flow direction
Pattern 2	○	✕	○	
Pattern 3	○	○	✕	
Pattern 4	○	○	○	

○ : operating   ✕ : stopped

Ventilation air flow direction →   JF: Jet fan

Fig. 10.6 Combinations of ventilation equipment in a tunnel
with jet fan longitudinal flow ventilation and dust collectors.

In the fuzzy air flow speed and concentration control to be discussed below, it is important to control not only the pollutant concentration but also the air flow speed inside the tunnel, so when the above air flow transport and standard number of ventilating machines computations are performed, a target value for the air flow speed over the roadway appropriate for the traffic volume load at that time is also computed.

10.3.3   Judgment regarding change of ventilating machine combination

The above - mentioned ventilation operation plan is based on the predicted traffic volume, so it does not necessarily match the actual traffic flow during transient periods, and cannot follow exact variations from one moment to the next. The operation plan might match the actual ventilation requirement averaged over an hour, but not at the moment a change is made. For example, at a time when the plan calls for the number of ventilating machines in operation to be reduced it is possible that just at the time the change is to be made a number of large trucks will come through, lowering the VI value. In such a case, it is desirable for the reduction in machines in operation to be delayed until after the trucks have passed through and the VI value has returned to normal.

Another possibility is that the ventilation operation plan prepared 1 hour in advance will call for the dust collectors in operation to be stopped, but after that the ventilation load again increases and it becomes necessary to restart the dust collectors that have just been stopped. This kind of operation leads to an increase in the frequency with which the equipment is turned ON and OFF, which is not good for the equipment. One possible way to deal with this problem is to compare the predictions made 1 hour in advance and 2 hours in advance, and use this information to avoid repeated turning of machinery ON and OFF.

For this purpose, this system is provided with a ventilating machine combination change timing judgment function. When it is time for the ventilating machine combination in operation to be changed according to the ventilation operation plan, this function judges the timing of the change, considering such factors as the load on each ventilating machine, the pollution condition in the tunnel and the ventilation operation plans made 1 hour in advance and 2 hours in advance. Consequently, although the operation plan is intended to be executed at 1 hour intervals, the actual changes might deviate from this timing.

When it is anticipated that operation according to the plan will make it difficult to keep the pollutant concentration at or below the allowable level due to insufficient ventilating machines in operation, this function can start machines that have been idle.

This function uses an expert system (production rules), and stores the conditions for switching and the switching method as a rule base. A sample rule is shown in Fig. 10.7.

> R1: If degree of pollution worsens (VI < a certain set
>   value, for example 50%),
>
> then do not execute planned reduction
> of number of ventilators in operation.

> R2: If excess ventilators are in operation (VI > a certain
>   set value, for example 70%),
>
> then do not execute planned
> increase of number of ventilators in operation.

> R3: If jet fan load is high (number of jet fans in
>   operation > a certain set value, for example 15 fans),
>
> then operate intake and exhaust fans at lowest
> air flow rate.

Fig. 10.7   Rules used in ventilating equipment
              combination switching judgment (examples).

10.3.4 Fuzzy air flow speed and concentration control

The above 3 functions:

1) traffic volume prediction,

2) ventilation operation planning,

3) ventilating equipment operation change judgment,

determine the combination of ventilating machines in operation, air flow transport and standard value of number of ventilating machines, mainly at long intervals such as 1 hour.

The fuzzy air flow speed and concentration control to be discussed here performs corrections to the ventilation machine operation air flow transport and number of ventilating machines at 5 minute intervals in accordance with sensor output; this takes the form of feedback corrections to ventilation operation plan values.

The purpose of this control is to keep the VI value steadily in the vicinity of the target value, but, as stated above, particularly in a tunnel with 2-way traffic and longitudinal flow ventilation, it is easy for the air flow over the roadway to fluctuate irregularly. Since reduction of air flow also results in reduction of the VI value, in addition to the VI value itself the air flow speed over the roadway is also controlled.

The VI target value is the value that will maintain the driving environment in the tunnel; the target value of air flow speed over the roadway is either the value according to the above - mentioned ventilation operation plan or a value set by an operator.

The principal operation quantity, in the case of a tunnel with a jet fan longitudinal flow ventilation system as shown in Fig. 1 or a tunnel with a jet fan longitudinal flow ventilation system with dust collectors as shown in Fig. 10.2, is the number of jet fans in operation. In a tunnel with jet fan longitudinal flow ventilation with dust collectors and vertical exhaust shafts, as shown in Fig. 10.3, the principal operation quantities are the number of jet fans in operation and the air flow transport in the vertical shafts. The response time for the effect of the dust collectors to show up in the VI value is long, so they cannot be used as a principal operation quantity in feedback control. However, in the type of system shown in Fig. 10.3, when there is a danger that feedback control of the air flow transport in the vertical columns will cause a large imbalance between the air flow transport in the vertical shafts and the air flow transport through the dust collectors, the air flow transport through the dust collectors is corrected, the corrections being coupled to the changes in air flow transport through the vertical shafts. Fig. 10.8 shows the basic concept of this control when control is applied principally to the jet fans. Suppose for example that visibility in the tunnel is poor and

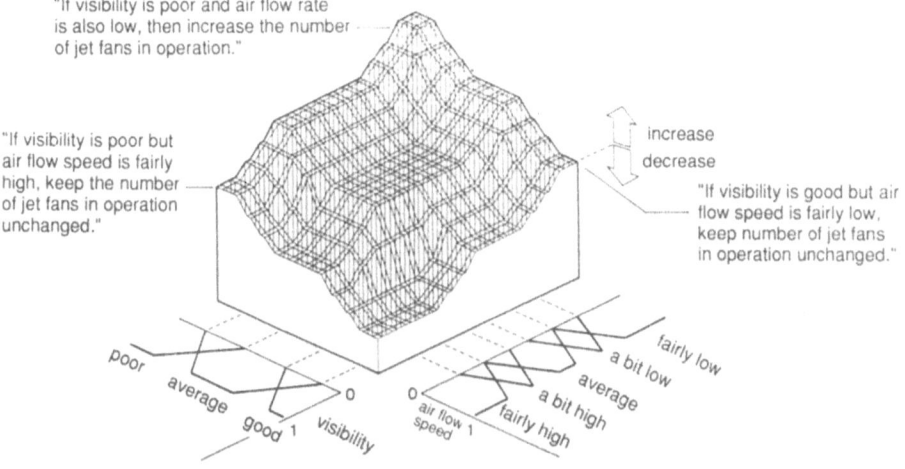

Fig. 10.8 Conceptual diagram of fuzzy air flow speed and concentration control.

air flow speed has dropped well below the target value. Then the number of jet fans in operation will have to be increased substantially. If, on the other hand, visibility is poor at a given time but the Air flow speed is fairly high, it can be expected that the visibility will quickly improve, so the number of jet fans in operation is not hastily increased (in order to save energy). Conversely, if visibility is good at a given time but air flow speed is low, if the number of jet fans in operation were to be decreased because visibility is good there would be danger of an instability causing the air flow to reverse, so changes are avoided in order to stabilize ventilation control. Fig. 10.9 shows a block diagram of fuzzy air flow speed and concentration control for a tunnel having jet fan longitudinal flow ventilation with vertical exhaust columns and dust collectors, as shown in Fig. 10.3.

A number of VI meters are installed in the tunnel. The VI value which is farthest below the VI target value is computed from the following formula, and that value is taken as the control deviation ΔVI of the VI value.

$$\Delta VI = \min(\Delta VI_1, \cdots, \Delta VI_6) \tag{3}$$

$$\Delta VI_i = VI_i - VI_{iref}(i = 1, \cdots, 6) \tag{4}$$

250   A ventilation control system using fuzzy control

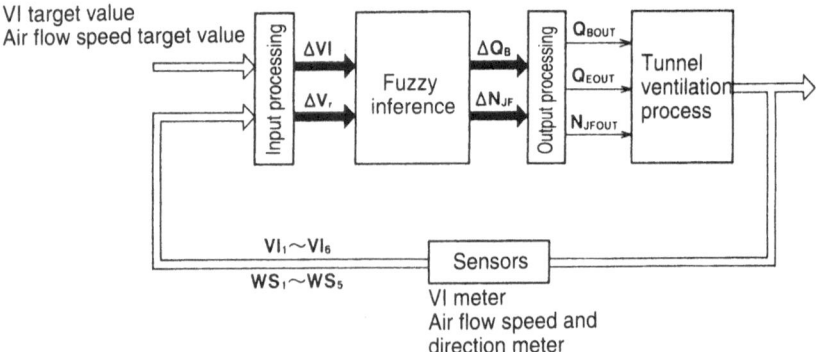

Fig. 10.9 Block diagram of fuzzy air flow speed and concentration control. For tunnel with jet fan longitudinal flow ventilation, vertical air intake and exhaust ducts, and dust collectors.

Here VI_i is the value (%) measured by the i'th VI meter; VI_{iref} is the target value (%) for the i'th VI meter.

In addition, a number of anemometers and wind vanes are installed. Facing toward Fig. 10.3, the air flow speed over the roadway from the left shaft opening (the upstream side shaft opening with respect to the ventilation air flow) to the vertical shaft and the air flow speed from the vertical shaft to the right side (downstream side) shaft opening are different; the average air flow speeds in the two intervals are computed. Then the air flow speed that is farthest below the target value is found from equation (5), and the control deviation ΔV_r of the air flow speed over the roadway is computed.

$$\Delta V I_r = \min(V_{r1} - V_{r1ref}, V_{r2} - V_{r2ref}) \tag{5}$$

Where:

V_{r1} is the average air flow speed (m/s) over the roadway in the upstream interval.

V_{r1ref} is the target air flow speed (m/s) in the upstream interval.

V_{r2} is the average air flow speed (m/s) over the roadway in the downstream interval.

V_{2ref} is the target air flow speed (m/s) in the downstream interval.

As shown in Fig. 10.9, the input variables to the fuzzy inference are the control deviation of visibility ΔVI and the control deviation of air flow speed over the roadway ΔV_r; the output variables are the the correction air flow transport in the vertical shaft ΔQ_B and the correction to the number of jet fans in operation ΔN_{JF}.

The fuzzy rules are described in an IF - THEN format. Some examples of rules which are used follow. In the following rules, Q_B is the air flow transport driven by the intake fans in the vertical shafts (the transport driven by the intake fans is taken as representative of the transport driven by both the intake and exhaust fans) and N_{JF} is the number of jet fans in operation.

Example 1:   IF $\Delta VI = NB_{VI} \& \Delta V_r = NB_{Vr}$
             THEN $\Delta Q_B = PM \& \Delta N_{JF} = PB$

(If the VI value and the air flow speed are fairly low, the transport driven by the intake and exhaust fans is medium and the number of jet fans in operation is substantially increased.)

Example 2:   IF $\Delta VI = PB_{VI} \& \Delta V_r = PB_{Vr}$
             THEN $\Delta Q_B = NM \& \Delta N_{JF} = NB$

(If the VI value and the air flow speed are fairly high, the transport driven by the intake and exhaust fans is medium and the number of jet fans in operation is substantially decreased.)

Example 3:   IF $\Delta VI = NM_{VI} \& \Delta V_r = Z_{Vr}$
             THEN $\Delta Q_B = PM \& \Delta N_{JF} = Z$

(If the VI value is low and the air flow speed is near the target value, the transport driven by the intake and exhaust fans is increased moderately, and the number of jet fans in operation remains unchanged.)

Example 4:   IF $\Delta VI = Z_{VI} \& \Delta V_r = NB_{Vr}$
             THEN $\Delta Q_B = Z \& \Delta N_{JF} = PM$

(If the VI value is near the target value and the air flow speed is fairly low, the transport driven by the intake and exhaust fans is kept constant, and the number of jet fans in operation is increased moderately.) Here we use the notation:

PB : positive big
PM : positive medium
Z :   zero
NM : negative medium
NB : negative big

Subscript$_{VI}$: Indicates a fuzzy expression relating to the VI value.

Subscript$_{V_r}$: Indicates a fuzzy expression relating to the air flow speed over the roadway.

10.3.5 Level control

Level control is a function provided to respond to increases of carbon monoxide concentration and sudden drops in the VI value. Control must be applied at even shorter intervals than in fuzzy air flow speed and concentration control. When the ventilating machines are operated by this control, they cannot be operated again for a certain time (the time needed to wait for an effect to occur). This time is on the order of 5 to 10 minutes.

10.3.6 Emergency control

Emergency control is provided to respond to sudden increases in pollutant concentration that cannot be dealt with by the above - mentioned fuzzy air flow speed and concentration control or level control. If the measured pollutant concentration remains in excess of a set danger level for a certain time, the ventilating machine air flow transport and number of ventilating machines in operation are increased in steps to return the concentration to normal as quickly as possible. This control is applied at intervals of tens of seconds. As in the case of level control, there is a prescribed time during which the system waits for an effect to occur.

10.4 Results of applying this system

The highway tunnel ventilation control system described above was applied in a group of tunnels on the Hokuriku Expressway starting with the Koshirazu Tunnel and the Ichiburi Tunnel, and in the Higo Tunnel between Yatsushiro and Hitoyoshi on the Cross - Kyushu Highway, and good results were obtained. Here the results obtained in the Higo tunnel are presented.

The Higo tunnel is 6,340m long; it is the 3rd longest expressway tunnel in Japan, after the Kan-Etsu Tunnel and Ena San Tunnel. It was opened to traffic in December 1989. At present it carries 2-way traffic; future construction of a separate tunnel for northward - bound traffic is planned. The traffic volume averages 10,000 vehicles per day. A diagram of the Higo Tunnel is shown in Fig. 10.10. The ventilation system is a jet fan longitudinal flow system with vertical intake and exhaust shafts and dust collectors, as shown in Fig. 10.3 above.

The membership functions and other control parameters used for the fuzzy air flow speed and concentration control were set in advance by simulation before the on-site test; when the actual on-site test was performed, the fluctuations of the intake and exhaust fan transport and the number of jet fans in operation tended to be too large. In particular, the intake and exhaust fan transport exhibited hunting. Therefore, adjustments were performed on-site, for the purpose of stabilizing the air flow.

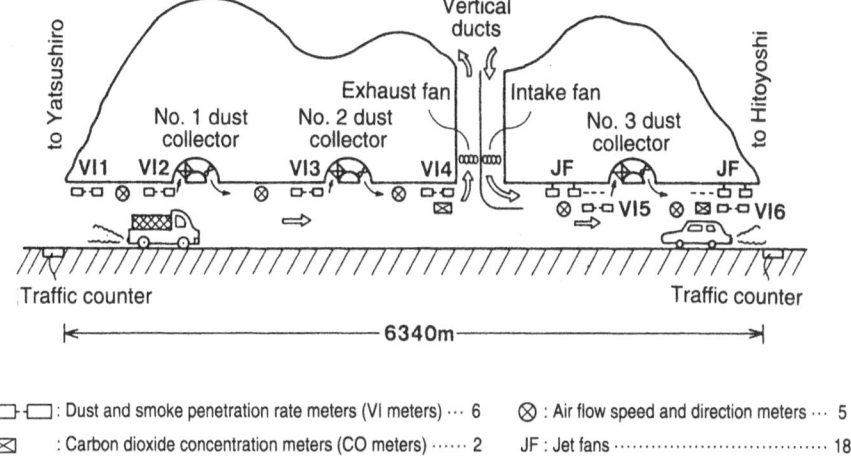

Fig. 10.10 Outline of system in Higo Tunnel.

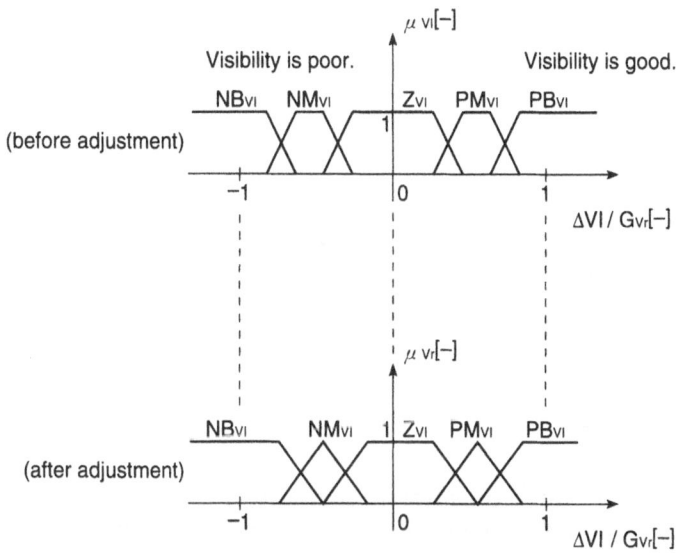

Fig. 10.11 Membership function with respect to input variable ΔVI.

The first thing that was tried in an attempt to stabilize the control was to reduce the control gain. If the VI target value is set a bit high (for example 60%), there is some leeway with respect to the allowable VI value (30%), so it was expected that even if the control response became slower as a result of reducing the gain, there would not be such a great problem. However, the VI target value in the Higo Tunnel is set to 50% to reduce the electric power required. Consequently, reducing the control gain resulted in too long a delay from the time the VI value fell until the time it recovered. Therefore, in addition to the gain, the fuzzy control membership functions were adjusted

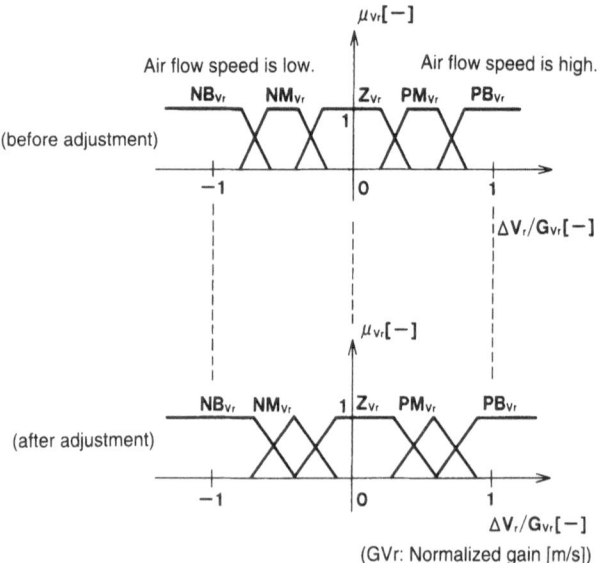

Fig. 10.12 Membership function with respect to output variables.

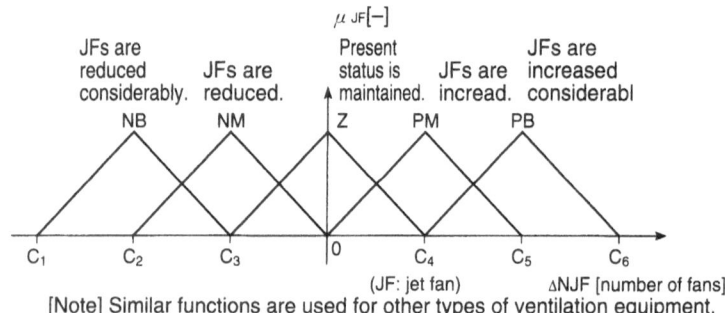

[Note] Similar functions are used for other types of ventilation equipment.

Fig. 10.13 Results of ventilation control in Higo Tunnel.

in an attempt to stabilize control without losing responsiveness when the VI value drops. The policy followed in adjustment was as follows.

1) If the VI value is above the target value, the gain is reduced to stabilize control.

2) If the VI value is near the target value, no change is made in ventilating machine operation (to avoid frequent turning of ventilating machines ON and OFF).

3) If the VI value is below the target value, the gain is controlled in a way that emphasizes responsiveness.

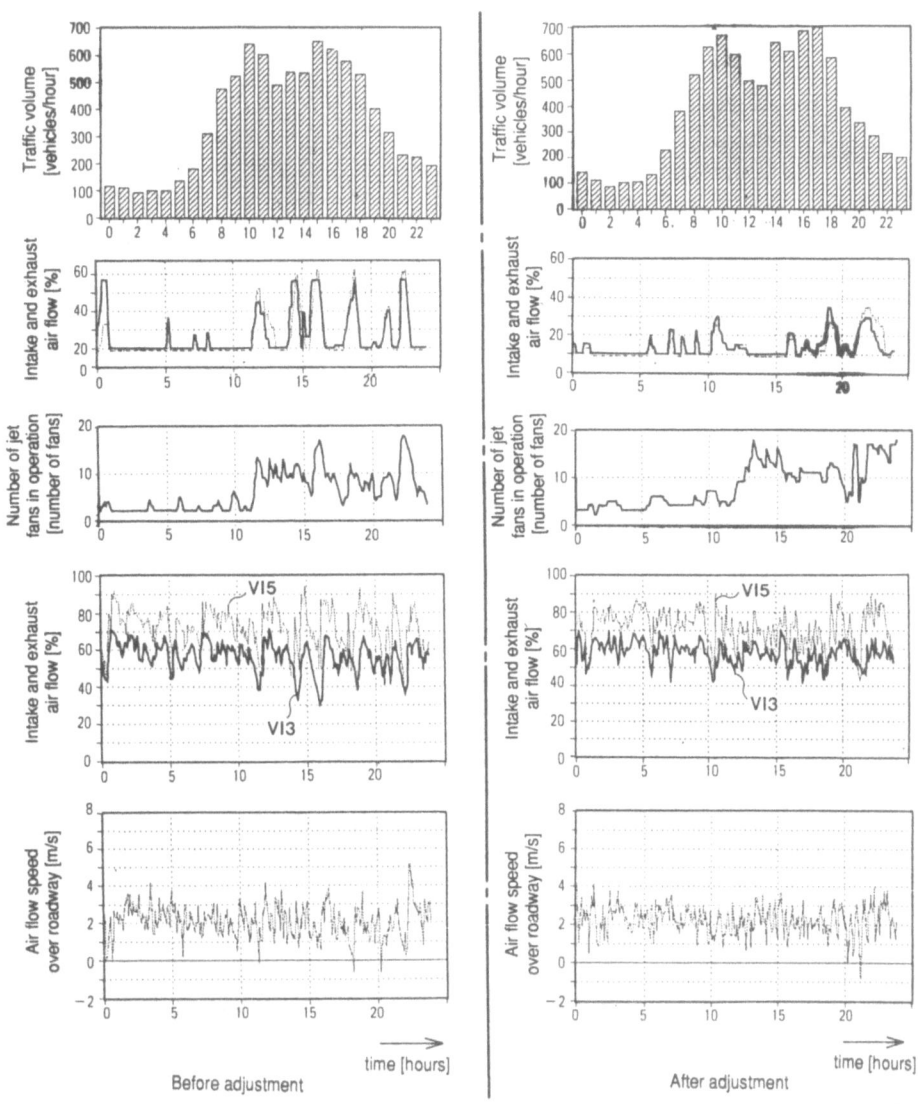

Fig. 10.14 Results of ventilation control in Higo Tunnel.

The membership function with respect to the input variables before and after adjustment are shown in Figs. 10.11 and 10.12. The membership function with respect to the output variables is shown in Fig. 10.13. As shown in these figures, the input and output variables are normalized. The membership functions after adjustment were shifted to the right a bit compared to those before adjustment; this is to speed the response up when the VI value and air flow speed are lower than the target values, and slow it down when they are higher than the target values. An example of the control results is shown in Fig. 10.14. The left - side chart shows the control condition before the membership functions were adjusted; the right - side chart shows the control condition after adjustment. It is seen that compared to the situation before adjustment, fluctuation of the intake and exhaust fan transports has been greatly suppressed. Fluctuations of the VI value were also smaller after the adjustment. Thus, stable control of the VI value was obtained by adjusting the membership functions.

10.5 Future problems

In the tunnel ventilation process, there is a long delay from the time the pollutant concentration increases until the increase is detected by the sensors. In addition, it takes a long time after control is applied to the ventilating machines until the concentration is restored to normal; the control response varies with conditions such as tunnel length but sometimes takes up to tens of minutes. Consequently, it is difficult to obtain adequate response with feedback control alone; it is necessary for effective feedback control to be developed. If the responsiveness is improved, the magnitude and frequency of pollutant concentration increases can be reduced, making it possible to lower the VI target value below its present level, leading to reduction of the amount of electric power required.

Tunnel process characteristics include secular and seasonal variations, so long - term data collection and analysis are necessary. In the Higo Tunnel, a system has been introduced that permits field data to be gathered by a personal computer at a remote location through a telephone line, making follow-up studies possible.

In addition, a problem remaining for the future development of tunnel ventilation control is to add a learning function to the control system so that it can respond to changes in process characteristics.

References

1. Watanabe T, Koyama T, Miyoshi M, Yoshino N, Aoki I, Kikuchi H (1990) "AI and fuzzy" -based tunnel ventilation control system. Int Conf Fuzzy and Neural Networks 1 : 71–75

2. Miyoshi M, Koyama T, Watanabe T, Yoshino N, Shinohara M, Aoki I (1991) Road tunnel ventilation control system. In : Proceedings of the International Fuzzy Systems Association fourth world congress, Brussels : pp137–140

3. Nagataki K, Koyama T, Miyoshi K, Shinohara M, Yoshino N, Iida K (1988) Highway tunnel ventilation control by fuzzy control (in Japanese). 1988 National Conference of Electrical Soc, No. 1577.

4. Nagataki K, Miyoshi K, Koyama T, Ezure H, Shinohara M, Adachi T (1989) First report on a tunnel ventilation operation system using artificial intelligence (in Japanese). 1989 National Conference of Electrical Soc, No. 1733.

5. Yoshino N, Miyoshi K, Adachi T (1989) Second report on a tunnel ventilation operation system using artificial intelligence (in Japanese). 1989 National Conference of Electrical Soc, No. 1734.

6. Aoki I, Koyama T, Miyoshi K, Yoshino N (1990) A tunnel ventilation operation system using artificial intelligence and fuzzy technology (in Japanese). Factory Automation 8 (2) : 23–27

7. Miyoshi K, Aoki I, Koyama T, Watanabe T, Shinohara M, Yoshino N (1990) A highway tunnel ventilation control system using fuzzy control (in Japanese). Automation 35 (4) : 93–97

8. Miyoshi K, Aoki I (1990) A highway tunnel ventilation control system using artificial intelligence and fuzzy control (in Japanese). Report of Electrical Soc Joint Study Group on Industrial Measurements and Control and Incorporation of Information Technology into Industrial Systems, IIC–90–13.

Chapter. 11

FUZZY CONTROL AND EXAMPLES OF APPLICATIONS

11.1 Introduction

Fuzzy theory is a theoretical framework having fuzzy sets, fuzzy logic and fuzzy measure at its core, which started with the proposal of the concept of fuzziness and its expression in the form of fuzzy sets by Prof. Lotfi A. Zadeh of the University of California at Berkeley in 1965[1]. When this theory is used, the subjective vagueness possessed by humans can be adequately expressed and processed [7, 12].

This theory was first used in such applications as automatic train operation and plant process control [2, 11], resulting in high quality, high efficiency and energy saving. Subsequently, it has found such additional applications as elevator group control, automobile control and household appliance control, in which a smooth relationship between a computer (or controller) and a human trying to operate the system is necessary [14], and applications such as finance and stocks in which a computer supports a human's judgment to assist decision making [13].

This chapter discusses trends in product markets and technology, and gives some examples of application of Hitachi's fuzzy systems.

11.2 Trends in markets and technology

11.2.1 Trends in markets

In the midst of great changes, such as the movement of cities toward 24-hour activity and globalization, markets are becoming more diversified and individualized, and there is a rapid shift of power from the manufacturer's logic to the user's logic. This is sometimes called the age of "light, thin, short and small"; this term arose as a reaction to heavy, thick, long and large. This is still manufacturer's logic. The user wants something that is fun to use, pleasant to look at, and convenient; to use another expression, the user

Table 11.1   Market and technological trends

Product needs	• Mass production: few types of products in large quantities • Heavy, thick, long, large to light, thin, short, small		•Flexible manufacturing: many types of products in small lots • Beautiful, good feeling, fun to use, creative (beautiful, good feeling, quiet, graceful)
Receiver of main benefit	• Maker's Profit		• User's Benefit
System	•Fixed type/calculation system • Closed Loop	⇒	• Flexible type/information system • Open Loop
Processing control system	• PID control (proportion + integration + differentiation) • Order type • Centralized system: main frame		• Knowledge-based control (AI, fuzzy) • Instruction type • Decentralized system: DB + network + WS

is seeking something that is individualized, beautiful, and has sensitivity, playfulness and imagination. That is to say, a product that is "beautiful, sensitive, playful and imaginative" will sell. As a result, we are in an age in which the demand for products fluctuates wildly, and the products have short lifetimes.

In order to keep up with the demands of this new age, it is necessary to develop control technology and information processing technology for flexible manufacturing systems which produce many types of products in small lots to respond to rapid market changes. Looking at this change in requirements from a systems point of view, previously the need was for control to reach fixed target values, and one was dealing with closed loop systems having fixed boundary conditions (this is called the good structure problem); in contrast, recently we have systems that are driven by information, the format is not fixed and moreover we are dealing with an open loop problem in which the boundary conditions are not clear (this is called the bad structure problem). In the bad structure problem, it is difficult to formulate a mathematical model, and problems must often be solved by using a skilled worker's knowledge and experience. In addition, whereas previously it was possible to solve problems with only knowledge of the natural sciences, we are now entering an age in which knowledge of the social sciences and humanities as well as natural sciences is required.

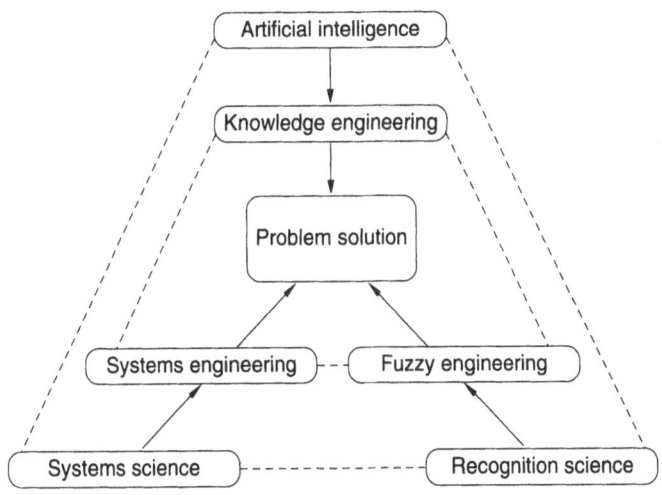

Fig. 11.1 Position occupied by fuzzy engineering.

Whereas the form of previous system products permitted treatment as good structure problems and embodiment as contralized type systems, the form of present and future system products requires dealing with bad structure problems and the trend is toward embodiment as down-sized decentralized systems. Decentralized systems consist of personal computers, work stations, local area networks (LANs), etc., and it has become important to process not only numerical data but also multimedia information such as natural images, graphics and knowledge data (Table 11.1).

11.2.2 Technological trends

The conceptual trend in problem solving has been from systems science to artificial intelligence (AI), and is now progressing toward recognition science. As for the approach to applying part of this learning to the world of industry, systems engineering, which is part of systems science, is used to solve the good structure problem, while knowledge engineering (KE), which is part of AI, and fuzzy engineering, which is part of recognition engineering, are used to solve the bad structure problem (Fig. 11.1). Systems engineering is suitable for deductive problems for which the formulation of a mathematical model is easy; AI and fuzzy engineering are suitable for inductive problems for which formulation of a mathematical problem is difficult but which can be solved using an experienced operator's knowledge.

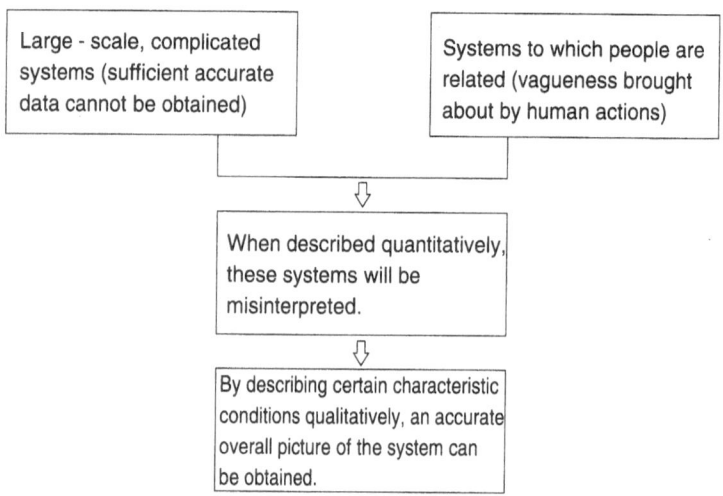

Fig. 11.2 How systems are understood in fuzzy engineering.

Bad structure problems to which fuzzy theory is applied include large-scale complicated systems and systems which involve human participation in some way. The large-scale complicated systems include many cases in which adequate accurate data cannot be obtained. Systems which involve human participation are characterized by including "vagueness" brought about by human actions. For this reason, there is a danger that if the system is described "quantitatively", the description will be inaccurate. However, by describing characteristic conditions "qualitatively", an accurate overall picture of the system can be obtained (Fig. 11.2). What makes it possible to express this "qualitative" sensitivity by a membership function as a degree of satisfaction is fuzzy theory. Note that the sensitivity of "vagueness" involves understanding and concepts which are common to the human participants, but this concept is often unintelligible to non-participants. Fuzzy theory is a subjective method of expression.

The application of fuzzy theory to control is fuzzy control. Seen from the point of view of control technology, "knowledge control" such as AI control and fuzzy control can be called 3rd phase control. 1st phase control is "classical control" typified by PID (proportion - integration-differentiation) control; 2nd phase control is "modern control" typified by state variable description. The next phase of "knowledge control", 4th phase control, is expected to lie in the direction of "intelligent control" typified by neurocomputing which is highly capable of learning and pattern recognition.

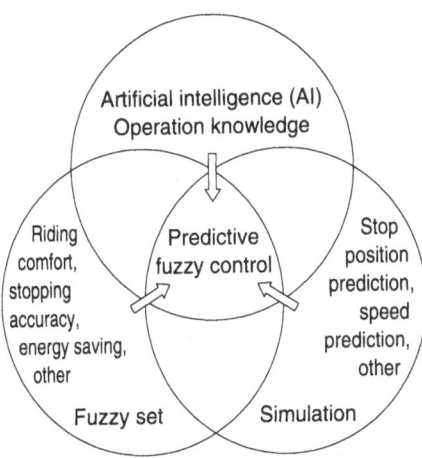

Fig. 11.3 The concept of predictive fuzzy control.

11.3 Skilled operator's operation and fuzzy control system

11.3.1 Skilled operator's operation

Conventional PID control systems are suitable for systems having tracking operation in which modeling of the object of control is easy, the dynamic range is narrow, and either the target value is fixed or a target value pattern is given and tracking is performed so that the controlled quantity acquires this value. However, PID control is difficult to apply to problems in which the characteristics of the object of control change with conditions and external conditions are dynamically changing; there are many control objectives; and moreover their weights vary with time. For such a problem, it often happens that a skilled operator can provide higher quality control.

A skilled operator handles a multi-purpose system by qualitatively judging the characteristics of the object system from past operating experience, and at the same time accumulates operating know-how as control knowledge which is then used to provide skillfully balanced control of the overall system.

11.3.2 Fuzzy control systems

Much control knowledge includes vagueness and is therefore suited for application of fuzzy theory. In addition, a skilled operator can perform control while looking well ahead, and avoid superfluous operations. These features have raised the technical problem of a new control system that performs fuzzy control while making predictions. The prediction function can be performed

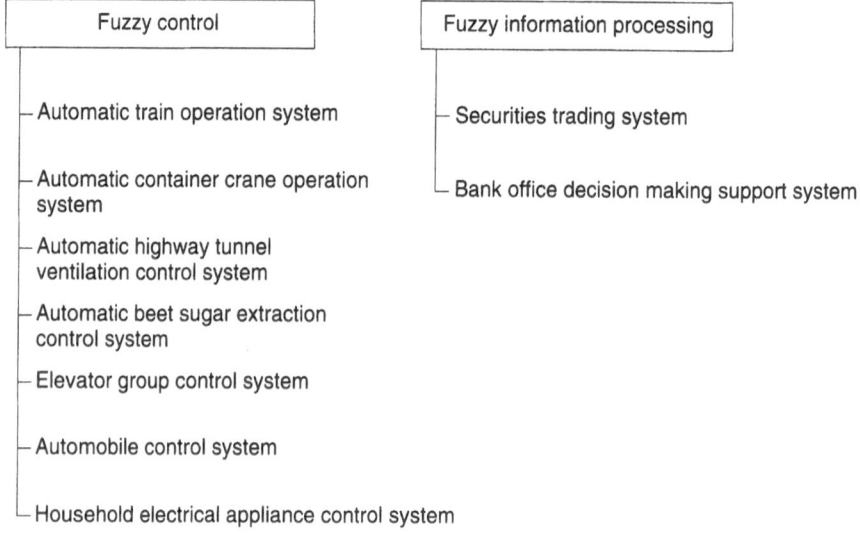

Fig. 11.4 Examples of fuzzy control application systems.

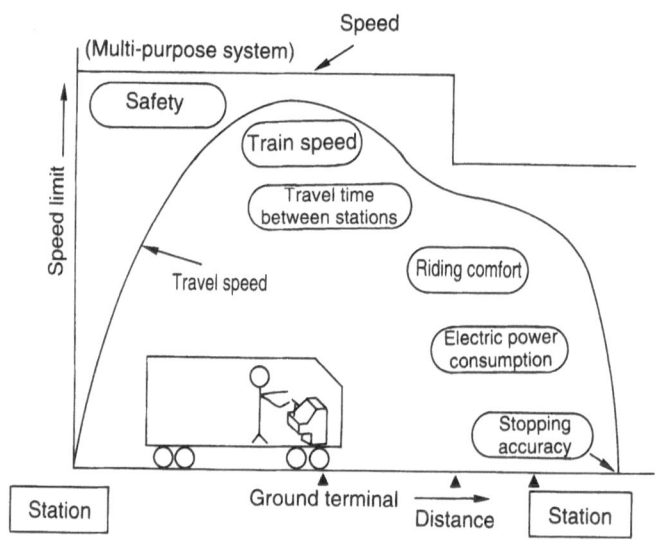

Fig. 11.5 Purpose of an automatic train operation system.

by simulation, so a predictive fuzzy control system has been developed for performing fuzzy control with simulation (Fig. 11.3). A predictive fuzzy control system has the following 3 characteristics. (1) The control knowledge can be expressed by a small number of IF $\cdots$ THEN $\cdots$ rules and a membership function that expresses the vagueness contained in those rules in functional form. (2) Developments in the somewhat distant future that will result from choices of the controlled quantities can be predicted by the simulator. (3) The optimum controlled quantity can be computed from a number of choices at high speed.

This control system has found a number of applications: to an automatic train operation, a container crane automatic operation system, a highway tunnel ventilation control system, a beet sugar extraction control system, an aluminum cold rolling control system, etc., giving control that maintains high quality, high productivity and low energy consumption. Subsequent to these applications, this control system then found new uses which require greater user friendliness such as an elevator group control system, an automobile control system, and household electrical appliance control systems; and, in the information processing field, a management strategy / tactical base decision making support system (Fig. 11.4).

11.4 Examples of applications of predictive fuzzy control systems

11.4.1 Application to an automatic train operation system

The purpose of this system is to obtain automatic operation on a par with that of a skilled operator; specifically this means multi-purpose operation that balances several requirements such as riding comfort, stopping accuracy and energy conservation (Fig. 11.5). The automatic train operation starts in response to a departure signal. The speed between stations is controlled so as not to exceed the speed limit, then at the next station the train is stopped accurately at the stop position through automatic application of driving control and a brake control command (notch). This automatic train operation system is called Fuzzy-ATO (Automatic Train Operation). The ATO operates the train automatically except when the ATC (Automatic Train Control) system, which operates the emergency brake to maintain safety, takes over (Fig. 11.6). The critical speed at which the ATC takes over is determined by the distance to the next train ahead; the ATO performs smooth automatic operation under that limiting speed.

The operating knowledge used by the ATO is described with 24 control rules for automatic operation such as "if it appears that the train will stop smoothly in the present notch condition, do not change the notch" and "if it appears that applying the brake a little bit harder will increase comfort and stopping accuracy, then apply the brake a little bit harder".

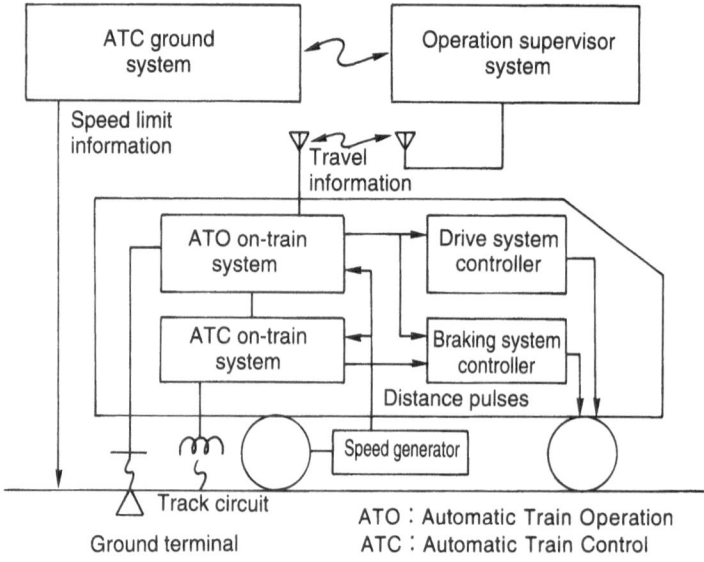

Fig. 11.6   Automatic train operation system composition.

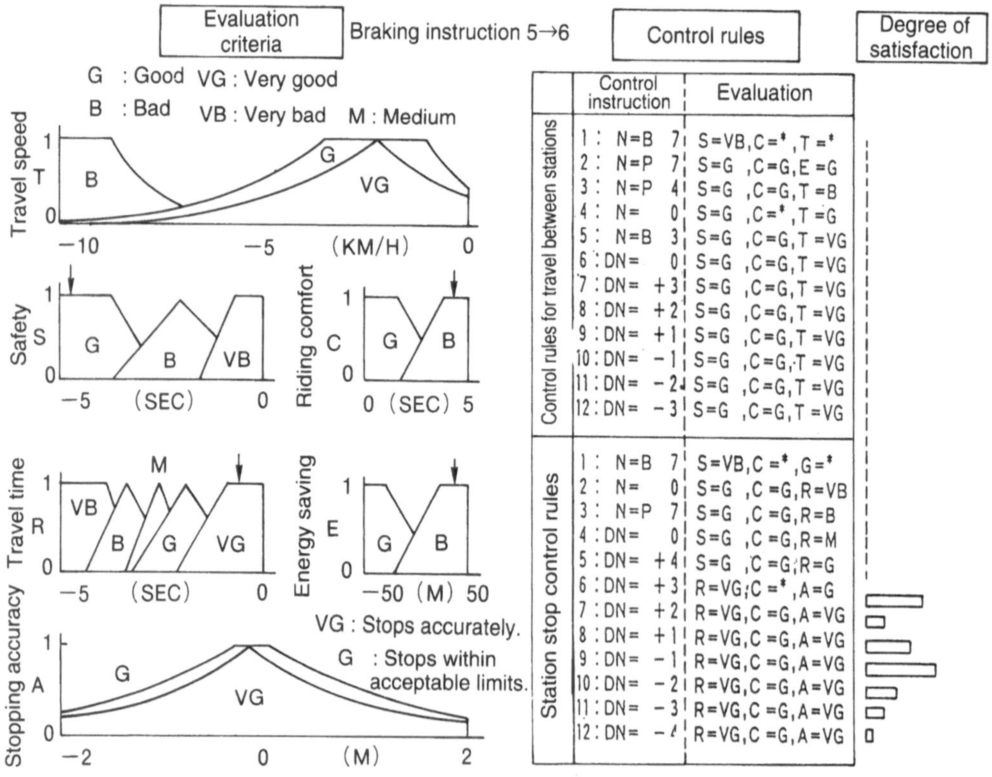

Fig. 11.7   Conditions for determination of control instructions by predictive fuzzy control.

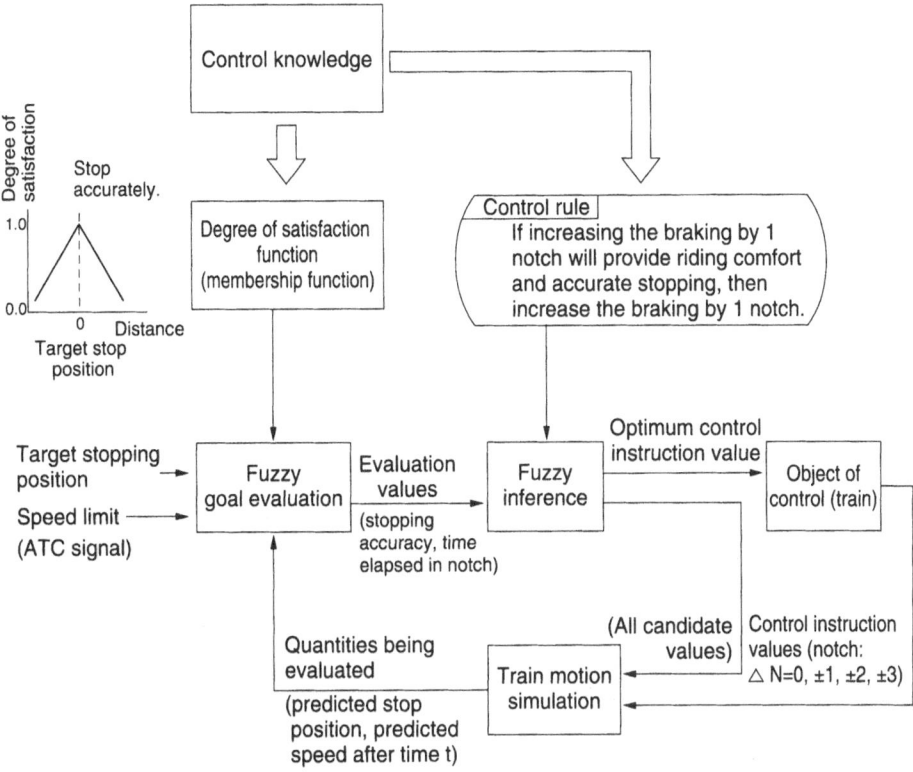

Fig. 11.8 Block diagram of predictive fuzzy control.

These control rules contain such vague expressions as "good riding comfort", "smooth stop", "stop accurately" and "a little bit harder"; these can be described by membership functions. This automatic train operation system has multi-purpose functions with 6 fuzzy indices; these multiple purposes are evaluated in real time and the optimum control commands determined (Fig. 11.7). The optimum control command (optimum notch) is determined by real time simulation prediction of the stop position corresponding to the (assumed) control command that can be selected by each control rule by the simulator which models the train's operating characteristics, followed by fuzzy inference based on these predicted values.

A block diagram of predictive fuzzy control is shown in Fig. 11.8, examples of its inference method and dynamic action in Fig. 11.9. In overall train operation, this system responds flexibly to changing conditions, incorporating inertial control involving for example shutting the notch off to save energy and maintain the required speed between stations while staying under the speed limit; at the same time during stop position control the speed is decreased according to the distance to the train ahead. A predictive fuzzy

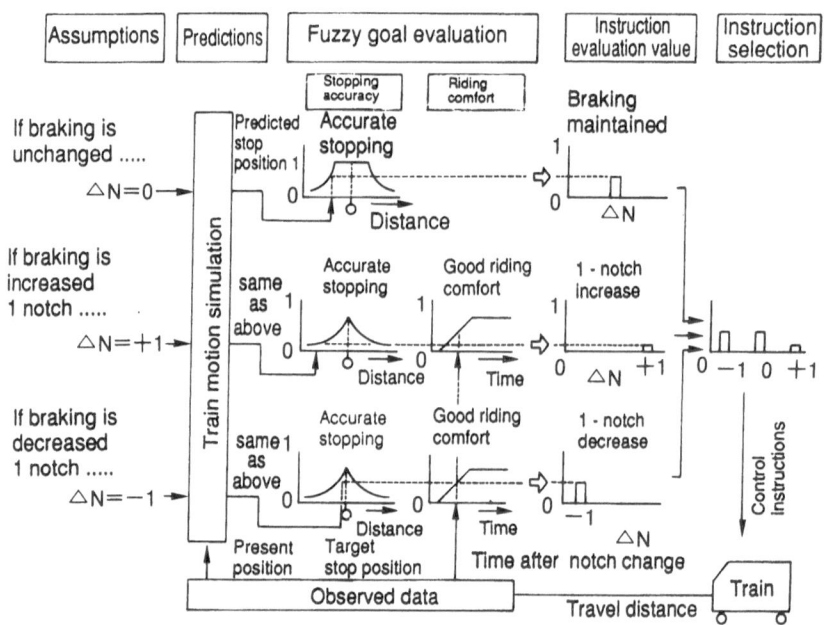

Fig. 11.9 Predictive fuzzy control inference process.

Fig. 11.10 Example of application of automatic train

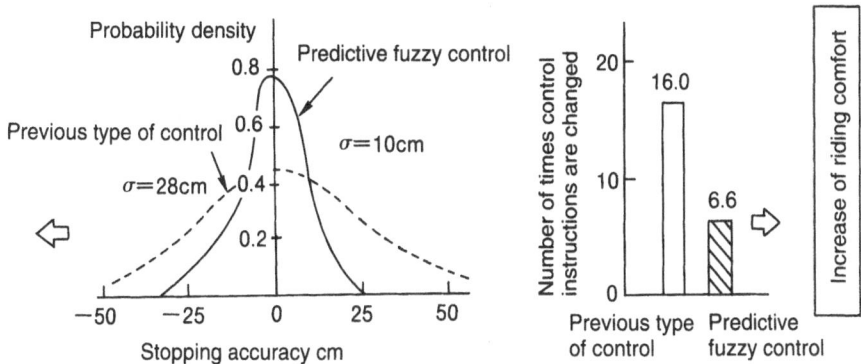

Fig. 11.11 Stopping accuracy simulation evaluation.

Fig. 11.12 Simulation evaluation of number of times control instructions are changed.

control system was put into actual use on the subway run by the Sendai Municipal Transit Authority in July, 1987 (Fig. 11.10). Before being put into use on the subway, automatic train operation by the predictive fuzzy control system was compared to a conventional PID control system by simulation. Since both AI and fuzzy engineering are forms of knowledge control, the problem is an open problem, reaches a solution or not one of whether it; the safety and optimalness of the system have not been theoretically proven. For this reason, evaluation by simulation before using the system is necessary. The results were of the following nature: (1) The standard deviation of the stop error was 1/3 (Fig. 11.11). (2) The frequency of notch variations was 1/3 (Fig. 11.12). (3) Simulated energy use was cut by 10%. It is clear that this system provided balanced control of riding comfort and energy saving. It can also be seen from the graphs of examples of actual runs with real trains that the system skillfully coped with changes in speed limit and incline grade; at the same time, in stopping at stations there were few changes of brake notch, and the operation provided good riding comfort (Fig. 11.13). Also, with regard to stopping accuracy, stopping tests done with 19-car trains running on all lines gave the following results after 11,395 stops (Fig. 11.14): (1) The average stop error was 3.57cm. (2) The standard deviation was 10.61cm.

11.4.2 Automatic container crane operation system

A container crane is a cargo handling machine that transfers shipping containers (a typical size would be 2.6m × 2.6m × 12.2m, weight 20 tons) between container vessel and trailer by lateral movement of a trolley on a crane track close to 30m above the ground and by raising and lowering a rope that hangs on the trolley. Fuzzy control that seeks to provide automatic operation equivalent to manual operation by a skilled operator has been applied to per-

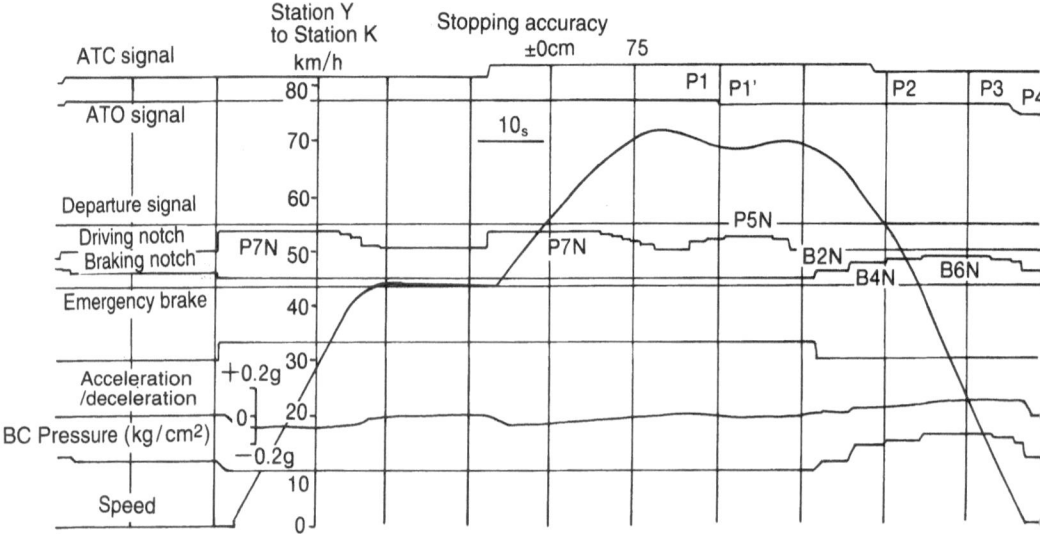

Fig. 11.13 Observed travel curves.

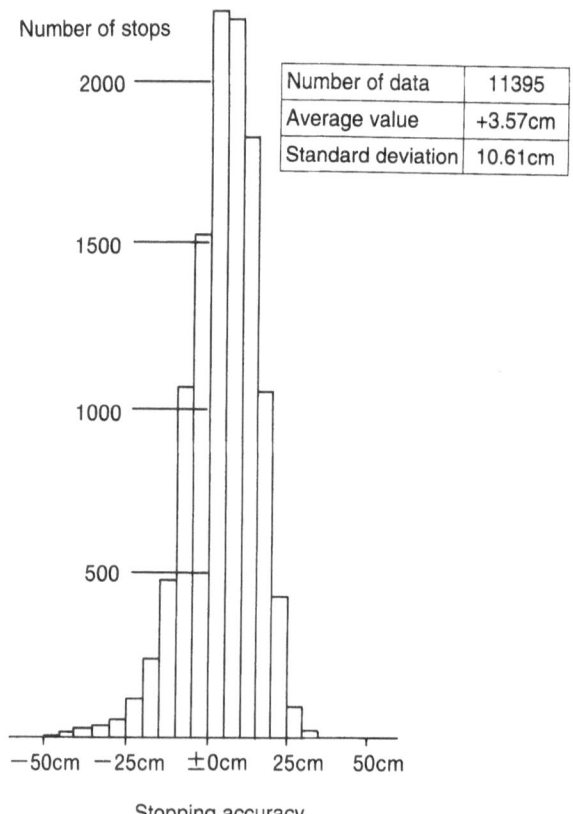

Number of data	11395
Average value	+3.57cm
Standard deviation	10.61cm

Stopping accuracy

Fig. 11.14 Observed stopping accuracy.

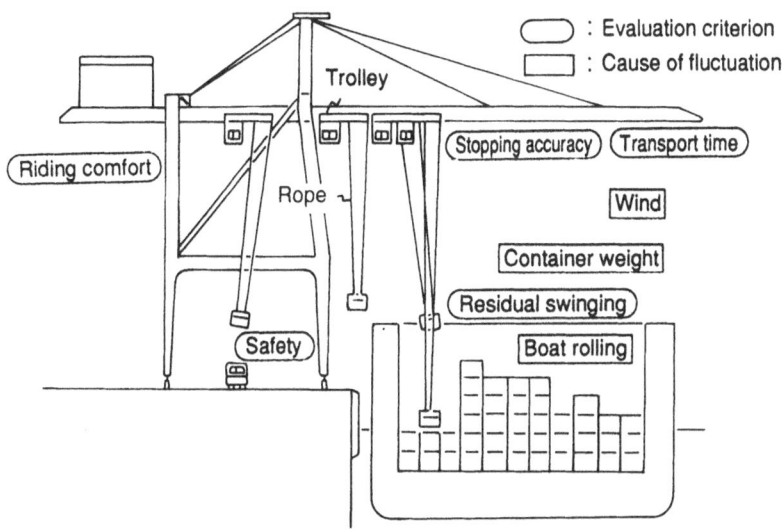

Fig. 11.15 Goals of container crane system.

form this cargo handling operation accurately and efficiently. The automatic operation makes it possible to bring a spreader, which is a metal device used to suspend the container, to the container to be handled as quickly and accurately as possible, and then to automatically transport the container that is grasped by the spreader to the target position. That is to say, it automates the process of minimizing the operation time, increasing the productivity of the cargo handling operation, and at the same time improving the trolley stopping accuracy and minimizing residual container swinging (Fig. 11.15). A container crane control system consists of the following two systems: (1) Lateral trolley speed control system (trolley system) (2) Rope wind-up speed control system (rope system). When a crane is operated manually, the operator watches the container's horizontal position and height, predicts various changes in his head, and gives target speed commands and brake commands to the trolley system and the rope system. The automatic system using fuzzy control automatically controls the crane by using the trolley position, rope length and shapes of obstacles as inputs, using a computer to predict various changes in lieu of an operator, and then outputs trolley system and rope system target speed commands and brake commands and trolley motor acceleration current settings. A functional block diagram is shown in Fig. 11.16.

In order to predict the container stop position and residual swinging, a numerical model that simulates the crane characteristics has been incorporated into the system. This model is used to predict the stop position and residual swinging for each possible set of operation commands in real time, permitting the optimum operation commands to be determined.

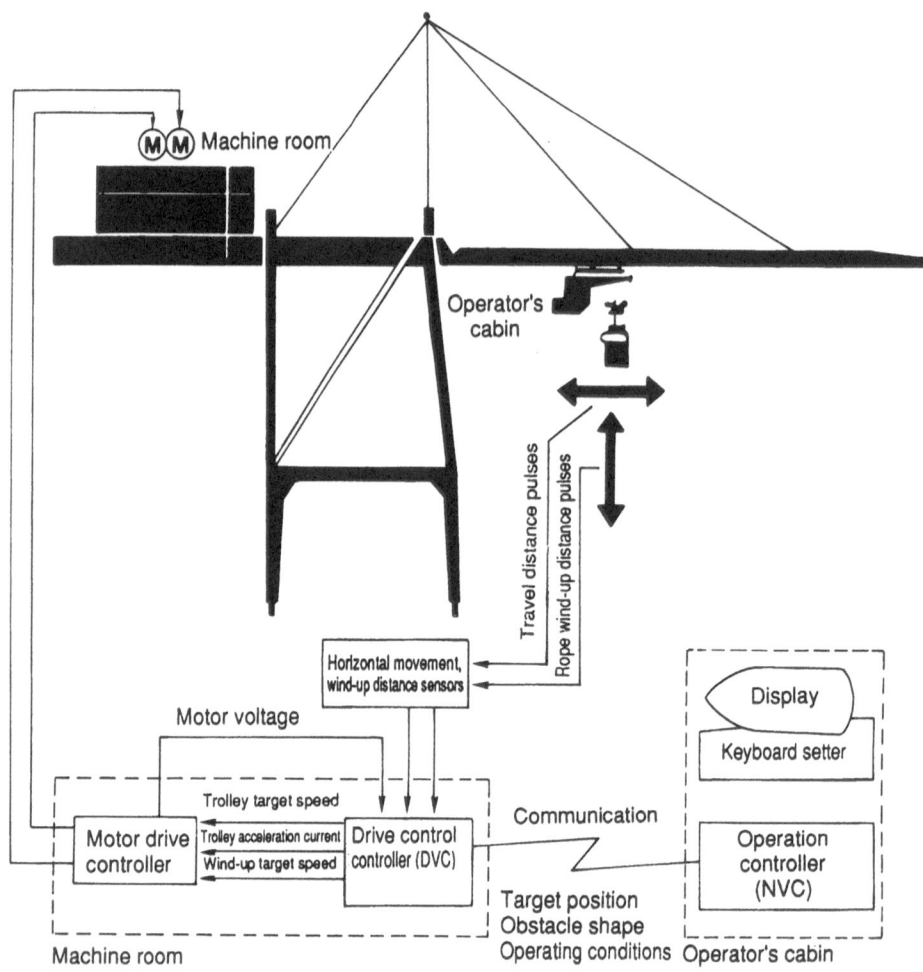

Fig. 11.16 Automatic container crane operation system.

The results of this automatic operation show considerable improvement over manual operation. The following results were obtained with an experimental crane that is a half-sized model of an operational crane (height 18m, span 50m, dummy container weight 6.45 tons). (1) The transport time averaged 47 seconds with a standard deviation of 1.0 second, compared to an average of 51.7 seconds and standard deviation of 4.3 seconds for manual operation. (2) The stop position accuracy was ± 5cm compared to ± 10cm for manual operation. (3) The residual swing was 5.8cm, compared to 10.8cm for manual operation. Subsequently, similar results were obtained with an actual operational crane, confirming that automatic operation results in increased productivity and good stopping accuracy.

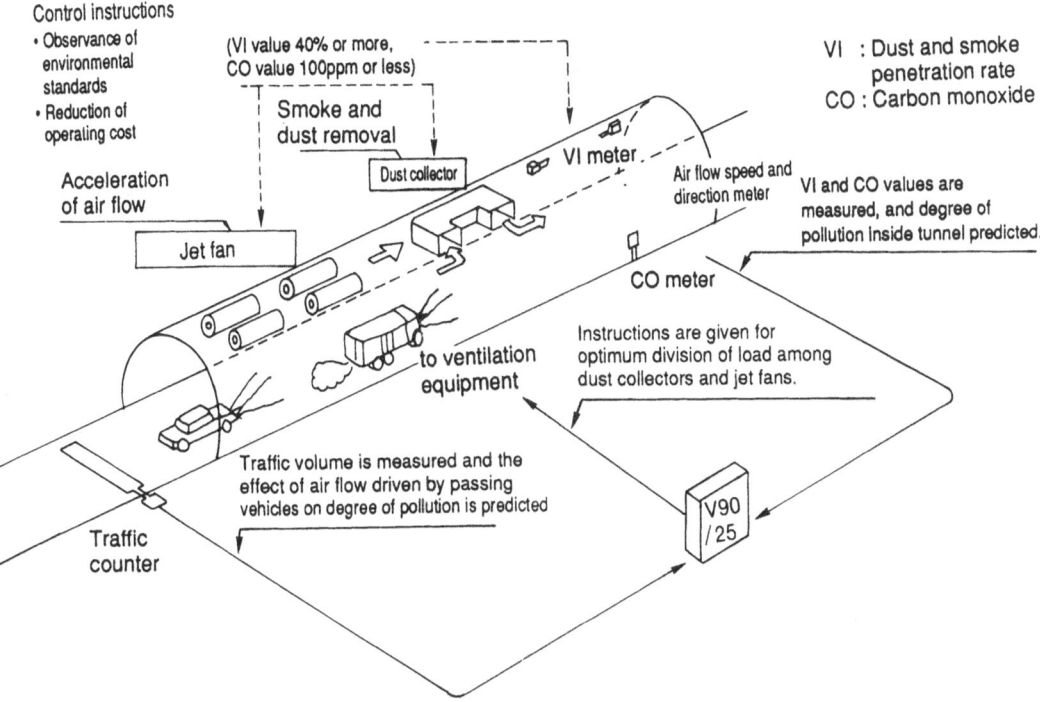

Fig. 11.17 Outline of a tunnel ventilation control system.

11.4.3 A highway tunnel ventilation control system

The purpose of tunnel ventilation control is to provide a safe and comfortable environment for the users (drivers). The necessary and sufficient amount of ventilation air flow for the traffic conditions in the tunnel at a given moment must be provided with minimum electric power consumption. In particular, increasing ventilation efficiency in long tunnels is important to control operating costs. The tunnel ventilation is optimized by controlling jet fans and dust collectors installed inside the tunnel (Fig. 11.17). The jet fans blow polluted air from inside the tunnel toward air exit ports. The dust collectors remove soot and smoke so that pollutant concentration inside the tunnel will not exceed the allowable level. Pollutant concentration inside the tunnel is measured by VI (Visibility Index) meters and CO (carbon monoxide) meters, while the numbers of large and small vehicles and their speeds are measured by traffic flow meters.

Ventilation control is based on this sensor information. The amounts of pollutants in exhaust gas, air flow driven by the vehicles and degree of pollution inside the tunnel are predicted, and optimized operation commands are given to the jet fans and dust collectors. "Optimum" means that pollutant

Input data
Inference processing
Output data
Traffic volume
Natural wind
Mechanically driven air flow
Degree of pollution (VI, CO)
Quantitative pollution prediction
Qualitative pollution prediction (fuzzy inference)
Pollution prediction from trend
Predicted values; overall judgment
Judgment on division of load among ventilation equipment
Control instructions
Number of jet fans
Air flow through dust collectors

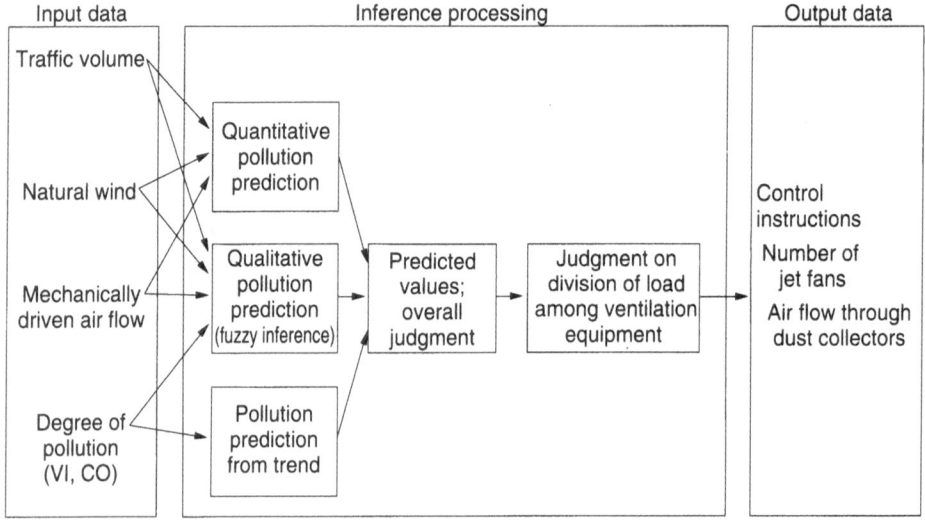

Fig. 11.18 Relationships among control functions in a tunnel ventilation control system.

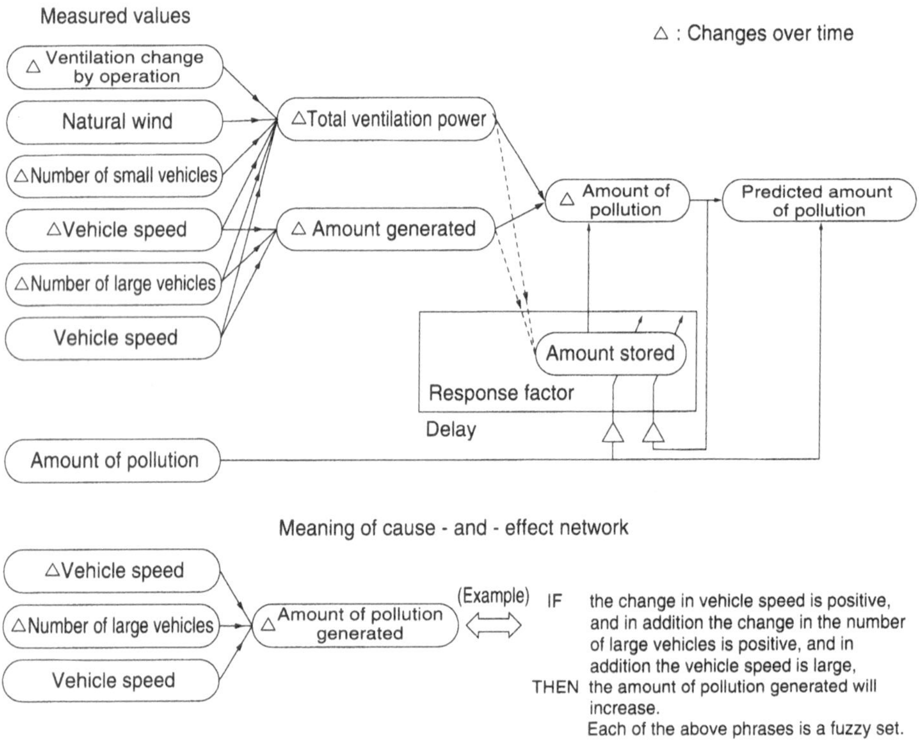

Measured values
△ : Changes over time
△ Ventilation change by operation
Natural wind
△Number of small vehicles
△Vehicle speed
△Number of large vehicles
Vehicle speed
△Total ventilation power
△ Amount generated
△ Amount of pollution
Predicted amount of pollution
Amount stored
Response factor
Delay
Amount of pollution
Meaning of cause - and - effect network
△Vehicle speed
△Number of large vehicles
Vehicle speed
△ Amount of pollution generated
(Example)
IF the change in vehicle speed is positive, and in addition the change in the number of large vehicles is positive, and in addition the vehicle speed is large,
THEN the amount of pollution generated will increase.
Each of the above phrases is a fuzzy set.

Fig. 11.19 A fuzzy model for predicting pollution inside a tunnel.

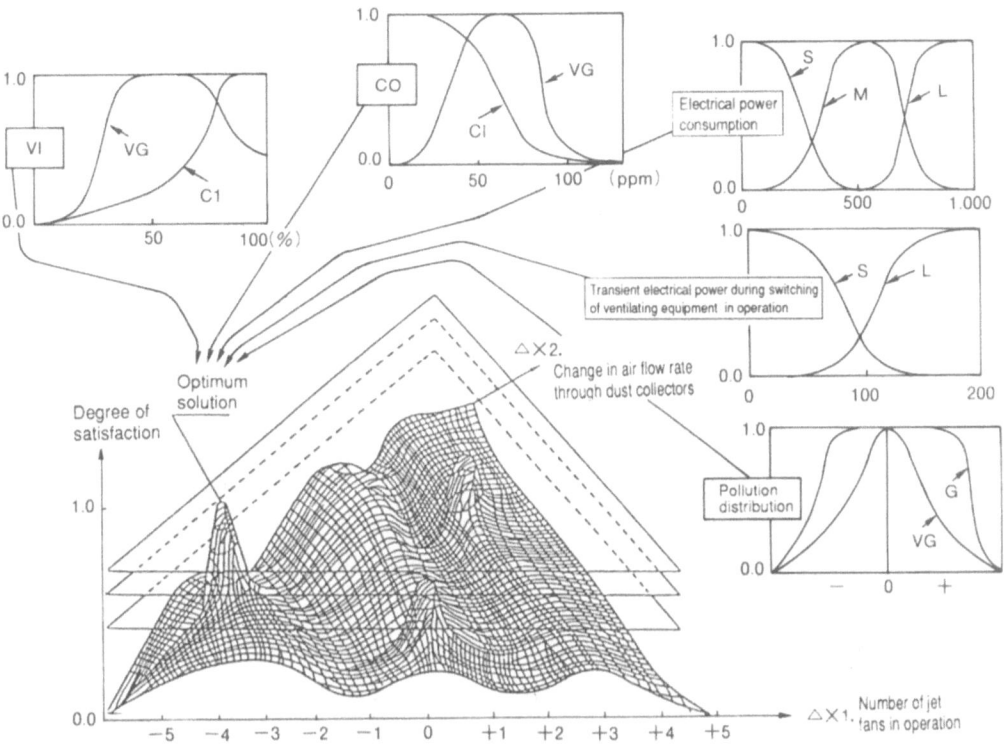

Note: Explanation of abbreviations VG: Very good, C1: Comfortable, S: Small, M: Medium, L: Large, G: Good

Fig. 11.20 Conceptual diagram of fuzzy inference for determining load distribution among ventilation equipment.

concentration is kept within the allowable limit, and at the same time electric power consumption is minimized. In the past, this control was performed using a quantitative numerical model; but the model failed to accurately account for a number of phenomena including turbulence inside the tunnel and emission of pollutants from vehicles, making it difficult to obtain optimum operation in which the electrical power consumption is minimized.

Since this is a process involving many elements which are difficult to quantify exactly, predictive fuzzy control was introduced to solve this problem. The control logic structure is shown in Fig. 11.18, and the pollution prediction model in Fig. 11.19. A conceptual diagram of the ventilation division of labor determination fuzzy inference is shown in Fig. 11.20. By introducing predictive fuzzy control, it was made possible to greatly reduce electric power consumption while keeping the degree of pollution within the allowable limit.

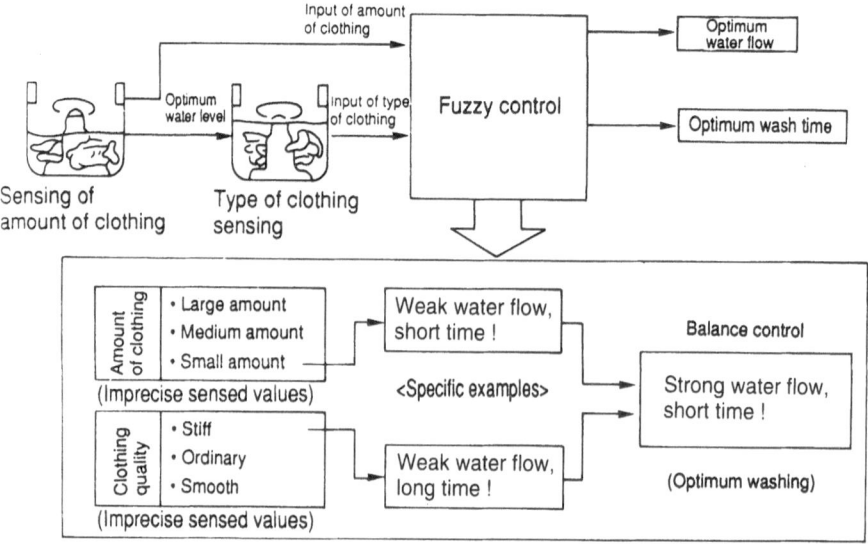

Fig. 11.21   Outline of a fuzzy fully automatic washing machine.

11.4.4   A fuzzy fully automatic washing machine

Fuzzy control technology has been moving from traffic, plant and process control into the areas of mass - produced products such as automobiles and household electrical appliances. In the midst of changes such as internationalization of urban activity and information usage, and extension of daily living activities around the clock, the household electrical appliance market has become more diversified and individualized, and we are rapidly moving into an age in which emphasis is shifting from the manufacturer's logic to the user's logic. Here we discuss a fuzzy fully automatic washing machine as a typical example of a fuzzy household electrical appliance. As living patterns change, washing machine usage is also changing as follows:

(1) From washing during daylight hours on a day with good weather to washing at any time of day or night without regard to weather.

(2) From washing clothes after they get dirty to washing them after wearing them once.

(3) From the housewife doing all the washing to the whole family sharing the responsibility.

For these reasons, the amount of clothes being washed and the frequency of washings are increasing, and it has become more important than before to wash clothes without damaging them and to simplify the operation. Against this background, the problem of optimizing washing machine control for a given type and amount of fabric has been formulated as the following specific technical problems.

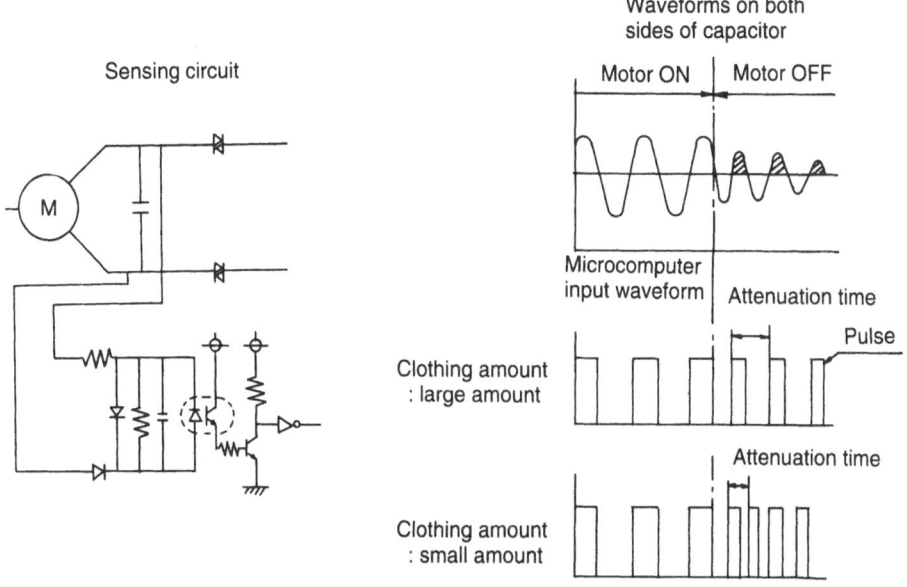

Fig. 11.22 Clothing amount and clothing quality sensor detection circuit and mechanisms.

(1) Development of fabric type and amount sensors (soft sensors that measure the reverse electromotive force of the motor).

(2) Development of one-button operation for the fully automated process (press one button and the machine does the rest).

(3) Development of control that optimizes the water flow strength and washing time according to the fabric type and amount (fuzzy control system for a fully automatic washing machine).

(4) Development of a man-machine interface to visualize the operation (liquid crystal display).

As in the case of the automatic train control system, the fully automatic washing machine control must allow for one button operation while balancing conflicting requirements such as minimizing damage to fabric, increasing washing power and minimizing time. For this reason, it was necessary to develop sensors for fabric type and amount, since this information is necessary for fuzzy control, and control technology to optimize the water flow strength and washing time based on that information (Fig. 11.21).

Fabric type and amount are determined by a 2-stage soft sensor (Fig. 11.22). First the fabric amount is measured by measuring the water level in the tank, running the motor a bit and then stopping it, and measuring the reverse electromagnetic force produced at that time by the fabric resistance.

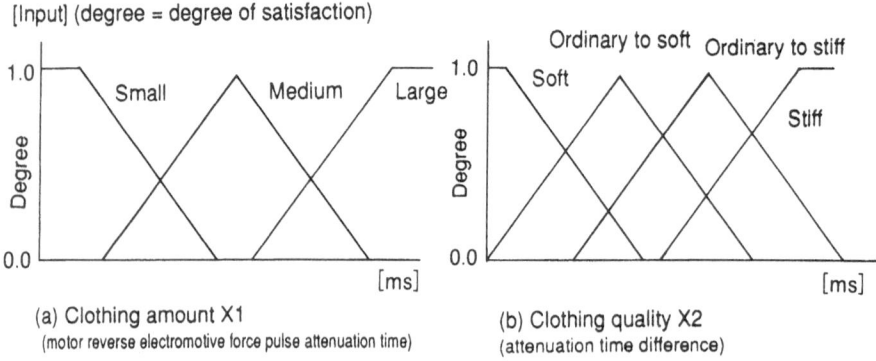

Fig. 11.23 Sensing input membership function.

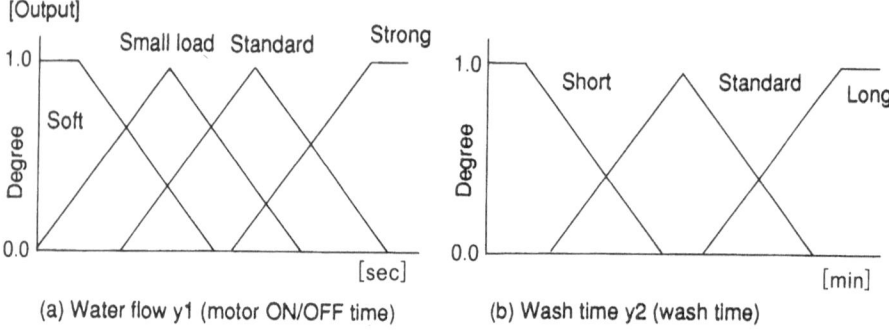

Fig. 11.24 Control output membership function.

The actual measurement involves converting the voltage waveform between the two terminals of a capacitor connected to the motor to a pulse, and measuring the attenuation time (T). To determine the fabric type, the water level that corresponds to the fabric amount (optimum water level) is determined and controlled, then it (fabric type) is estimated by a similar method, and the pulse attenuation time difference ((DELTA)T) is measured.

In practical application of this system, the motor reverse electromotive force pulse attenuation time and its time difference, which are related to the fabric amount and type respectively, are used as fuzzy measurement inputs (Fig. 11.23), and fuzzy inference is used to determine control commands, the fuzzy control outputs being water flow strength and washing time (Fig. 11.24). This has resulted in a fuzzy fully automatic washing machine "quiet machine with one button operation" system which optimizes the water flow, washing time, rinsing time and spin wring-out time in accordance with the fabric amount and type input information (Fig. 11.25).

Fig. 11.25 A fuzzy fully automatic washing machine that can be operated by pushing one button.

11.5 Future expectations

Really sophisticated applications of fuzzy theory still remain to be developed, and there are great expectations for it. In the past, most applications were to machine systems, mainly control, but in the future many additional applications are expected, including application to man-machine systems such as diagnosis of facilities, various types of optimized planning and management, decision making support, information search and image and voice recognition; also application to human systems such as human behavior models, human relations models, various types of assessments and market models (Table 11.2). In addition, by combination of fuzzy research with other fields such as neuro research processing of comprehensive and intuitive knowledge will be made easy, opening the way to application to higher level problems (Fig. 11.26).

	Social, management	Information processing	Control
Machine systems	• Safety and maintenance systems • Production control	• Image and voice recognition • Character recognition	• Train control • Automobile control
Man systems	• Direction and behavior models • Demand model	• Organic inspections • Product selection	
Man-machine systems	• Management decision making support • Development planning	• CAD/CAI • Diagnosis of facilities	• Automotive equipment • Household electrical appliances
Miscellaneous	• Risk analysis • Earthquake prediction	• Fuzzy LISP • Fuzzy PROLOG	• Inference chips • Inference engines

Table 11.2   Trends in fuzzy research

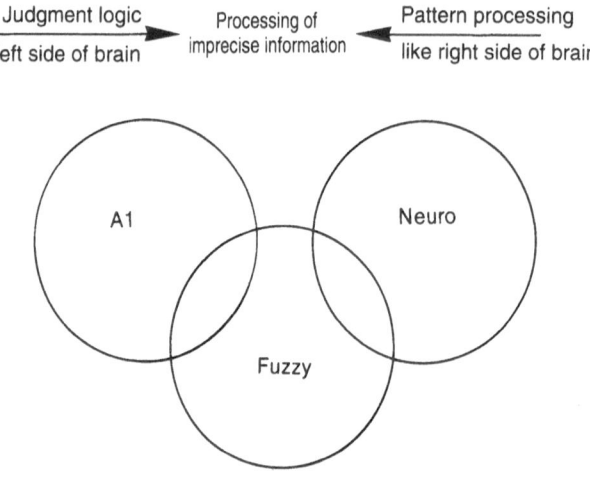

Fig. 11.26   The relation among fuzzy, neuro and artificial intelligence.

11.6   In conclusion

This chapter has discussed fuzzy theory, predictive fuzzy control and examples of its application, and technology application trends. Application of fuzzy control to other bad structure problems for which computerized control has until now been difficult will perhaps make automatic operation equivalent to operation by a skilled worker feasible in those areas as well. Applications of fuzzy theory are expanding, from control problems to information processing problems and to man - machine systems that involve stronger relations with human beings. At present very active research is being conducted on fuzzy chips, fuzzy computers and fuzzy software, and at the same time fuzzy technology is being actively applied to the mass production market including automobiles and household electrical appliances. I believe that this trend of applications to wider fields can be expected to accelerate.

References

1. Zadeh LA (1965) Fuzzy sets. Information and Control 8 : 338–343

2. Matsumoto K (1988) Fuzzy control and its applications (in Japanese). Computer Technol 18 (4) : 2–7

3. Matsumoto K (1990) From "light, thin, short and small" to "beautiful, sensual, quiet and graceful" by fuzzy (in Japanese). Electronics : 62–65

4. Matsumoto K (1990) Application of fuzzy control to an automatic train operation system (in Japanese). J Textile and Machinery Soc 43 (7) : 19–26

5. Yasunobu S (1988) Application of fuzzy theory to a real system (in Japanese). J Jpn Machinery Soc 91 (836) : 25–30

6. Yasunobu S (1986) Development of a system for automatic container crane operation by predictive fuzzy control (in Japanese). Collection of Papers of the Jpn Measurement and Automatic Control Soc 22 (10) : 60–65

7. Sugeno M (1988) Fuzzy control. Nikkan Kogyo Shimbun Sha

8. Nagataki, Kotsuji, Yashiro, Inoue (1988) Application of knowledge engineering to tunnel ventilation, and its effect (in Japanese). Electrical Soc Res Report TER 88 (27) : 35–41

9. Sakurai T, Maekawa O (1988) Fuzzy control of a sugar extraction process (in Japanese). Automation 33 (6) : 55–59

10. Yahiro M, Inoue (1988) A concentrated expressway monitoring control system (in Japanese). Hitachi Rev 70 (5) : 45–50

11. Funahashi (1988) Highway tunnel ventilation control using quantitative inference (in Japanese). Papers from Jpn Measurement and Automatic Control Soc, 5th Industrial Systems Symposium pp.23–28

12. Terano T (1987) Introduction to fuzzy systems (in Japanese). Ohm Sha.

13. Hirota K (1989) Fuzzy control and trends in its application (in Japanese). Measurement and Control 28 (11) : 28–33

14. Matsumoto K (1990) A fuzzy fully automatic washing machine (in Japanese). J Jpn Fuzzy Soc 2 (4) : 40–44

Chapter. 12

APPLICATION OF FUZZY THEORY TO HOME APPLIANCES

12.1 Introduction

The "Aisaigo (Loved Wife) Day Fuzzy" fully automatic washing machine, which went on sale in February 1990, triggered a flood of applications of fuzzy technology to all kinds of household electrical appliances and video equipment throughout the rest of 1990. "Fuzzy" won the gold medal as the outstanding new word of 1990, and Japan was engulfed in a "Fuzzy boom".

This is perhaps because the application of fuzzy, a leading technology that is a quantum jump beyond previous control by microprocessors and sensors, gives people the satisfaction of being able to do work on a par with a skilled professional with only simple operation. By applying fuzzy to control, the electrical appliance manufacturers were able to use the know-how concerning their products that they had accumulated over the years and incorporate features that users had wanted but had been difficult to design. This is a major reason that fuzzy was such a big hit.

Clearly, what is required of a manufacturer is to understand what the user is really after and provide it in a timely manner.

Meanwhile, viewed from the viewpoint of technology, as the technology becomes higher and higher level, and the systems and networks spread out wider and wider, will it not become necessary for the machines to become human? That is to say, they will have to be kind and warm - hearted toward people, blend into living space, and provide people with a comfortable and pleasant life. We believe that for this to be possible, it will become important for the manufacturers to stand in the users' shoes and become users them-selves when making products that are driven by electronics (this is called human electronics).

Fuzzy control involves a number of control rules which are expressed by fuzzy sets (fuzzy rules) in the format IF ..... THEN ..... . It is necessary to find the degree to which an actual situation matches the IF ..... part of each rule, and then apply control that matches that degree of matching.

Words that contain quantitative meanings are included in both the IF
..... part (condition part) and the THEN ..... part (operation part) of a
rule. The appeal of fuzzy control is that if the know-how consisting of the
experience and intuition of veterans and experts accumulated over the years
is incorporated without change into quantitative expressions, then skillful
control equivalent to that of expert workers is obtained.

The ability to incorporate both simple operation and expert know - how
into products, combined with the latent possibility of making products that
match users' desires in the future, makes the development of fuzzy theory
very important as a basic technology for bringing about the human electronics
referred to above.

However, there are technological problems in the application of fuzzy
theory to household electrical appliances. These include simplifying fuzzy
inference, lightening the load on the microprocessor as much as possible, and
determining how membership functions should be tuned in fuzzy inference.

In this chapter, we first explain inference simplification methods and
tuning methods, which become important in the application of fuzzy theory
to products.

Next, as specific examples of the development of fuzzy theory, we take up
washing machines and vacuum cleaners as examples of household electrical
appliances, and video cameras as an example of video equipment.

12.2 Fuzzy inference simplification methods and tuning methods

One way to simplify fuzzy inference is to have the operation part give
real numbers instead of fuzzy quantities [1,2].

In control that uses fuzzy inference, it is necessary to adjust the fuzzy
variables and number of rules so that the control system acts in accordance
with the desired specifications. This adjustment process is called tuning. One
tuning method is to use a neural network learning function [3].

12.2.1 Simplification of fuzzy inference

The simplified fuzzy inference proposed by Ichihashi et al (referred to
below as simplified fuzzy inference) uses real numbers as the operation parts
of fuzzy control rules; it is said that the control results are better than those
obtained from ordinary fuzzy inference [4].

A simplified fuzzy inference control rule is expressed in the format shown
in expression (1), with $e1$ and $e2$ as the input variables and u as the output
variable.

Here PB (positive big) and NS (negative small) are fuzzy variables, and f_i is a real number.

$$\text{IF } e_1 \text{ is } PB \text{ and } e_2 \text{ is } NS \text{ THEN } u \text{ is } f_i \tag{1}$$

When input data (e_1', e_2') are input, from expression (1) equation (2) is obtained; $\bigwedge$ is the min operator.

$$\mu_i(f_1) = \mu_{PB}(e_1') \bigwedge \mu_{NS}(e_2') \bigwedge 1 \tag{2}$$

There are several control rules of the form (1), so the final inference result obtained from all of the equations of form (2) is obtained from equation (3).

$$u^* = \frac{\sum \mu_i(f_i) \cdot f_i}{\sum \mu_i(f_i)} \tag{3}$$

Thus, by using real numbers as the operation parts, simplified fuzzy inference makes it possible to reduce the number of adjusted parameters and thus makes tuning easier; at the same time, the inference operation can be simplified.

12.2.2 Tuning by means of a neural network

However, when it comes to creating fuzzy rules for control that involves more complicated input and output and/or complicated judgment, at present there are many cases in which it takes several days, or even is not possible for humans. In addition, the more complicated the input information, the more difficult it becomes to define the membership functions used in the rules.

One way of performing such tuning is by means of a learning function using a neural network. Here we explain neural network drive type fuzzy inference [3].

This method makes it possible to apply fuzzy control to complicated systems with many input variables, for which it was not previously possible to design control systems by human labor alone.

Neural network drive type fuzzy inference is a fuzzy inference method based on the concept of fuzzy modeling [5]. In fuzzy modeling, fuzzy control rules such as the following are used.

Rule 1:

If the temperature (x_1) and humidity (x_2) in the air conditioning duct are high, then the degree of valve opening is:

$$y = 5x_1 + 2x_2 + 1.5 \tag{4}$$

Rule 2:

If the temperature (x_1) and humidity (x_2) in the air conditioning duct are medium, then the degree of valve opening is:

$$y = 3x_1 + 1.8x_2 + 1.2 \tag{5}$$

The inference rules of equations (4) and (5) are described using fuzzy quantities such as "high" and "medium". Neural network drive type fuzzy inference automatically determines the membership values of these fuzzy variables, and can identify the structure of the equation that expresses the input/output relationship in the operation part.

In this method the fuzzy inference rules are expressed as follows.

$$R^s; \quad \text{IF} \qquad X = (x_1, x_2, \cdots, x_n) \text{ is } A^s$$
$$\text{THEN} \quad y^s = NN_s(x_1, x_2, \cdots, x_m), s = 1, 2, \cdots, r \quad m \leq n \tag{6}$$

Here A^s expresses the fuzzy set in the condition part; NN_s is a neural network model that expresses the relation between the inputs $(x_1, x_2, \cdots, x_n)$ and the output (y^s) when k - layer back propagation is used. Here the scale of the back propagation type neural network model is expressed by the symbol "k - layer" $[u_0 \times u_1 \times \cdots u_k]$. $u_0, u_1 \cdots$ and u_k are the numbers of neurons in the input layer, hidden layers and output layer. A 4 - layer $[3 \times 2 \times 2 \times 2]$ neural network is shown in Fig. 12.1.

Suppose that input and output data $(x_{i1}, x_{i2}, \cdots, x_{in}, y_i)i = 1, 2, \cdots, N)$ have been obtained. The fuzzy inference result $y_i{}^*$ can be found from equation (7) after substituting input data $(x_{i1}, x_{i2}, \cdots, x_{in})$ into inference rule (6). Here $\mu_A{}^s(x_{i1}, x_{i2}, \cdots, x_{in})$ is the membership function for the fuzzy set A^s ; $ey_i{}^s$ is the inferred value obtained from $y^s = NN_s(x_1, x_2, \cdots, x_n)$.

$$y_i{}^* = \frac{\displaystyle\sum_{s=1}^{r} \mu_A{}^s(x_{i1}, x_{i2}, \cdots, x_{im}) \cdot ey_i{}^s}{\displaystyle\sum_{s=1}^{r} \mu_A{}^s(x_{i1}, x_{i2}, \cdots, x_{im})}, i = 1, 2, \cdots, N \tag{7}$$

Next, we explain algorithms for the purpose of formulating inference rules in neural network drive type fuzzy inference. A diagram of an algorithm is shown in Fig. 12.2. The operation value $y_i{}^*$ for the given inference rule and inference data is obtained by the following procedure.

[Step 1] Determine the input variables $x_1, x_2, \cdots, x_n$ which are related to the output y. Then assume that input and output data $(x_{i1}, x_{i2}, \cdots, x_{in}, y)i = 1, 2, \cdots, N$ have been obtained.

[Step 2] The input and output data are divided up into r (parts) $R^s, s = 1, 2, \cdots, r$. This division corresponds to the number of fuzzy inference rules. Here the input and output data for each R^s are expressed as $(x_{i1}{}^s, x_{i2}{}^s, \cdots, x_{in}{}^s, y_i^s)$, $i = 1, 2, \cdots, N^s$.

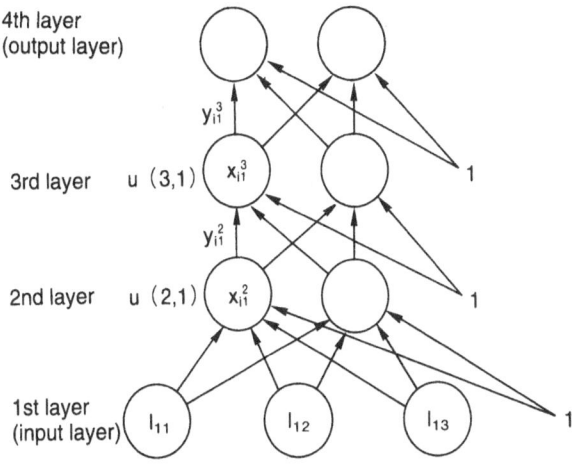

Fig. 12.1 A 4-layer error back-propagation type neural network.

[Step 3] Using the error back-propagation learning model NN_{mem} in Fig. 12.2, the form of the condition part membership function is determined.

An example of determination of the condition part fuzzy quantity membership values in the case of 2 condition part input variables (x_1, x_2) is shown in Fig. 12. 3. In Fig. 12. 3 a 3-layer error back-propagation learning model is used; $(x_{i1}{}^s, x_{i2}{}^s)$ are assigned to the input layer, and the degree of belonging to each rule R^s is assigned to the output layer as a value 0,1. The number of learning iterations is on the order of 1,000. After learning is performed, when new data other than the initially given input data are given, the membership values of those data for the condition part fuzzy sets can be obtained.

[Step 4] $NN_1, NN_2, \cdots, NN_r$ in Fig. 12.2 are used to determine the structure of the operation part. Each error back-propagation learning model NN_s describes an inference rule of operation part, and that structure is taken to be M - layered $[k \times u_2 \times \cdots \times u_{M-1} \times 1], k = n, n - 1, \cdots, 1$. The optimum model for each NN_s is selected.

First, k is set equal to n and the input data $(x_{i1}{}^s, x_{i2}{}^s, \cdots, x_{ik}{}^s), i = 1, 2, \cdots, N^s)$ are assigned to the input layer of the error back-propagation learning model NN_s. The output data are assigned to the output layer of NN_s. The set of input variables assigned to the input layer is expressed as follows.

$$\Lambda_s = \{x_1, x_2, \cdots, x_j, \cdots, x_k\} \tag{8}$$

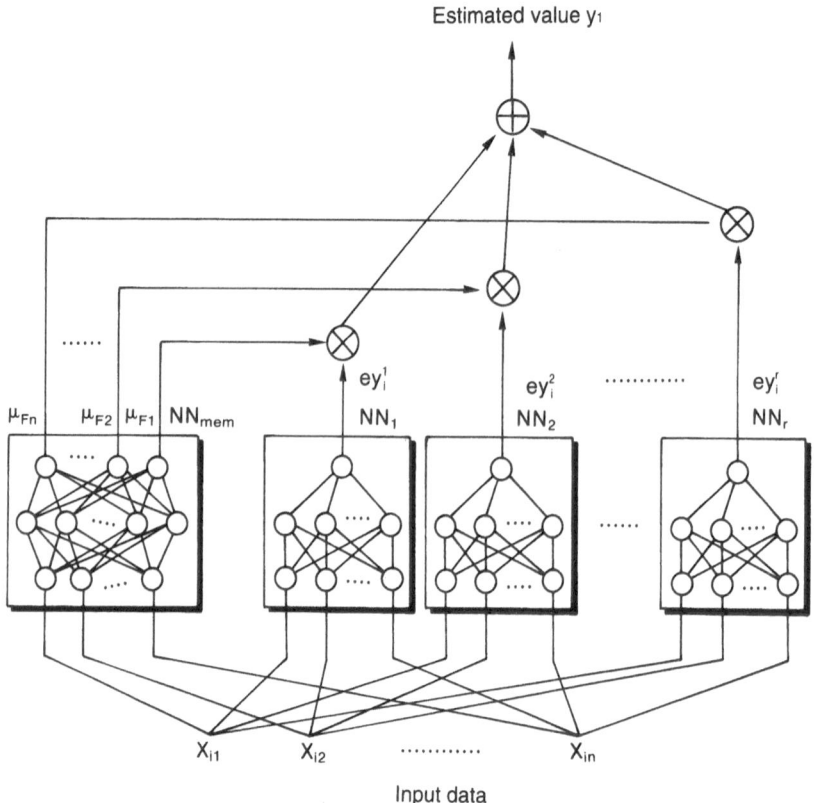

Fig. 12.2 Neural network drive type fuzzy inference configuration.

After the NN_s error back-propagation learning, the inferred values $ey_i{}^s$ for the input data $x_{i1}{}^s, \cdots, x_{ik}{}^s$ are obtained. The number of learning iterations is on the order of 3,000. The mean squared error between the output data $y_i{}^s$ and the estimated values $ey_i{}^s$ is calculated, and then an evaluation value $\Theta_k{}^s$ for determining the input variables is found.

$$\Theta_k{}^s = (\sum_{i=1}^{N^s}(y_i{}^s - ey_i{}^s)^2)/N^s, s = 1, 2, \cdots, r \tag{9}$$

In order to investigate the strength of the connection of the input variable x_j to the output variable y, the input variable x_j is temporarily removed from the set of input variables Λs. The input data $x_{i1}{}^s, \cdots, x_{ij-1}{}^s, x_{ij+1}{}^s, \cdots, x_{ik}{}^s$, $i = 1, 2, \cdots, N^s$ with the input variable x_j removed are assigned to the input layer of the M - layered structure $[k - 1 \times u_2 \times \cdots \times u_{M-1} \times 1]$ of the error back-propagation learning model; the output data y_i are assigned to the output layer, and learning takes place. After the learning, the estimated values

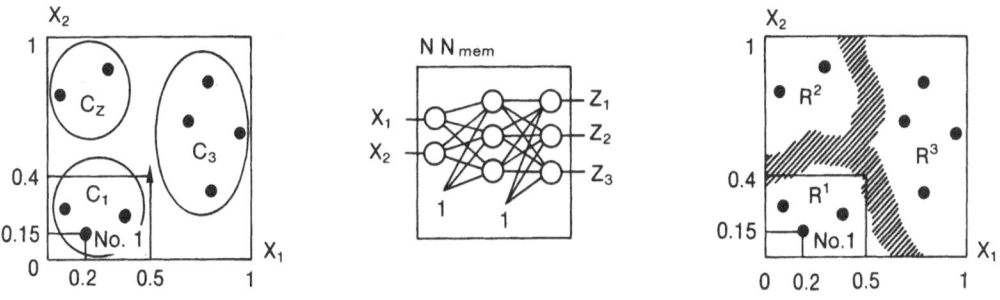

(a) Conceptual diagram of membership function shape determination

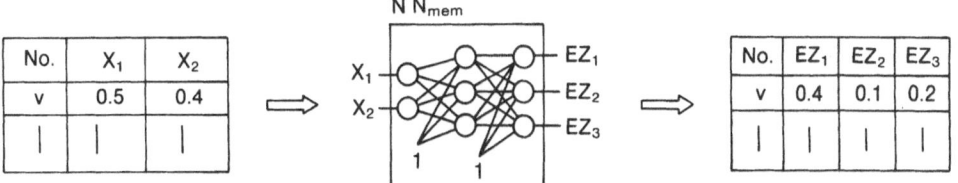

(b) Learning method with respect to No. 1 data

(c) Example of membership value with respect to data other than learning data

Fig. 12.3 Determination of the condition part membership function.

$ey_i^{s\prime}$ for the input data $x_{i1}^s, x_{ij-1}^s, x_{ij+1}^s, \cdots, x_{ik}^s$ are obtained. The mean squared error between these estimated values $ey_i^{s\prime}$ and the output data y_i^s is calculated, and the evaluation value Θ_{k-1}^{sj} for input variable determination is found.

$$\Theta_{k-1}^{sj} = (\sum_{i=1}^N (y_i^s - ey_i^{s\prime})^2)/N^s, s = 1, 2, \cdots, r \qquad (10)$$

Similar operations are performed for the input variables other than x_j, and the evaluation values $\Theta_{k-1}^{s1}, \Theta_{k-1}^{s2}, \cdots, \Theta_{k-1}^{sj}, \cdots, \Theta_{k-1}^{sk}$ are found. Then the evaluation value Θ_{k-1}^{sc} with the smallest value among the evaluation values is found.

$\Theta_{k-1}{}^{sc}$ and $\Theta_k{}^s$ are compared, and the set Λ_s of input variables assigned to the input layers of each error back-propagation learning model NN_s is altered as follows.

$$\Lambda_s = \{x_1, x_2, \cdots, x_{c-1}, x_{c+1}, \cdots, x_k\} \cdots \Theta_{k-1}{}^{sc} < \Theta_k{}^s \qquad (11)$$

$$\Lambda_s = \{x_1, x_2, \cdots, x_k\} \cdots \Theta_{k-1}{}^{sc} \geq \Theta_k{}^s \qquad (12)$$

If the number of input variables can be decreased, k is varied among $n-1, n-2, \cdots, 1$, and step 4 is repeated until equation (12) is satisfied. Once equation (12) is satisfied, the procedure for reducing the number of input variables has been completed.

[Step 5] The estimated values $y_i{}^*$ are obtained from the following equation.

$$y_i{}^* = \frac{\displaystyle\sum_{s=1}^{r} \mu_A{}^s(x_{i1}{}^s, x_{i2}{}^s, \cdots, x_{in}{}^s) \cdot ey_i{}^s}{\displaystyle\sum_{s=1}^{r} \mu_A{}^s(x_{i1}{}^s, x_{i2}{}^s, \cdots, x_{in}{}^s)}, i = 1, 2, \cdots, N \qquad (13)$$

Here $ey_i{}^s$ are the estimated values obtained from the optimum error back-propagation learning model in step 4.

This tuning method becomes especially effective as the number of input and output quantities increases. When the number of rules is small, tuning by human "cut and try" is possible. However, in control involving more complicated input and output, and in control involving more complicated judgments, tuning can take a number of days, or it can even become impossible for humans to describe the rules. If tuning is done automatically by this method the design takes only a small amount of time, and it becomes possible to design fuzzy control involving many input variables that has been impossible to do by human labor alone.

12.3 Application to electrical appliances

In the household work and cooking field, it is often necessary to handle data that cannot be expressed quantitatively, for example, whether there is "a lot" or "a little bit" of dirt, or whether the heat applied should be "strong" or "weak". In such a situation, humans apply their past experience to understand and judge how to control a washing machine, vacuum cleaner or cooking appliance.

Fuzzy technology can be considered well - suited to such daily tasks which humans do without giving much thought to them, and to control operations requiring long years of know - how, such as flame control when the instructions for cooking state that "at the beginning simmer over medium heat.....".

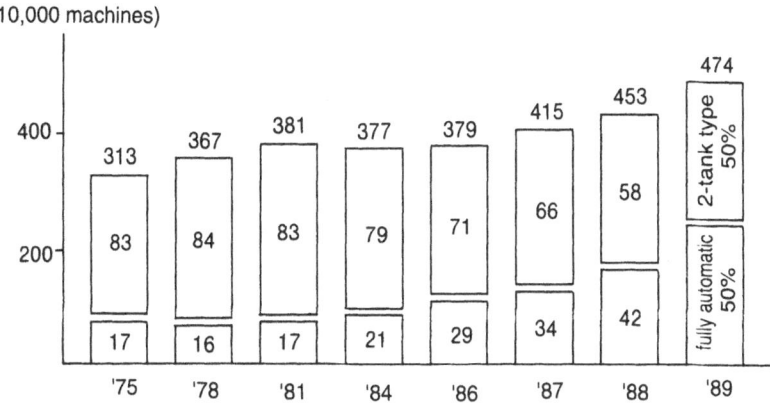

Fig. 12.4 Trend in total demand for washing machines, by year.

Meanwhile, social changes are taking place; Japan's population is aging, and more and more females are entering the work force. For this reason, it is necessary to have products that elderly people, men and children can use safely and easily.

It is desirable to have appliances that require one to just press one button, and it will perform an operation as skillfully as, or in some cases more skillfully than, a human would.

The introduction of electronics into household electrical appliances started about 1970. About 10 years later microprocessors began to be used; most of today's products contain microprocessors.

Since it is possible to respond to social needs by applying fuzzy control in the programming of microprocessors, fuzzy has found an unprecedented range of applications in this field and its use has spread rapidly.

12.3.1 A fuzzy fully automatic washing machine

Many people are replacing their washing machines, most of them switching to fully automatic. Nationwide (Japan) sales in 1988 set a record of 4,530,000 machines, and since then sales have continued to increase. Since 1989 the proportion of fully automatic washing machines has exceeded 50% (Fig. 12.4). As the spread of fully automatic washing machines suggests, the time that housewives spend doing the laundry is decreasing; in particular, working wives have far less time available than do full - time housewives. There is now greater awareness of the need to streamline the washing operation and reduce the amount of labor involved; the need for fully automatic washing machines that will do a good careful job of washing and require only simple operation is increasing.

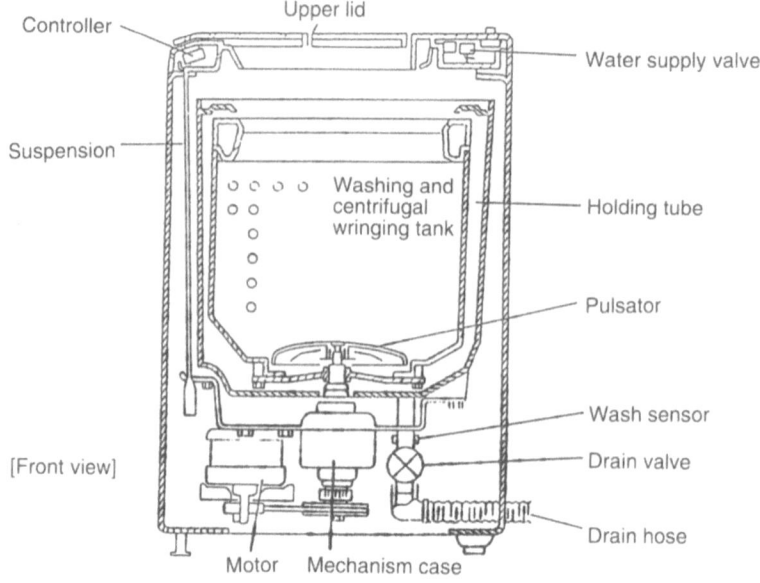

Fig. 12.5 Composition of a washing machine.

In order to respond to this need a fuzzy washing machine was developed, based on the following concept. (1) It washes skillfully, and anybody can operate it easily. (2) It is of large capacity, capable of washing 5kg, and takes up only little space. (3) It is of new design, matching the interior living environment.

Of these conditions, it is (1) to which fuzzy is relevant; in order to do the wash skillfully, operation is automated with fuzzy control based on the basic principle of laundry that "when dirt has been removed, washing is stopped". A wash sensor that indirectly detects the dirtiness of the laundry by detecting the turbidity of the wash water with an optical sensor [6, 7] provides input, then fuzzy inference [8] is used to set a suitable washing time. This way of using a sensor plays an important role in characterizing fuzzy appliances.

The wash sensor

The wash sensor consists of a light emitting diode and a photo transistor; these are installed near the drain valve on opposite sides of the drain pipe (Fig. 12.5). The intensity of the light from the light emitting diode is converted to a voltage by the photo transistor; that level is read by a microprocessor to measure the transmissivity of the wash water to light. The light emitting diode unit of the wash sensor has the configuration shown in Fig.

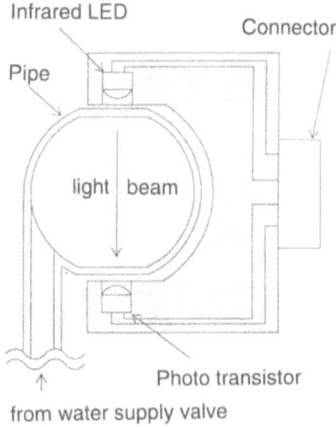

Fig. 12.6 The wash sensor.

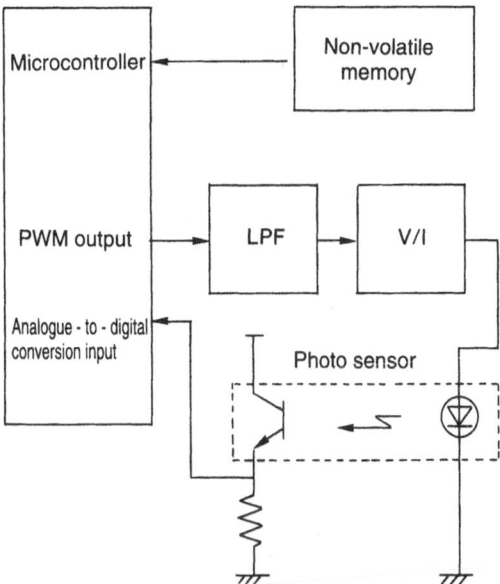

Fig. 12.7 Wash sensor control circuit.

12.6; it can be installed as is in the drain pipe. A block diagram of the control circuit is shown in Fig. 12.7; a PWM signal from the microprocessor is passed through a low pass filter, then converted from voltage to current and used to control the intensity of light emitted by the light emitting diode.

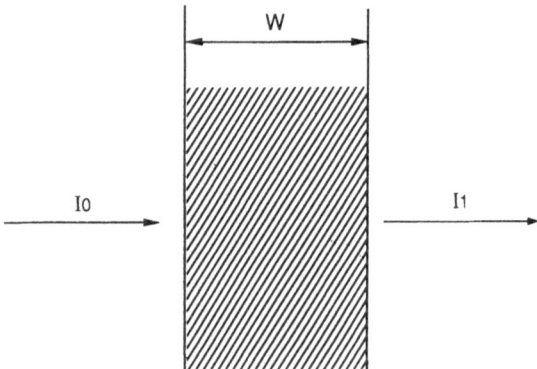

Fig. 12.8   The principle of detection of degree of penetration.

The light emission intensity is set initially so that it will have a specified level in clean water that has just entered the washing machine.

The principle of transmissivity detection is the same as the principle of a spectrometer; transmission of light through the liquid of unknown concentration is compared with transmission through a liquid of known concentration. In Fig. 12.8, monochromatic light enters a liquid of thickness w; the incident light intensity is I_0 and the transmitted light intensity is I_1. Then, from Lambert - Beer's Law, the transmissivity is given by equation (14). Here k_1 is the absorbance of the known liquid.

$$I_1/I_0 = e^{-k_1 \cdot w} \tag{14}$$

When the light enters a different liquid, the transmitted light intensity changes to I_2 and the transmissivity changes according to equation (15). Here k_2 is the absorbance of the unknown liquid.

$$I_2/I_0 = e^{-k_2 \cdot w} \tag{15}$$

If we let I_0 be the intensity of light transmitted through clean water and I_2 the intensity of light transmitted through the wash water, their ratio is given by equation (16).

$$I_2/I_1 = e^{-(k_2-k_1)w} \tag{16}$$

If the emitted light output is controlled so that it is always I_0 in clean water, the intensity of the output voltage from the light receiving element is proportional to the transmitted light intensity, so, letting V_1 and V_2 be the light receiving element output voltages in the clean water and wash water, respectively, their ratio is given by equation (17).

$$V_2/V_1 = e^{-(k_2-k_1)w} \tag{17}$$

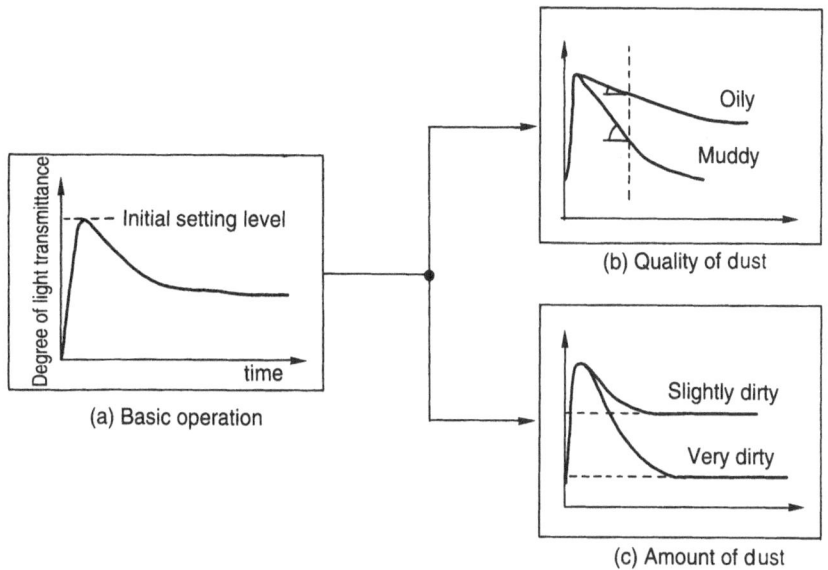

Fig. 12.9 Wash sensor output change.

Taking the logarithms of both sides gives:

$$\ln(V_2/V_1) = -(k_2 - k_1) \cdot w \tag{18}$$

That is to say, the logarithm of the ratio of the output voltage V_2 (transmissivity) for wash water to the output voltage V_1 for clean water is proportional to the change of absorbance, permitting the dirtiness of the wash water to be measured.

The change of the output of this wash sensor over time is shown in Fig. 12.9.

When washing is started, the dirt in the clothes is gradually washed out and the wash water becomes dirty, causing the transmissivity to light to decrease. The rate of decrease of the transmissivity depends on the quality of the dirt, being fast for "muddy dirt" and slow for "oily dirt". This is because "muddy dirt" is removed easily by the mechanical force of the water flow produced by rotation of the pulsator, while "oily dirt" is not adequately removed until the detergent takes effect. When most of the dirt has been removed the transmissivity approaches a state of saturation. The transmissivity at this time is lower the greater the dirtiness of the clothing, and vice versa.

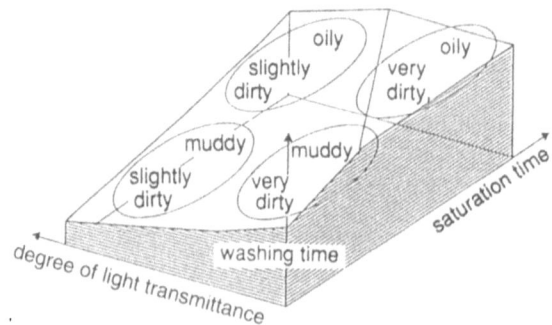

Fig. 12.10 Wash sensor input/output image.

Fig. 12.11 Fuzzy inference.

(Washing time)

	very dirty	medium dirty	slightly dirty
muddy	T1	T2	T3
oily	T4	T5	T6

Fig. 12.12 Rule table.

Fuzzy inference

From the rate at which the transmissivity measured by the wash sensor decreases (the time required to reach saturation) and the level at the time of saturation, the dirtiness of the wash is determined quantitatively from the areas shown in Fig. 12.10. The optimum relation between dirtiness and wash time is determined experimentally from several sample points. However, in actual practice, while there is no limit to the number of degrees of dirtiness (wash sensor output patterns), collecting detailed experimental data for all such levels is nearly impossible. In addition, due to the nature of the washing machine wash mechanism the relation between dirtiness and wash time is

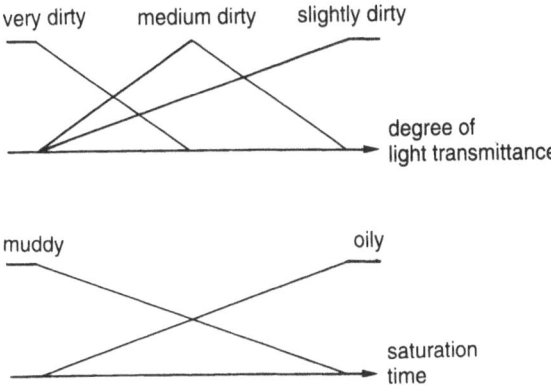

Fig. 12.13 Membership functions.

nonlinear, and it is difficult to obtain a unique numerical formula relating them.

Therefore fuzzy inference is used to determine the wash time.

Fuzzy inference processing is based on several rules expressed quantitatively; the nonlinear curve is approximated by interpolation.

Fig. 12.11 shows a diagram of fuzzy inference. From the time T until the transmissivity reaches saturation, which is related to the quality of the dirt, and the output level V at that time, which is related to the amount of dirt, the remaining wash time T is determined by fuzzy inference.

The following 6 fuzzy rules, for example, are used. "If the wash is very dirty and the dirt has an oily quality, make the wash time very long." "If the wash is only slightly dirty and the dirt has a muddy quality, make the wash time very short."

Fig. 12.12 shows the fuzzy rule table (T1 to T6 are wash times), and Fig. 12.13 shows the membership functions. Simplified fuzzy inference [2] is used as the inference method in order to simplify the inference operations, reduce the amount of memory required, and simplify the parameter tuning.

Rule creation and parameter tuning are performed using laboratory data and a skilled launderer's know-how.

Fig. 12.14 shows the input and output characteristics which indicate the results of fuzzy inference after tuning; the ordinate is wash time.

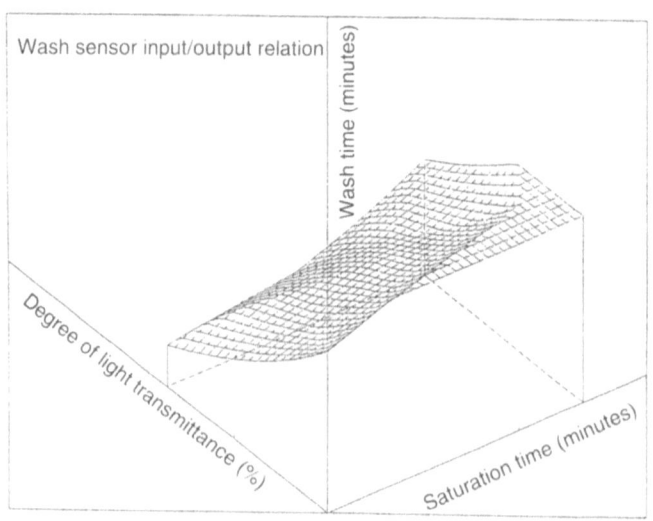

Fig. 12.14   Fuzzy inference input/output diagram.

Characteristics of a fuzzy washing machine

By using fuzzy theory, the complicated relation between the degree of dirtiness of the wash and the wash time can be expressed, making it possible to perform careful washing that corresponds to the degree of soiling.

In addition, a wash weight sensor determines the amount of the wash immediately before water supply starts to determine the appropriate water level, so that the appropriate wash time can always be determined, without the fuzzy inference being thrown off by differences in the amount of the wash.

There are both solid and liquid detergents; the effect of these on the transmissivity is taken care of by correcting a scaling factor that is one of the input variables.

It is necessary to consider differences in water quality among different parts of the country. In particular, the hard water in Okinawa could conceivably cause a metal soap film to adhere to the wash sensor which is installed inside the drain pipe; correcting the scaling factor is also an effective means of dealing with this problem. In addition to this electrical correction, a mechanism is included to clean the wash sensor with clean water after the washing is completed.

Thus, the fuzzy fully automatic washing machine came into being not merely through introducing fuzzy inference into the microprocessor program, but through introducing appropriate sensors and electrical and mechanical technology.

Table 12.1 The status of housewives' cleaning work (according to survey by our company)

	full-time housewives	working housewives	self-employed housewives
weekdays	32.2	21.7	20.4
Sundays	26.4	33.3	17.8

	weekdays	Sundays
careful	25.8	26.4
ordinary	58.7	39.1
simple	15.5	34.5

The fuzzy fully automatic washing machine, which performs optimum washing based on the amount and degree of soiling of the wash, gives the kind of detailed control that fits human desires, and economizes on time and electricity without damaging the fabric by overwashing or leaving anything inadequately washed.

12.3.2 A fuzzy vacuum cleaner

With the spread of wool carpets and increased concern about cleanliness and health, vacuum cleaners have recently been becoming more functional, with the suction power being increased dramatically; 5,600,000 units were sold in Japan in 1989. As shown in Table 12.1, the modern housewife who is

busy with work and hobbies wants to complete cleaning and other housework speedily and efficiently, and at the same time must be able to clean the increasing diversity of floors in modern homes.

The fuzzy vacuum cleaner development concept consists of the following 4 points. (1) Anybody can use it easily, and moreover it picks dust up completely. (2) It can be operated easily without feeling resistance from the floor surface. (3) It leaves the user with a satisfied and secure feeling of having cleaned the floor completely. (4) It is quiet so that cleaning can be done even at night.

In order to make this concept a reality, it is necessary for the suction power to be automatically controlled in accordance with the amount of dust and the floor surface material. To do this, fuzzy inference is performed with the amount of dust passing through the suction hose, as measured by an optical sensor, as input; the inference then sets the suction power corresponding to that amount.

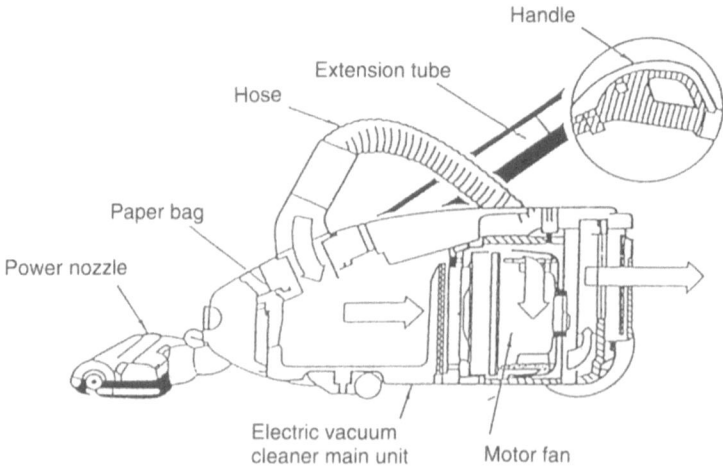

Fig. 12.15 Construction of a vacuum cleaner.

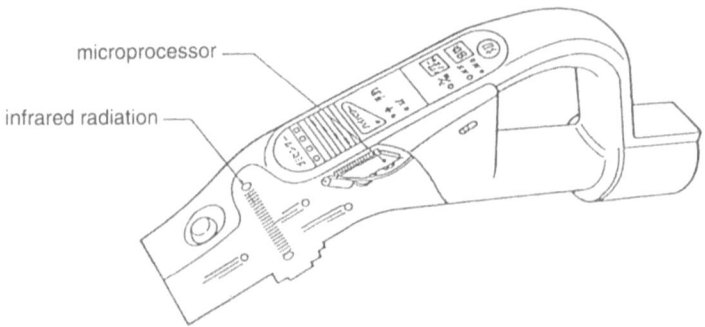

Fig. 12.16 Enlarged view of the handle.

Composition of the dust sensor

A conceptual cross - sectional diagram of a fuzzy vacuum cleaner is shown in Fig. 12.15. An enlarged view of the user's operation section is shown in Fig. 12.16; the cleaner incorporates a dust sensor, microprocessor and control circuit.

A block diagram of the control circuit is shown in Fig. 12.17. Dust on the floor is sucked up by the force of the motor - driven fan inside the main unit; it then passes through the dust sensor in the handle and is collected in a paper bag in the main unit.

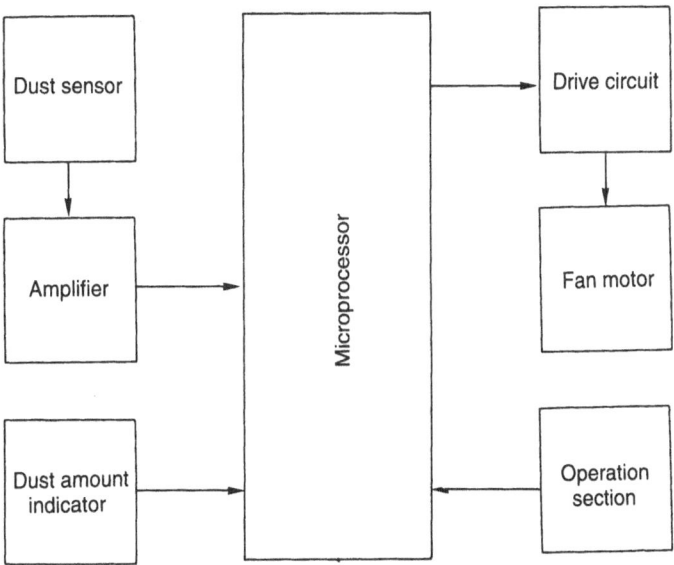

Fig. 12.17 Control circuit.

The dust sensor shown in Fig. 12.16 is inside the pipe handle on which the controls are located. A photo transistor is mounted opposite a light emitting diode (infrared type); infrared rays are emitted in a beam. When dust passes through, some infrared rays are cut off causing the amount that reach the photo transistor to decrease. The varying component is extracted and amplified, then a waveform is formed and converted to pulses. The amount of dust is judged by counting these pulses. Specifically, the cumulative number of pulses in 0.1 second intervals is taken to indicate the amount of dust. If the surface of the dust sensor becomes dirty, the transmissivity of infrared rays will decrease, but by detecting the differentiated change the measurement can be made insensitive to the dust. In addition, the sensor has the configuration described below so that no problems arise in practice. (1) The air path where the dust sensor is installed is partially restricted to change the air flow so that less dust will adhere to the sensor. (2) The light emitting diode and photo transistor are encased in hard transparent acryl holders to protect them from scratching. (3) The dust sensor is installed near an opening so that it can be cleaned. The time variation of dust amount that can be detected by this dust sensor is shown in Fig. 12.18. As cleaning proceeds the amount of dust decreases, but whereas this decrease is fast in the case of a tatami mat or wooden floor, it is slow in the case of a wool carpet. This is because it is easy for dust to be picked up off of a tatami mat or wooden floor but hard for it to be picked up off of a wool carpet. Thus, by measuring the time variation of the rate at which dust is picked up, the floor surface material can be judged.

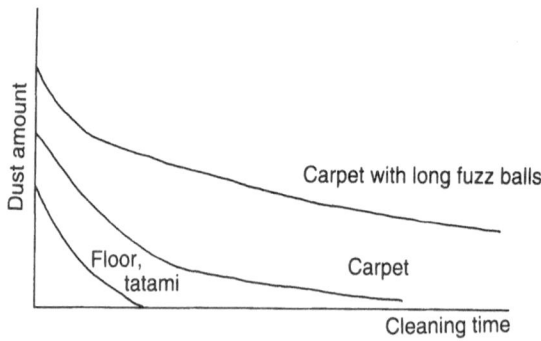

Fig. 12.18   Dust sensor output variation.

Fig. 12.19   Fuzzy inference.

Fuzzy inference

When cleaning is done, if the suction power is too strong the suction nozzle will stick to the floor, making it difficult to handle. Conversely, if it is too weak it will be hard to pick dust up. For this reason, there is an optimum suction strength for a given floor material and amount of dust. In order to optimize the tradeoff relationship between suction power and ease of handling, providing a vacuum cleaner that is easy to use, the suction power is determined using fuzzy inference. Fuzzy inference is performed as shown in Fig. 12.19. The suction power is determined from the cumulative amount of dust detected in a unit of time and the floor surface condition (material).

Inference is performed with 15 rules such as the following. "If there is a great deal of dust and the floor surface has the properties of a wool carpet, make the suction power very strong."

The fuzzy rule table is shown in Fig. 12.20 (P1 to P15 in the figure are levels of suction power), and the membership functions are shown in Fig. 12.21. Simplified fuzzy inference, with real number type operation parts [2], is used in order to simplify the inference operations, reduce the amount of memory used and simplify the parameter tuning. By doing this, fuzzy infer-

(Suction power)

	Carpet	Tatami	Wooden floor
Very strong	P1	P2	P3
Strong	P4	P5	P6
Ordinary	P7	P8	P9
Weak	P10	P11	P12
Very weak	P13	P14	P15

Fig. 12.20 Rule table.

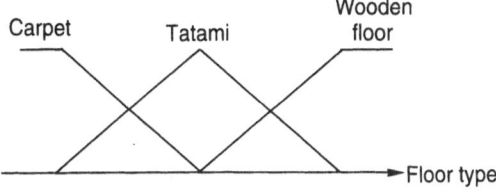

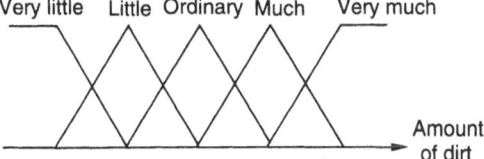

Fig. 12.21 Membership functions.

ence is performed by software operations on a 4-bit 1-chip microprocessor, with a memory capacity on the order of 0.8Kbytes.

Rule creation and parameter tuning are done by "cut and try" based on data from monitoring in ordinary households and the developer's cleaning know-how.

A diagram of input/output characteristics, showing the results of fuzzy inference after tuning, is shown in Fig. 12.22.

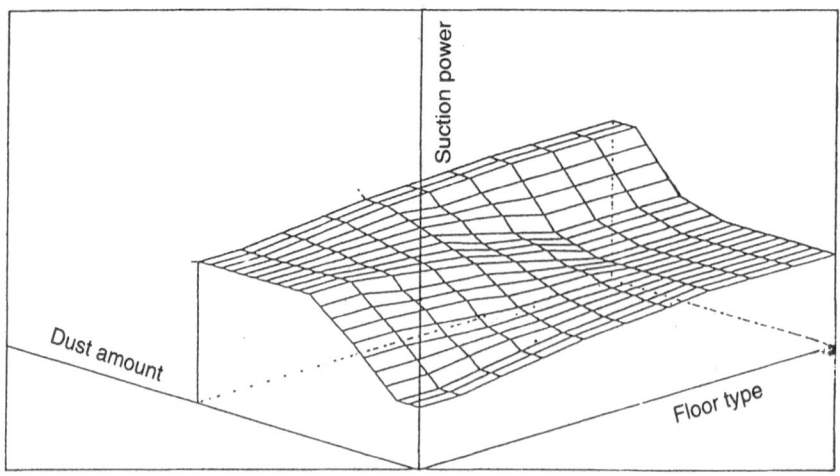

Fig. 12.22   Fuzzy inference input/output graph.

Characteristics of a fuzzy vacuum cleaner

By using fuzzy inference, the optimum suction power for a given amount of dust and floor surface material can be determined, eliminating the problems caused by a suction power which is too strong so that the suction nozzle sticks to the floor, or, conversely, a suction power that is too weak so that some dust is not picked up. In addition, the completion of cleaning is indicated by a lamp on the control section, reducing the incidence of incomplete cleaning in places where it is difficult to see such as deep down in the wool hairs and under beds.

12.3.3   Future development of fuzzy household electrical appliances

There was a great boom in fuzzy household electrical appliances in 1990, but it is necessary to continue research and development work with the aim of continuing the development of this field if fuzzy is not to end up being a passing fad.

For example, present fuzzy household electrical appliances incorporate know - how based on the manufacturer's point of view into their rules and membership functions; future developments will have to be based more on the user's point of view, reflecting the user's desires and subjective feelings.

In order to derive the greatest benefit from fuzzy, it is not sufficient merely to apply fuzzy technology in a straightforward manner. To use a human being as a comparison, no matter how intelligent he is, unless he has good eyes, a good nose, and good hands and feet, his intelligence will be wasted. Research and development on peripheral technologies, such as sensor technology, microprocessor technology and man - machine interfaces, must proceed hand - in hand with that on the fuzzy technology itself.

Development of housework and cooking software, and research into daily life are also important in applying fuzzy theory to household electrical products. The intuition, skills and know-how possessed by human beings are indispensable in research on housework and cooking software; fuzzy theory is merely useful as a technique for incorporating the results into devices. It is necessary to do more thorough research on which the applications are to be based. Until now it has been sufficient merely to incorporate microprocessors and various sensors into household electrical products and then to make them intelligent, but in the future fuzzy will have to be used to make them evolve into being more user - friendly, with human qualities based on the user's point of view.

12.4 Application to video equipment

Among the various types of video equipment, applications of fuzzy theory to TV and VTR are particularly active. Perhaps the leading single area of fuzzy application is to video cameras.

The first applications of fuzzy to video cameras were to automatic focusing and automatic diaphragm control [11]. Recently there is great demand for video cameras that are smaller and lighter, and more convenient to use; it is sales of this type of video camera that have been primarily responsible for the explosive spread of video in recent years.

The smaller and lighter video cameras become, the easier it is for camera movement to become a problem. When zooming up to take closeups, even a small amount of camera vibration is magnified on the screen. In addition, video cameras are being used in increasingly diverse ways, and there is demand for a camera that can be used even in a vibrating environment such as inside a car or a train. In these increasingly light weight video cameras, a camera movement correction function has become indispensable. Fuzzy theory is used in this camera movement correction function. This technology is explained below.

The camera movement correction function

Previous technology involved using an angular velocity sensor to detect camera motion, which was then corrected for by mechanically moving the pickup optical system. However, such a correction mechanism is quite large, and is therefore unsuitable for a small, light weight video camera. Therefore technology for correcting for camera motion entirely by image processing

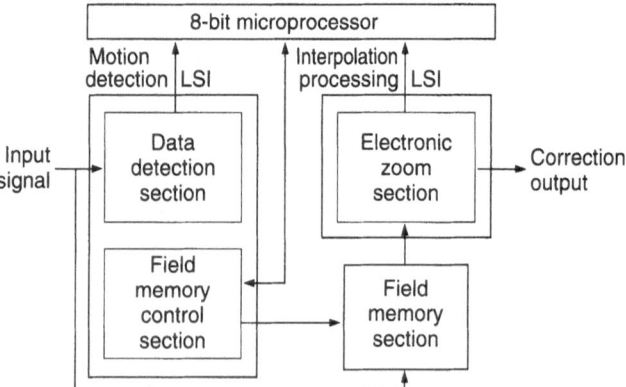

Fig. 12.23 Block diagram of camera vibration correction function.

was developed, making it possible to incorporate a camera motion correction function even into a compact video camera. This involves detecting the amount of camera motion by digital signal processing, then using fuzzy to decide whether or not the motion is motion of the subject, and finally fuzzy is applied again to correct the image for camera motion.

A diagram of the camera movement correction function is shown in Fig. 12.23. The system consists of a motion detection LSI, an interpolation processing LSI and field memory, and an 8-bit microprocessor.

The principle of camera movement correction is shown in Fig. 12.24. First the motion detection LSI compares the

present field image signal to the image signal 1 field previously, and the amount and direction of image movement due to camera motion (the motion vector) are found. Next, the field memory read-out position is controlled, the screen frame moved antiparallel to the motion vector, the image signal extracted and the camera movement corrected for. If the image signal is output directly the edge of the TV screen will be cut out, so the image signal is enlarged to fill the TV screen by the interpolation processing LSI.

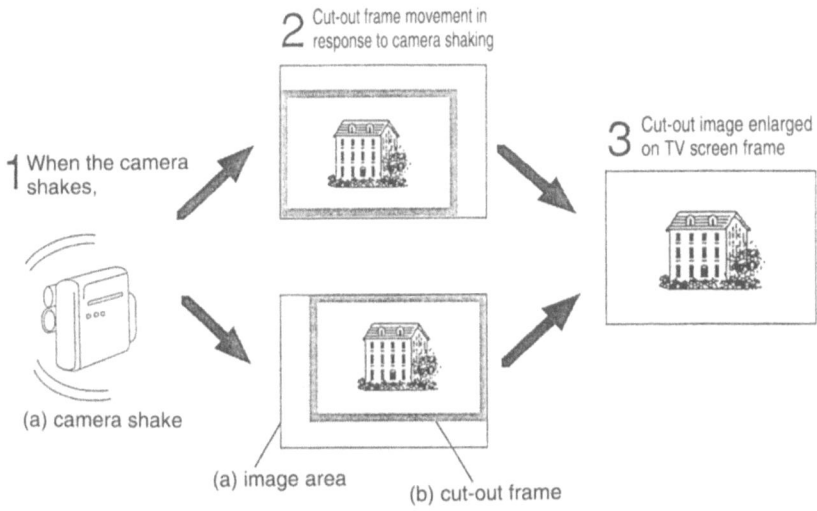

Fig. 12.24 The principle of camera vibration correction.

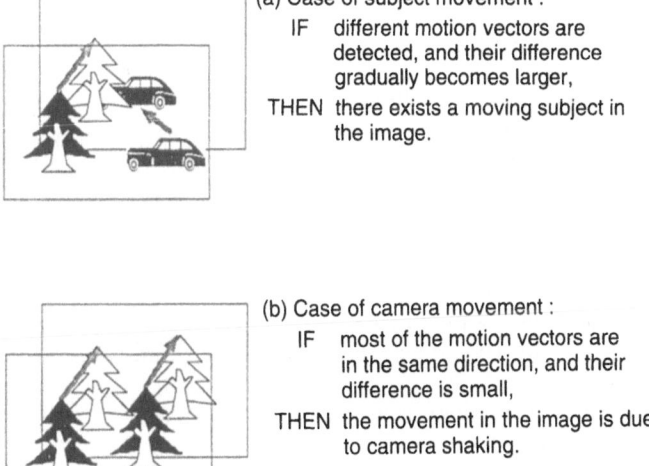

Fig. 12.25 Fuzzy rules.

Fuzzy inference

In correction for camera movement by image processing, the amount of movement is detected from the image signal, so it is very difficult to distinguish between actual camera movement and movement of the subject. The human eye can easily decide which of the two is moving, but it is very difficult to do the same thing with a machine. Fuzzy inference is used to distinguish one from the other.

One example of the fuzzy rules used to distinguish camera movement from subject movement is shown in Fig. 12.25. (a) is the rule for judging that the subject has moved; in this case the individual motion vectors are scattered in various directions. (b) is the rule for judging that the camera has moved; the motion vectors are all in nearly the same direction and also of the same magnitude. By using fuzzy rules such as these, it can be judged accurately whether movement in the image is due to camera movement or subject movement.

A human can easily judge whether image movement is due to camera movement or subject movement. This human judgment has been reproduced by using fuzzy theory.

Thus, even without the use of highly sophisticated image processing and recognition technology, by skillfully combining digital technology and fuzzy, judgment comparable to that of a human can be obtained with a computer on the level of a microprocessor. This is the great appeal of fuzzy.

12.5   In conclusion

In the last few years intelligent functions have been incorporated into household electrical appliances by using microprocessors and various sensors; the introduction of fuzzy has accelerated progress toward creating user - friendly products.

Let us summarize the characteristics of fuzzy. (1) Subjective concepts and imprecise ideas such as meanings of words can be handled. (2) Judgments similar to human macro judgments can be performed. (3) Fuzzy is suitable for expressing human thought processes (inference and sensitivity).

In the future, technology that is friendly to people, close to people and close to the heart will become important.

In the process of bringing this about, progress in research on flexible artificial intelligence - based fuzzy, research on combination with neurocomputer technology, and research on human interfaces that harmonize man and machine, will make it possible to incorporate actions, judgments and thought that are near - human into a wide variety of devices and systems. Products using fuzzy will respect users' subjective desires and be introduced into a greater variety of products.

By using a computer that can make various judgments while learning like a human, the appearance of a human type robot is perhaps not impossible.

Machines will perhaps become able to say things like "Is this what you are trying to say?"

Perhaps the appearance of such machines will free humans from the present need to learn machine operation.

We want to pursue the development of "human electronics" by seeking harmony between advanced fuzzy technology and respect for humans that gives a feeling of security to the user.

References

1. Maeda M, Murakami S (1987) Follow up control of an automobile with fuzzy logic (in Japanese). Collected Papers of the 3rd Fuzzy Systems Symposium. Osaka, 1–3 June, 1987, pp 61–66

2. Ichihashi H, Tanaka H (1988) PID–Fuzzy hybrid controller (in Japanese). Collected Papers of the 4th Fuzzy Systems Symposium. Tokyo, 30–31 May 1988, pp 97–102

3. Hayashi I, Nomura H, Wakami N (1990) Acquisition of inference rules by neural network driven fuzzy reasoning (in Japanese). J Jpn Soc Fuzzy Theory and Systems 2 (4) : 585–597

4. Mizumoto M (1988) A method to improve fuzzy inference (in Japanese). Collected Papers of the 4th Fuzzy Systems Symposium. Tokyo, 30–31 May, 1988, pp 91–96

5. Kang GT, Sugeno M (1987) Fuzzy modeling (in Japanese). J Soc Instrument and Control Engineers 23 (6) : 650–652

6. Nonaka H, Oiwa Y, Shinchi Y (1984) Fully automatic washing machine with sensors (in Japanese). National Technical Report 30 (5) : 606–112

7. Yamashita H, Abe S, Kondo S, Kiuchi M, Imahashi H (1990) A washing process monitoring sensor for fully automatic washing machine using fuzzy inference (in Japanese). In : Proceedings of the 9th sensor symposium, pp 163–166

8. Kondo S, Kiuchi M (1990) Introduction to the Aisaigo Day Fuzzy (NA-F50Y5) fully automatic washing machine (in Japanese). J Jpn Soc Fuzzy Theory and Systems 2 (3) : 112–114

9. Kamogawa T, Haruki T (1990) Fuzzy control (in Japanese). J Inst Television Engineers Jpn (ITEJ) 44 (9) : 1196–1202

10. Uomori K, Morimura A, Ishii H, Sakaguchi T, Kitamura Y (1990) Automatic image stabilizing system by full-digital signal processing (in Japanese). IEEE Trans Consumer Electronics 36 (3) : 510

11. Morimura A, Uomori K, Ishii H, Sakaguchi T, Kitamura Y (1991) A purely electronic image movement correction system (in Japanese). ITEJ Tech Reports 15 (7) : 43–48